Drug Delivery Systems for Musculoskeletal Tissues

Sumit Murab
Editor

Drug Delivery Systems for Musculoskeletal Tissues

Springer

Editor
Sumit Murab
School of Biosciences & Bioengineering
Indian Institute of Technology Mandi
Mandi, Himachal Pradesh, India

ISBN 978-3-031-55652-4 ISBN 978-3-031-55653-1 (eBook)
https://doi.org/10.1007/978-3-031-55653-1

This Springer imprint is published by the registered company Springer Nature Switzerland AG
The registered company address is: Gewerbestrasse 11, 6330 Cham, Switzerland

Paper in this product is recyclable.

Dedicated to
My Parents
Mr. & Mrs. Suraj Bhan and Pushpa Murab

Preface

Musculoskeletal tissues are hard to heal if damaged by disease or trauma. If damaged they cause pain and hinder regular movements of our bodies. Drug delivery systems that can be used for healing and regenerating these tissues need to be specialized to bear physiological loading on these tissues while helping in controlled release of drug or biomolecules for healing/regenerating the tissue. Localized drug delivery to specific tissues is of significant interest to the biomedical community. Drug delivery specifically to musculoskeletal tissues have its distinctive challenge of withstanding physiological loading applied to these tissues apart from the unique physio-chemistry prevailing in the diseased tissues. The book discusses about different musculoskeletal tissues including bone, cartilage, ligament, tendon, muscle, and musculoskeletal tissue interfaces and the disease pathologies affecting them. Each section further discusses the existing treatment modalities available for the treatment of these diseases, including surgical techniques and their disadvantages. The main focus is on the existing drug delivery platforms that are available or under development that can provide either symptomatic relief from pain/infection or help in regeneration of these degenerated tissues through controlled delivery of drugs/biomolecules. The book also discusses in detail the mechanical properties and support that these systems provide to the damaged musculoskeletal tissues and how long-term healing/regeneration is possible.

All of the contributors to the book are active researchers in the field and have brought new insights through their research into this book. I would like to extend my sincere thanks to all of them for contributing to this book. I would like to thank my students, especially Ms. Ankita Negi, who helped me in the final steps of completing the book. I would also like to express my sincere gratitude towards all my Gurus who have shaped me as a researcher. A big thanks to my family and friends who have been a constant support throughout.

Mandi, Himachal Pradesh, India Sumit Murab

Contents

Chapter 1
Introduction to Musculoskeletal Tissues and Drug Delivery Systems

Ankita Negi, Pankaj Sharma, Vedante Mishra, and Sumit Murab

Contents

A. Negi · P. Sharma · V. Mishra
School of Biosciences & Bioengineering, IIT Mandi, Mandi, Himachal Pradesh, India

S. Murab (✉)
School of Biosciences & Bioengineering, IIT Mandi, Mandi, Himachal Pradesh, India

Indian Knowledge System and Mental Health Applications Centre, IIT Mandi, Mandi, Himachal Pradesh, India

BioX Centre, IIT Mandi, Mandi, Himachal Pradesh, India

Advanced Materials Research Centre, IIT Mandi, Mandi, Himachal Pradesh, India
e-mail: sumitmurab@iitmandi.ac.in

S. Murab (ed.), *Drug Delivery Systems for Musculoskeletal Tissues*,
https://doi.org/10.1007/978-3-031-55653-1_1

1 Musculoskeletal Tissues

The musculoskeletal system is a diverse set of specialized tissues, which is responsible for the structural shape, support, stability, and mobility of the human body. Bone, cartilage, ligaments, skeletal muscles, and tendons are all components of the musculoskeletal system. Among musculoskeletal tissues, muscle moves the skeletal bones, ligaments connect the bone, tendons bind the muscle to the bones, and articular cartilage functions as a shock absorber and acts as a lubricant between the bones, which makes up a freely movable joint (Fig. 1.1). The biomechanical and homeostatic processes of the human body, including hematopoiesis and the storage of carbohydrates, are largely dependent on the musculoskeletal tissues. At the same time, these tissues are also affected by various diseases and degenerative processes. The most common cause of disability worldwide is lower back pain, which is a musculoskeletal illness. Generally, musculoskeletal tissues become less functional as people age, which causes changes in their structure and function. Thus, the skeletal muscle loses its strength, bones become fragile, cartilage loses its resilience, and the elasticity of ligaments is reduced. Especially, in an aging population, musculoskeletal conditions become more prominent.

The following sections will discuss the anatomy of different musculoskeletal tissues from a microscopic and macroscopic perspective as well as changes in these tissues with specific pathological characteristics.

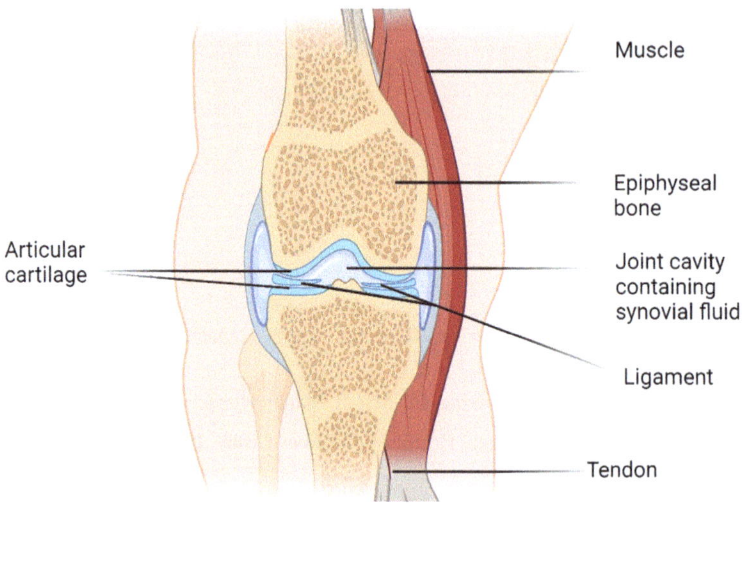

Fig. 1.1 Musculoskeletal system comprises of bones, cartilage, tendons, ligaments, and muscles

1.1 Types of Musculoskeletal Tissues

1.1.1 Bone

In vertebrates, the skeletal system provides a structural support system to the body that can resist mechanical forces [1]. The skeletal framework is formed by bone tissue. Bone is a natural nanocomposite between inorganic minerals, organic matrix, and cellular components [2]. The inorganic part mainly consists of calcium phosphate salts in the form of hydroxyapatite crystals. In addition to this, bone contains carbonate, phosphate, magnesium, and fluoride in minor quantities. In contrast, collagen type I makes up most of the organic matrix. Along with this, proteoglycans, glycoproteins, and various growth factors, such as osteocalcin and osteonectin, are also present in the bone matrix [3]. The cellular component involves osteoclasts, osteoblasts, and osteocytes. Among them, osteoblast cells are involved in bone formation and mineralization, whereas osteoclasts are responsible for bone resorption. Osteoblasts and osteocytes are derived from mesenchymal stem cells, while osteoclasts are derived from hematopoietic stem cells closely related to monocytes/macrophages.

The hierarchical structural arrangements of bone components provide mechanical properties to the bone [4]. On a macroscopic level, bones also have different densities in different parts—the inner region is made up of cancellous or spongy bone, which has a porous network of bony spicules called trabeculae, while the exterior is made up of hard cortical or compact bone [1, 4]. The microscopic bone structure of both cortical and trabecular bone is formed of structural units called osteons. In cortical bone, osteon is referred to as the Haversian system, which is tubular in shape. The Haversian canal is present in the central part of the tube, which contains blood, lymphatic vessels, and nerves enveloped by a concentric layer or lamellae. In cortical bone, osteon plays an important role in the mechanical properties where each osteon functions as a fiber that prevents fracture by deformation [5]. Furthermore, there are two types of lamellae: outer circumferential lamellae which are present next to the periosteum, and inner circumferential lamellae which lie next to the endosteum. Interstitial bone, which is made up of the remains of earlier, partially resorbed osteons, fills the space between adjacent osteons and is visible as short, arching layers. Osteons are referred to as bone packets in trabecular bone and are saucer-shaped stacks or layers of lamellae. A cement line divides adjacent packets in trabecular bone. On a microscopic level, both types of bone show tissue with either a woven or lamellar structure. This classification is based on the arrangement of collagen fiber. Woven bone has randomly arranged collagen fibers having osteoblast and osteoprogenitor cells alongside. This type of bone is present majorly in the developing fetus, which is later replaced by lamellar bone in adults, whereas lamellar bone has a well-organized arrangement of collagen fibers [1, 5].

Different parts of the skeletal system develop from different embryogenic cell lineages [6]. The dentin of teeth, connective tissue, and parts of the bones and cartilages of the face and anterior skull are all produced by neural crest cells that migrate from the dorsal edges of the closing neural tube into this area of the skull. The posterior portion of the skull's bones and cartilage are produced by prechordal mesodermal cells. The cellular source of axial skeletal is the paraxial mesoderm (somites), while the appendicular skeleton is made up of lateral plate mesodermal cells [7]. The process of skeletogenesis involves the migration of mesenchymal cells derived from these embryonic lineages to the locations of the future bones. Thereafter, condensed high cellular density mesenchymal cells either differentiate into chondrocytes or directly into osteoblasts. The axial skeleton, appendicular skeleton, and posterior portion of the skull all develop through endochondral ossification, where the initial cartilage present during the fetal and postnatal stages is gradually replaced by bone in the later stages of growth. In contrast, the membranous neuro- and viscerocranium, as well as a portion of the clavicle, all develop through intramembranous ossification where osteoclast cells produce bone [6]. Endochondral ossification is a complex process where a developing long bone has two areas of growth cartilage: the growth plate and the articular epiphyseal growth cartilage, which drive the extension of the primary and secondary centers of ossification, respectively (Fig. 1.2). Chondrocytes in growth cartilage undergo hypertrophy in which they increase in size dramatically and secrete extracellular matrix, which eventually becomes mineralized. This results in the elongation of bone. The transverse septa of the cartilage matrix surrounding the hypertrophic chondrocytes are destroyed as they die, leaving the vertical septa largely intact but allowing the invasion of blood vessels, osteoclasts (multinucleate bone-resorbing cells), precursors of osteoblasts (bone-forming cells), and bone marrow cells. The differentiating osteoblasts utilize the leftover cartilage matrix as a scaffold for the deposition of bone matrix, and the osteoclasts help remove the cartilage matrix [8].

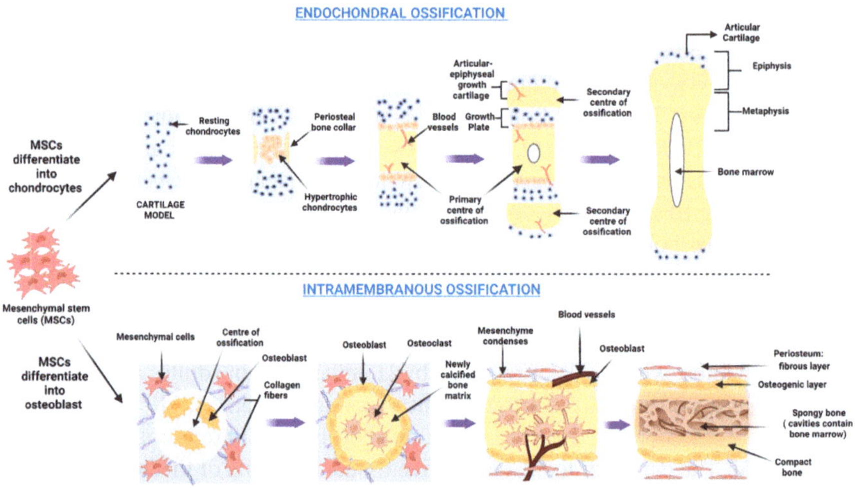

Fig. 1.2 Bone development: endochondral and intramembranous ossification

1.1.2 Cartilage

Cartilage is a connective tissue with a high resistance to compression, tension, and shear forces. It is located in various body parts such as ribs, the lower part of the nose, the auricle of the ear, and synovial joints. In addition to this, most of the skeletal system is cartilaginous in early fetal life, which is replaced through a complex process in the later stage of bone development. This process is called endochondral ossification. Cartilage is mainly composed of collagen type II, elastin, water, a varying amount of collagen type I, and a variety of proteoglycans with low-density chondrocytes [9]. The molecular organization of cartilage is described as the intermingling of collagen fibrils with electrically charged proteoglycans stabilized by inorganic ions. Water tends to move toward higher ion concentration, thus creating osmotic pressure [10]. Being an avascular tissue, it receives nutrients from surrounding connective tissues through diffusion. In contrast, in the early embryonic stage, cartilage is characterized by high-density cellular cartilage with a narrow matrix septum between the cells. Cartilage mainly grows in two ways: (1) interstitial growth, in which chondrocytes, mature cartilage cells, divide and multiply, and (2) appositional growth, in which cells of the inner layer of the perichondrium develop into chondroblasts and start laying down cartilage matrix [11].

Based on their molecular composition, they are classified as hyaline, elastic, and fibrous cartilage [11–13]. Differences in molecular composition give rise to differences in biomechanical properties and appearance in each type of cartilage. Among all, hyaline cartilage is the most prominent form, which includes nasal, some laryngeal, tracheobronchial, and most common articular cartilage, which is present at the end of long bones [13]. Articular cartilage is characterized by a dense network of approximately 50% collagen fiber and 35% proteoglycan. Nonstructural proteins, glycoproteins, and other minerals and lipids make up the final 10%. Chondrocytes represent around 10% of the total volume of articular cartilage. Generally, it is divided into five distinct zones. As the depth increases from the superficial to the deep zone, cellular, collagen, and proteoglycan content increases. In the superficial zone, flat morphology cells are arranged parallel to the articular surface, whereas globoid-shaped chondrocytes are present perpendicular to the articular surface. Tidemark divides Zone III from calcified cartilage Zone IV. In calcified cartilage, cells are embedded in the calcified matrix, and it is known to connect the uncalcified cartilage with the subchondral bone [12, 14–16]. Chondrocytes respond to mechanical stimuli in a zone-specific manner, and hence, the zonal arrangement is crucial for cartilage activity. Although it is a relatively resilient tissue, it has limited healing potential because of its avascular nature, low cellularity, and absence of undifferentiated cells. Lack of vascularization prevents cartilage from responding to injury normally, which involves the production of a fibrin clot, which prevents perfusion of inflammatory cells to the injured site, followed by undifferentiated cells that would build a new matrix and aid in tissue regeneration. In addition to collagen 1, fibrocartilage has a high amount of type I collagen, contains proteoglycans in small quantities, and has high cellular density. Chondrocytes and fibroblasts are the two main cell types found in fibrocartilage. It is mainly located in the annulus fibrosis of intervertebral discs, some regions of tendons and ligaments, and the pubis symphysis [13]. It is the strongest type of cartilage because it has denser and well-organized collagen fibers than hyaline cartilage. Fibrocartilage provides flexibility to the tissue and is resistant to compressive stress. The extracellular matrix of elastic cartilage contains elastic fibers, which give the tissue its elastic properties. It is mostly located in the external ear's auricle, the auditory tube, the epiglottis, and a portion of the larynx. Elastic cartilage is not a significant part of the musculoskeletal system.

1.1.3 Muscle

The most dynamic and largest organ of the human body is the skeletal muscle, which comprises approximately 40% of the total body weight and contains 50–70% of all body protein. Muscle is mainly characterized by high water content (70%), protein (20%), and other substances such as fats, carbohydrates, and minerals (5%) [17]. Tissue function is reflected in its structure; skeletal muscles have a well-defined arrangement of muscle fibers (cells of muscle tissue, also known as myofibers) to generate force and produce body movement.

Each skeletal muscle is formed by thousands of muscle fibers covered by connective tissue sheaths where an individual bundle of muscle fiber is known as fasciculi and each muscle fiber is made up of several myofibrils containing multiple myofilaments [17, 18]. The epimysium, perimysium, and endomysium are the three anatomical components of the connective tissue in the muscle, where the whole muscle is surrounded by epimysium, fascicles are surrounded by perimysium, and the innermost single muscle fiber is surrounded by a cell membrane called as sarcolemma, which is overlaid by endomysium connective tissue [19]. In muscle, elastin fibers are present in the majority along with collagen. Myofibers show characteristic properties of contractibility and conductibility, which in turn convert chemical energy into mechanical energy. Under the microscope, each muscle fiber is elongated, cylindrical, and multinucleated and has a series of alternating dark (A band) and light bands (I band) in internally arranged myofibrils, hence giving the striated appearance to skeletal muscle. Myofibrils have cytoskeletal elements arranged into thick and thin filament structures, hence forming sarcomeres, the basic unit of skeletal muscle (Fig. 1.3).

Actin and myosin are the two myofilaments (proteins) that are most prevalent in a single fiber. The primary molecular motor is myosin, and 11 different sarcomeric myosin genes have been identified in mammals along with the related protein products (some of which may not be produced in humans). There are other regulatory proteins such as the calcium-dependent troponin complex (including troponins C, I, and T) and tropomyosin that are associated with the actin filament and play very important roles in the activation process that leads to myofilament sliding and force generation. Titin and nebulin are two additional proteins that support the physiological and mechanical characteristics of muscle. Titin, a large elastic protein, connects myosin and the Z disc (defines the lateral boundaries) of the sarcomere and aids in stabilizing and aligning the thick filament. Nebulin, on the other hand, is combined

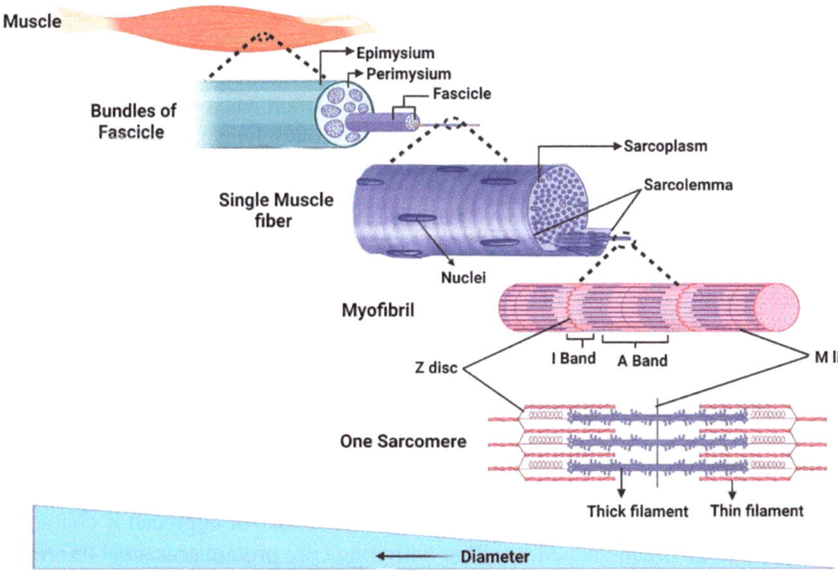

Fig. 1.3 Hierarchical structure of skeletal muscle

with other proteins to form actin-containing thin filaments. These proteins affect the passive tension and stiffness properties of individual cells, support sarcomere integrity, and may be important for myofibril assembly and cell signaling. Actinin is one of the proteins found in the Z disc of the sarcomere, which also acts as an attachment site for the actin myofilament. The additional cellular components in the sarcoplasm of muscle fibers include the transverse tubular system (T tubule), the sarcoplasmic reticulum, and a mitochondrial network. T tubule system plays an important role in the transmission of the nerve action potential to the interior of the cell, while the sarcoplasmic reticulum is the storehouse of calcium, which releases upon muscle activation. Sarco/endoplasmic reticulum Ca^{2+}-ATPase (SERCA) and calsequestrin are two classes of proteins found in the sarcoplasmic reticulum that help to maintain calcium homeostasis. After muscle activation, the former is in charge of calcium reuptake into the cisterna, and the latter loosely binds calcium inside the sarcoplasmic reticulum [12, 17, 20, 21].

1.1.4 Intervertebral Disc

Intervertebral discs are located between the vertebrae in order to provide flexibility to the spine. It primarily functions as a shock absorber, transferring load and dissipating energy under a load. The anatomical structure of the intervertebral disc (IVD) is divided into three parts—the ***nucleus pulposus (NP)*** in the center, surrounded by the ***annulus fibrosus (AF)***, and the ***two cartilaginous endplates*** located superiorly and inferiorly, which allow their attachment with vertebrae [22]. The intervertebral disc has no vascular or nerve supply. Both NP and AF differ in structural, cellular, and molecular composition, which elucidates differences in their biomechanical properties. The core part is the nucleus pulposus, which has a hydrogel-like structure as it holds a high amount of water and large proteoglycans, loosely held by an irregular network of type II collagen and noncollagenous protein elastin fibers. Other collagens are also present in small amounts such as types VI, IX, and XI [23]. AF is known to provide tensile strength to IVD by controlling and maintaining the osmotic pressure produced by NP. AF contains collagen fibers of Type I and Type III, a small amount of type II collagen, proteoglycans, and non-collagenous protein elastin arranged in a series of 15–25 concentric lamellae. Based on structure, AF is further divided into two parts—outer and inner AF where inner AF has a high amount of Type II collagen in comparison to outer AF. Water content decreases in NP as we age, whereas in AF, it remains the same throughout life. During spinal compression, IVD resists the applied load by generating hydrostatic pressure on the NP, which is maintained by AF by creating tension [24]. Aggrecan is a major proteoglycan present in the NP region of IVD. The structure of aggrecan is characterized by the bottlebrush-shaped structure with long core protein present at the center where chondroitin sulfate (CS) and keratan sulfate GAG chains are covalently connected to it. The highly anionic GAG content provides the osmotic characteristics required to maintain NP height and turgor against compressive pressures. The IVD aggrecan has more sulfated keratan (a highly hydrated sulfated glycosaminoglycan). Therefore, it offers a greater ability to hydrate. It interacts with specific proteins like aggrecan, versican, lymphatic vessel endothelial HA receptor-1, tumor necrosis factor (TNF)-inducible gene-6 protein, and thrombospondin as well as membrane receptors like CD44, HA-mediated motility receptor, and toll-like receptor 4/2. HA is a widely expressed glycosaminoglycan (GAG) found in the extracellular matrix (ECM) and intracellular environment. It also contributes to cancer, apoptosis, inflammation, cell migration, and morphogenesis [24, 25].

There are predominantly big vacuolated notochord cells (NCs) and small, chondrocyte-like nucleopulpocytes (NPCs) in the NP, which have a low cell density ($3000/mm^3$) [26]. The NPCs primarily synthesize the ECM of the NP, which is substantially distinct from hyaline cartilage. On the other hand, AFCs have a density of $9000/mm^3$ and are primarily made up of fibroblast cells [27]. The cartilaginous end plate (CEP) joins the vertebral bodies to the IVD via a hyaline cartilage structure. The CEP is made up of chondrocytes and hyaline cartilage cells that create type II collagen and PGs to carry nutrients and metabolites to the avascular IVD. The CEP serves as the blood supply's terminus, making it an important transportation hub. In order to protect the vertebral bodies, the CEP also takes the strain from the IVD [28]. Thus, these three components work together to create a special anatomical structure that preserves the unity of the IVD and preserves its microenvironment's homeostasis and IVD function.

The development of the intervertebral disc starts from an embryonic structure called the notochord [22, 29]. It is a structure around which axial development occurs. Mesenchymal somites are temporary, spherical-shaped structures that are aggregated around the notochord to form a continuous perichondral tube. Thereafter, somites differentiate into sclerotomes, which form connective tissues of the axial skeleton: vertebrae, cartilaginous end plates, annulus fibrosis, tendon, and ligaments. The sclerotome is temporary embryonic tissue made up of pluripotent mesenchymal stem cells. In addition to this, notochord gives rise to NP.

1.1.5 Ligaments and Tendons

Ligaments and tendons are closely related dense fibrous connective tissues that are essential for the stability and mobility of the musculoskeletal system. In contrast to ligaments, which attach bone to bone and stabilize joints, tendons connect muscle to bone and enable body movement by transmitting tensile stresses and storing elastic energy [30, 31]. Generally, both are designed to withstand longitudinal tensile stress and strain. Tendons are inelastic collagenous soft tissues that contain spindle-shaped cells called tenocytes [31]. These are sparsely arranged and sandwiched between parallel layers of collagen type I. Tendons undergo different levels of stress depending on the muscle they support, and as a result, their tensile strength varies significantly. Ligaments, which are considered to be dense connective tissue, resemble tendons in terms of morphology. They are mostly made up of fibroblasts encased in type I collagen layers. The fact that ligaments connect one bone to another means that they encounter traction stresses from a variety of angles, which gives their collagen fibers a more varied directional range than tendons. Compared to tendons, ligaments are shorter and wider, and their extracellular matrix contains more growth substance rather than collagen. However, tendons and ligaments have dynamic properties. They can heal after injury and adjust their tensile strength in response to activity or immobilization [32].

Tendons come in a variety of shapes and sizes. Positional tendons play a major role in the muscular control of joint position and have distinct sites of osseous attachment, such as the tendon of origin of the biceps brachialis. The sheath that surrounds positional tendons frequently contains fluid to allow for low-friction sliding motion. Positional tendons that undergo a sudden shift in direction as they traverse a joint include wrap-around tendons (such as the digital extensor tendons) [33]. A fibrous retinaculum typically stabilizes the segment of a wrap-around tendon that crosses a joint, and the tendon itself may contain fibrocartilage that offers resistance to the compressive pressures that concentrate at the site of the joint. Tarsal extensors and carpal flexors, which store energy, are adapted to carrying weight and maintaining posture and have higher elastin concentrations than other tendon types. Aponeuroses, which resemble flattened sheets, connect muscles to other fascial structures often (e.g., insertional aponeuroses of the abdominal wall muscles). There are two types of ligaments. Collagen-rich thickenings of the joint capsule are known as capsular ligaments, such as the collateral ligaments of a majority of joints [34]. Within synovial compartments, intra-articular ligaments (like the femoral head's round ligament) cross joints. A thin epiligament that fuses with the periosteum at the ligament's origin and insertion surrounds the intra-articular ligaments

[35]. Ligaments and tendons have significant anisotropy and viscoelasticity. Due to their dynamic behavior, tendons and ligaments may be able to detect changes in mechanical load and coordinate their reaction to change the extracellular matrix's (ECM) composition. When placed in tension, they have considerable strength and stiffness, but when loaded in compression, they buckle easily. Although tendons are routinely subjected to 4% strain during in vivo physiological loading which causes microscopic tearing of collagen fibers, in vitro testing has shown that tendons may be elastically stretched up to 14% [33, 36].

The normal physiology of the musculoskeletal system tissues can get deregulated both due to diseases and trauma. The next section will discuss the various pathological conditions that these tissues are prone to.

1.2 Understanding Pathologies of Musculoskeletal Tissues

The growing population has become more vulnerable to noncommunicable diseases such as musculoskeletal problems. The burden of musculoskeletal-related problems has been rising globally due to an unhealthy lifestyle and the increasing age of the global population [37]. These disorders lead to mental and behavioral illnesses in society, hence impacting the global economy and health. Musculoskeletal diseases or injuries are referred to as pathological conditions that affect bone, cartilage, muscle, or other connective tissue. The most common and severe musculoskeletal disorder includes osteoarthritis (OA), rheumatoid arthritis (RA), and gout [38].

Bone strength and integrity are maintained by the balance between the osteoblast, which is responsible for bone formation, and the osteoclast, which is responsible for bone resorption. This phenomenon is in a balanced state in younger adults, whereas, in older age or a diseased state, bone resorption exceeds bone formation. This condition leads to local or systemic bone loss known as ***osteoporosis***. Osteoporotic bone eventually loses its integrity and mechanical property, hence becoming susceptible to fracture. It is prevalent in postmenopausal women due to a decrease in estrogen levels. The rise in osteoclast numbers is driven by the elevated level of cytokines, such as RANK (receptor for activator of nuclear factor-kB), TNF-a (tumor necrosis factor-a), interleukin-1(IL-1), IL-6, IL-11, M-CSF (macrophage-colony stimulating factor), and prostaglandin E [39], which in turn affect the level of estrogen. Other than estrogen deficiency, pathological disorders that cause bone loss include multiple myelomatosis, hyperparathyroidism, and hyperthyroidism. Apart from this, stress fractures, bone breaking, and inflammatory disorders are caused by the overstressing of bones as a result of trauma or injury. Bone has been found to be the most common site for ***metastasis***, particularly in the case of breast and prostate cancer, but it is the less common site for primary cancer. Bone sarcoma is also a prevalent disease that primarily affects patients between the ages of 10 and 20 and is characterized by the production of osteoid and bone by malignant spindle cells. Although it can occur everywhere in the body, this type of tumor typically forms in the metaphyseal areas of the distal femur, proximal tibia, and proximal humerus. An ***osteosarcoma*** develops radially to form a mass that resembles a ball. After that, it enters the bony cortex and contracts the surrounding muscles to create the "reactive zone," a pseudo capsular layer where nodules (also known as "satellites") that are microscopic expansions of the original mass may also develop [40]. Osteoporosis, bone metastases, and osteosarcoma can all be treated with medications that block osteoclastic activity or promote bone growth. The current medication therapies work on the grounds that either osteoclast development is suppressed (e.g., by estrogens or selective estrogen receptor modulators) or osteoclast activity is suppressed (e.g., by bisphosphonates).

Osteoarthritis (OA) is the most prominent form of arthritis characterized by the degeneration of articular hyaline cartilage, which mainly affects multiple joints such as the knee, hip, lumber facet joints, and temporomandibular joints [41]. OA also causes structural changes in the subchondral bone, the synovial membrane, and the synovial fluid. It is the greatest cause of disability in the elderly population, which affects more than 10% of the population aged 65 and older [41]. Histological analysis of osteoarthritic tissue shows fibrillated articular surface, loss of cartilage structure, presence of chondrocytes in clusters, and proteolysis of extracellular matrix. Different factors such as cytokines and growth factors influence chondrocyte phenotypes and cartilage matrix structure. Among the cytokines, IL-1β and TNF-α are the main inflammatory mediators contributing to the progression of osteoarthritis. Both IL-1β and TNF-α induce the synthesis of MMPs, aggrecanases, inducible nitric oxide synthase (iNOS), and cyclooxygenase 2 (COX-2) and reduce the synthesis of tissue inhibitor of metalloproteinases (TIMP). However, growth factors, such as the transforming growth factor (TNF), fibroblast growth factors (FGF), and bone morphogenetic proteins (BMP), regulate cartilage homeostasis. Another form of arthritis is ***rheumatoid arthritis,*** which is classified as an autoimmune disease characterized by chronic systemic inflammation, leading to the progressive destruction of cartilage and synovial joints. Cytokines such as TNF-α, IL-6, IL-1β, and IL-17α are responsible for the progression of the disease [42]. Analgesia is the most common treatment against OA and RA as both share common symptoms such as joint pain and stiffness, which includes paracetamol, NSAIDs, and opioid medication. However, there are various side effects associated with these medications. Injectable medications, such as corticosteroids and hyaluronic acid, are also commonly used, but they have only momentary benefits [43]. There are disease-modifying antirheumatic medications (DMARDs) that act as second-line treatment to slow and stop the progression of the joint deformity. Hydroxychloroquine, sulfasalazine, methotrexate are the most common DMARDs [44]. Unfortunately, there is no approved standard pharmaceutical medication for the treatment of OA. The majority of the disease-modifying OA medications (DMOADs) are currently being studied. Based on current knowledge of the pathophysiology of OA, the activity of DMOADs is directed at cartilage preservation.

Another form of inflammatory arthritis is ***Gout*** caused by the deposition of uric acid crystals in the joints. ***Chondrocalcinosis*** is another cartilage-related disease that resembles gout or chronic arthritis. It involves the deposition of calcium pyrophosphate dihydrate (CPPD) crystals in joints. In cartilaginous tissues, CPPD crystals mostly develop in the pericellular gaps around chondrocytes. A small amount of inorganic pyrophosphate and ATP is released from chondrocytes into pericellular spaces where ATP dissociates to form inorganic pyrophosphate. Calcium and inorganic pyrophosphate react to form CPPD crystals. Inflammatory responses are generated when CPPD crystals release from cartilage into joint spaces [45].

About 637 million people worldwide suffer from ***low back pain (LBP),*** which is the most prevalent musculoskeletal-related problem [46]. The major reason behind LBP is ***Intervertebral disc degeneration***, which also contributes to other spinal conditions such as ***disc herniation, spondylosis, and lumbar spinal stenosis***. Earlier, it was thought that continuous mechanical load and aging were the only causes of IDD, but recent epidemiologic studies have concluded that genetic factors are the main dominating risk factor for the occurrence of IDD along with environmental factors. Genetic defects are involved in differential ECM turnover, thus determining the size and mechanical strength of IVDs and the onset of IDD. These genes mainly involve *ACAN, CLIP, COL1A1, COL9A2, COL11A1, GDF5, IGF1R, IL-1, IL-6, MMP2, MMP3, MMP9, SKT, THBS2, and VDR.* Some recently identified genes that are also involved include PARK2 and CHST3 [47]. In the early stages of disc degeneration, the biochemical composition of the ECM changes in the NP as type II collagen, proteoglycan (PG), and water content decrease, leading to decreased axial stresses in the AF. Cellular death can result in low cellularity and phenotypic alterations in cells, both of which impair the cells' capacity to synthesize ECM and disrupt the homeostasis of the ECM. The production of ECM-degrading enzymes including matrix metalloproteinases (MMPs) and aggrecanases also rises with ageing and degeneration [48]. Phase II is referred to as the mild degeneration phase and is characterized by increased levels of inflammatory mediators in the disc. As degeneration progresses, AF, NP, and immune cells in the discs, such as macrophages, neutrophils, and T cells, produce high levels of inflammatory cytokines including interleukin (IL)-1β, IL-6, IL-17, IL-2, IL-8, IL-10, IL-4, tumor necrosis factor (TNF)-α, interferon-γ (IFN-γ), and various chemokines such as C–C chemokine ligand 20 (CCL20) and C–C chemokine receptor 6 (CCR6) [49]. At a later stage or severe degeneration phase, the disc loses its biomechanical function as a result of structural alterations. Fissures begin to form in the AF, causing the NP to extrude, allowing nerve growth and vascularization in the disc, resulting in discogenic pain. Disc cells and the infiltrating immune cells like CD68+ macrophages, neutrophils, and T cells (CD4+ and CD8+) continue to release cytokines and neurogenic factors including nerve growth factor (NGF) and brain-derived neurotrophic factor (BDNF) [50]. In general, nonpharmacological therapies are used in daily life which include exercise, massage, acupuncture, physical therapy, or other alternative therapies with insufficient proof of efficacy. The most frequently prescribed medications given to IVDD patients for pain alleviation and increased function are nonsteroidal anti-inflammatory medications [51–53]. In addition to NSAIDs, oral steroids, opioids, analgesics, antidepressants, gabapentinoids, and muscle relaxants are also prescribed regularly [24].

The health of the muscle fiber is determined by a dynamic equilibrium between the rate of synthesis and breakdown of contractile proteins. A change in this equilibrium results in significant changes in the composition, structure, and functionality of the muscle fiber and is caused by both internal and external factors, such as inflammation, oxidative stress, muscle disuse, and disease [54]. The IGF1-Akt/PKB-mTOR signaling pathway is the main signaling system that controls muscle mass and protein synthesis [55]. One of the primary causes of increased muscle bulk is increased activation of IGF1-AKT-mTOR signaling pathway activation, resulting in satellite cell (SC) activation and muscle hypertrophy. Clinically, hyperplasia, or a rise in the quantity and size of myocytes, is a sign of ***hypertrophy***. When a muscle is "exercised," it becomes clinically strong and healthy; this is seen in athletes and conditions like ***myotonia congenita*** [56]. In addition to the traditional IGF1-AKT-mTOR route, it has been demonstrated that SC activation via G-protein coupled receptors, specifically via the subunit G-i2 and through myostatin inhibition, induces hypertrophy in healthy muscles. It has been shown that aerobic exercise alters protein metabolism both acutely and chronically and causes skeletal muscle growth. ***Muscle atrophies*** occur when the balance between protein synthesis and degradation shifts increasingly in favor of protein degradation in response to a variety of stimuli, such as bacterial and viral infections, exposure to pro-inflammatory cytokines, mitogens, growth factors, and oxidative and biomechanical stresses. The muscle that is atrophying is identified by loss of muscle mass coupled with a decrease in the cross-sectional area of the muscle fiber, the volume of the muscle, and the amount of protein in the muscle. For a variety of reasons, including illness, injury, and prolonged periods of immobility, muscles can undergo atrophy (e.g., limb immobilization or even space flight). Based on numerous environmental and cellular stimuli, four primary proteolytic systems, including lysosomal (autophagic), proteasomal, calpains, and caspases, activate and contribute to atrophy. The ubiquitin-proteasome system has been the subject of numerous studies that aim to systematically decode the gene expression profiles for protein degradation in the atrophying muscle. Initiation factor eIF3f and myosin chains are the two primary substrates for atrogin-1 and MURF-1 and two ubiquitin E3 ligases of the ubiquitin-proteasome system, respectively. In general, they are regarded as "master regulators" of muscle atrophy [57]. These two regulators are able to regulate muscle expression, which is dependent on the movement and activity of the FOXO transcription factors, which in turn are regulated by the AKT pathway. In addition to upregulating the expression of the two important E3 ligases, MURF1 and atrogin1, pro-inflammatory factors, particularly those like interleukin (IL) 1 and TNF, cause muscle atrophy by signaling through the well-established pathways of p38MAPK and NFκB [58, 59].

2 Drugs for Musculoskeletal Pathologies

There are several diseases associated with musculoskeletal tissues that can be acute or chronic and should be treated to prevent any severe or fatal outcomes. Depending on the severity of the disease, there are several methods for their treatments, some of which are physiotherapy, drug therapy, and radiotherapy. The most common approach for the majority of pathological conditions is drug therapy. It can be employed for a broad spectrum of actions, such as the prevention of inflammation, bacterial infection, and pain as well as the suppression of tumors. Although it is the most common approach for treatment, there are several aspects that should be considered before the administration of drugs into the body. The extent of success for a drug depends on its composition, dose, interaction with the cells in its passage, absorption rate, action time, and by-product formation. The route of drug administration and the dose is determined by the results produced after the diagnostics. Drug administration can be via oral, topical, intravenous, intramuscular, or subcutaneous routes depending upon the type and severity of the disease. In general, routes of administration can be systemic or targeted. Although systemic drug delivery has been in use for time immemorial, it has several side effects and leads to post-treatment complications. These can be avoided with targeted delivery of drugs, and also, the quantity of the administration can be decreased, thereby preventing unintentional interaction with nontargeted cells and tissues.

The musculoskeletal system includes bone, cartilage, muscles, tendons, and ligaments, which are responsible for the framework, shape, movement, and locomotion of the body. The characteristic symptoms of musculoskeletal diseases include restricted mobility, pain, and/or distortion of the tissue shape. Apart from these, tissue-specific and disease-specific symptoms are also expressed, which help in assessing the disease and type of drug prescription. Targeted delivery of drugs with a suitable carrier for the specific disease is being studied for the treatment of all types of diseases. Here, we have categorized the drugs as low-molecular-weight drugs, stem cells, peptides, and therapeutic nucleic acids (Fig. 1.4). Apart from the conventional drug administration, several new biomaterial-based drug delivery approaches are discussed as well.

2.1 Low-Molecular-Weight Drugs

This category of drugs has a molecular weight in the range of 100–1000 Da and can be administered through various routes, thereby enabling their use for both systemic and targeted delivery. Patients with rheumatic diseases have common symptoms of pain and stiffness and are generally prescribed analgesics of nonsteroidal origin, called Nonsteroidal Anti-Inflammatory Drugs (NSAIDs). They are used for pain relief and can be employed in any musculoskeletal disorder and have a similar mechanism of action, which involves the inhibition of cyclooxygenase proteins

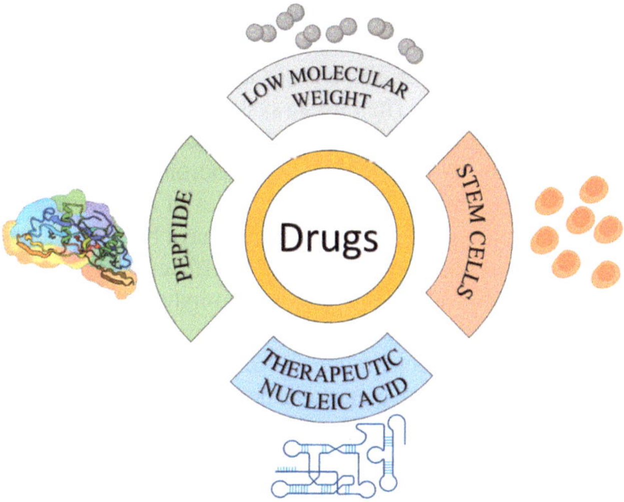

Fig. 1.4 Classification of drugs into different categories used for treating musculoskeletal pathologies

(i.e., COX-1 and COX-2), responsible for the formation of prostaglandins, the causative agent of inflammation. In most cases, COX-2 is responsible for inflammation. Earlier, they were administered as ointments on the skin or taken orally. In either case, their efficacy is quite low, and thereby, heavy doses were suggested for the patients, further leading to toxicity in the body. New target-specific delivery methods are being developed to prevent these issues. Meloxicam is one such NSAID that is given to patients with chronic arthritis but generally develops side effects such as abdominal pain, dyspepsia, and flatulence. This can be cured if there are controlled and target-specific releases of the molecules, as done by Yongli et al. with pH-sensitive hydrogel embedded with microsphere particles of meloxicam and block polymer for oral delivery. The system was developed while considering the acidic pH of the stomach, which causes side effects and results in low efficacy of the drug. This method allows maximum release of the nanoparticles and the drug into the small intestine for better absorption [60]. The process of development of these delivery particles is complex, and the concentration of the drug enclosed in a capsule cannot be precisely controlled. A simpler and modified method has been developed for delivering naproxen (another NSAID). This method is used for the development of methoxy poly (ethylene glycol)-poly(ε-caprolactone) (mPEG-PCL) diblock copolymer-based micelles and conjugated with naproxen. This system is not only pH sensitive but also has enhanced permeability and retention [61]. Although several other systems use different NSAIDs, one of the major problems concerning these systems is that there are not many COX-specific drugs that can be used for loading into these systems.

Bisphosphonates are the most commonly used drugs for the treatment of bone diseases such as osteoporosis, osteolytic lesions, multiple myeloma, and Paget's disease. Its molecular weight lies in the range of 200–300 Da and is mostly given to patients via the oral or intravenous route. These conventional routes are systemic drug delivery methods; however, these drugs are also being administered through targeted drug delivery methods. Their use in systemic drug delivery was quite direct, where they either help in resorption or control calcification-decalcification of bones. Etidronic acid, risedronic acid, clodronic acid, alendronic acid, zoledronic acid, and pamidronic acid are some of the common examples of bisphosphonates. Although these were administered orally to the patients and have been in use for a long time, there are adverse consequences. First, they are not efficiently delivered to the desired site, and second, they often lead to bisphosphonate-induced necrosis (BIONJ) in the mandible and/or maxillofacial bones. Therefore, targeted delivery of these drugs can help in the treatment of bone ailments [62]. In a recent study, alendronate is being used for the treatment of metastatic bone cancer in conjugation with curcumin. In this study, the anti-cancer potential of curcumin and the binding affinity of bisphosphonates resulted in high bone turnover. The system comprises pluronic F127–alendronate conjugated nanoparticles, which are then converted into nanosize micelles with the core made up of curcumin molecules [63].

Doxorubicin is another low-molecular-weight drug (~532 Da) that is being extensively used for the treatment of cancer as it intercalates the DNA and inhibits topoisomerase enzyme but has serious side effects like cardiotoxicity [64]. Targeted delivery of this drug can reduce post-treatment complications; hence, Pham et al. illustrated a delivery system using graphene oxide nanosheets functionalized with alendronate. They focused their study on breast cancer metastasis to the bone since bones are highly susceptible to cancer cells leading to osteolytic lesions and show high concentrations of bisphosphonates around lesions when treated with the same. Using this information, a hybrid delivery system was developed with graphene oxide nanosheets conjugated with alendronate to maintain its accumulation at the target site. The graphene oxide-alendronate carrier was loaded with doxorubicin, and the system was stabilized by polyethylene glycol (PEG). This system helped in local countering of the acidic environment of the metastasis [65]. A more target-specific and smart method has been demonstrated by Bagheri et al., where they used nanoparticles loaded with doxorubicin and capped with ATP and MUC1 aptamers. ATP aptamers make the particles sensitive to high ATP concentration in bone cancer tissue and MUC1 aptamer was added as a cancer marker [66]. A multiple-drug delivery system was developed by Hecheng et al., for treating osteosarcoma. This system was made up of thermosensitive hydrogel of triblock polymer poly(L-lactide-co-glycolide)-poly (ethylene glycol)-poly(L-lactide-co-glycolide), i.e., PLGA-PEG-PLGA and loaded with doxorubicin (DOX), cisplatin (CDDP), and methotrexate (MXT). This combination exhibited enhanced tumor inhibition of osteosarcoma in a rat model [67]. Like DOX, CDDP (MW 301 Da) also has anti-tumor activity and can target multiple intracellular sites to induce tumor cell death. It can induce reactive oxygen species (ROS) release, inhibit DNA synthesis by forming bifunctional adducts, and activate the p53 gene, which, on expression, results in cell cycle arrest and apoptosis [68, 69]. It makes cells susceptible to tumor necrotic factor (TNF)-induced apoptosis by binding and aggregating TNF-related apoptosis-inducing ligand (TRAIL) receptors and activating the extrinsic pathway of apoptosis. Intravenous administration of cisplatin was inefficient as 65–95% of the drug would bind to the proteins in blood plasma, resulting in its deactivation [68]. There are other methods to deliver this combination of anti-tumor drugs; one such method is using tricalcium phosphate beads and hydroxyapatite matrix scaffold loaded with these drugs using the freeze gelation technique. This method demonstrated that co-delivery can prolong the release and synergistic effects of both drugs [70].

2.2 *Therapeutic Nucleic Acids*

These are charged high-molecular-weight nucleic acids or closely related compounds used for the treatment of specific diseases. They are broadly classified into DNA therapeutics and RNA therapeutics, while their action mechanism includes the suppression of certain endogenous DNA and RNA. Nucleic acids have enormous functional diversity, high specificity, and low toxicity and are suitable for therapeutic use, but as they are a novel concept, the approval for clinical use is complicated. Major clinical research efforts are underway to develop suitable delivery methods for these nucleic acids [12, 71]. There are several therapeutic nucleic acids commercially available such as *Vitravene* (for the treatment of cytomegalovirus retinitis) and *Nusinersen* (for the treatment of spinal muscular atrophy) [72].

DNA-based nanomaterials are stable, flexible, stimuli-responsive, and easy to modify as per the need [73]. In arthritis, the cyclooxygenase-2 (COX-2) gene is activated forming the COX enzyme that catalyzes the formation of prostaglandins, resulting in inflammation and pain. There are methods available for its suppression, but the impact is short-term. COX-2 antisense oligonucleotides gapmer sequences or COX-2 siRNA, when injected with modified HA/chitosan hydrogel, showed long-term suppression of the COX-2 gene in vitro [74]. Bone regeneration can be promoted by osteoblast differentiation. For this, gene-activated matrices were found to be useful, which have TGF-β1 DNA incubated on chemically linked bifunctional RGD-containing lysine-rich nonviral peptide sequence. RGD motif facilitates cell binding, and the lysine-rich motif facilitates TGF-β1 DNA binding to the biomaterial. Results showed the differentiation of bone marrow stromal cells to osteoblasts [75].

Damage in articular cartilage is one of the important features of osteoarthritis, and RNA interference (RNAi) has been researched extensively for its treatment. This involves the editing of targeted genes by short-interfering RNA (siRNA) delivered through various biomaterials. The function of the introduced siRNA is to bind with the endogenous mRNA from the faulty gene ("Gene Knock-down"), thereby reducing the rate of degeneration of the articular cartilage. The advantage of using RNAi therapy is that there are minimum or no side effects unlike other chemical inhibitors used for treatment. For example, the activity of miR-23a-3p promoted cartilage regeneration, which was loaded into stem cell extracellular vesicles and

Table 1.1 List of miRNAs that promote osteogenic and chondrogenic differentiation

Type	miRNA	Target Gene	Function	References
Osteogenic	miR-10b	SMAD2	Suppresses Smad2 protein and promotes osteogenic differentiation	[77]
	miR-877-3p	SMAD7	Promotes TGF-β1 mediated osteoblastic differentiation by silencing Smad7	[78]
	miR-200c	Sox2	Suppresses Sox2 leading to upregulating of Wnt activities	[79]
	miR-26a	GSK-3β	Activates Wnt signaling and promotes osteogenic differentiation	[80]
	miR-19b	Smurf2	Suppresses Smurf2 by Wnt signaling pathway	[81]
Chondrogenic	miR-9	NF- κB1	Promotes chondrocyte proliferation and prevents apoptosis	[82]
	miR-23b	Protein Kinase A (PKA)	Suppresses PKA signaling to promote chondrogenic differentiation	[83]
	miR-92a-3p	WNT5A	Suppresses WNT5A expression and inhibits cartilage degradation	[84]
	miR136-5p	ELF3	Promotes proliferation and migration of chondrocytes by decreasing ELF3 expressions	[85]
	miR-140	NF-κB	Inhibits inflammation and stimulates chondrogenesis by down-regulating activators of the NF-κB pathway	[86]

delivered with GelMA/nano clay for enhanced mechanical and biological performance. This system provides sustained release of miRNA-rich extracellular vesicles and thereby increases the regeneration of cartilage [76]. Similarly, there are many other miRNAs that can be targeted to promote osteogenesis, chondrogenesis, and myogenesis, some of which are presented in Table 1.1. These microRNAs are generally loaded onto polymers or metallic structures for their targeted and sustained release at the site of defect/injury. Generally, the polymer is the choice for the loading of these microRNAs because of their biocompatibility, mechanical strength, and non-toxicity, e.g., polycaprolactone (PCL) and poly-lactic acid (PLA). MicroRNA therapy is also used for treating pathologies in other components of the musculoskeletal system.

Tetrahedral Framework Nucleic Acid (TFNAs) is another new regenerative medicine technology widely researched as a drug delivery system. These nanostructured specialized DNA polyhedrons were first prepared by Turberfield et al., via a single-step process [87]. The structure consists of four designed single-stranded oligonucleotides, self-assembled at a 60° angle to each other (Fig. 1.6). The reason for their self-assembly is the complementarity rule proposed by Watson & Crick, as each ssDNA consisted of three segments that were complimentary to the segments of the other three ssDNA. Along with the precise control over the programmability of these frameworks, it has a few other traits as well, which include entering the cell without any resistance and nonimmunogenicity [88]. Several studies have demonstrated their various applications; for example, they promote cell proliferation and migration [89–91], osteogenic and chondrogenic differentiation [90, 92–94], decrease cellular ROS and pro-inflammatory cytokines, and reduce apoptosis [95, 96]. Their uptake is facilitated by caveolin present on the membrane, and then, they are internalized into lysosomes with the help of tubulin [88]. The ability of TFNAs to penetrate the cell membrane makes them a suitable drug carrier. Strong et al. loaded Wogonin, a natural flavonoid that has anti-inflammatory properties, inhibits matrix degradation, and promotes expression of chondrogenic genes in osteoarthritis as an intercalating agent onto the programmed TFNAs [97]. This resulted in the lowering of the inflammatory response, increase in collagen-II and aggrecan expression, and delayed progression of osteoarthritis in a rat knee model [92]. These were found to be suitable for the regeneration of damaged skeletal muscle, which was demonstrated by high-level PAX-7 gene (Paired box 7) expression, which is required for cell proliferation and myogenesis [98].

Bisphosphonate-induced osteonecrosis of the jaw (BIONJ) is a drug-related disease of the maxillofacial bone. One of the causes of the BIONJ is that bisphosphonates suppress osteoclast differentiation and maturation, thereby leading to a weakened load-bearing capacity of the facial bones (when bisphosphonates are orally administered). Although their prevention method is already discussed in the previous section that includes targeted delivery of bisphosphonates on the site, in case of patients already suffering from BIONJ. Precisely designed TFNAs demonstrated differentiation and maturation of the osteoclast cells via the Wnt/β-catenin signaling pathway [99] and also favored angiogenesis and polarization of macrophages to M2 phase (BIONJ promotes M1 polarization of the macrophages) [100]. As mentioned earlier, the anti-inflammatory response is also one of the major properties of TFNAs, which was further enhanced by loading curcumin in this system. In this composite, TFNAs' role was to enhance the uptake of the curcumin molecules by the cells and their sustained release [101]. TFNAs can be further modified with aptamers to target tumors and suppress their growth. Aptamer-TFN inhibits the cell cycle of tumor cells without affecting the normal cells [102].

2.3 *Peptide Drugs*

Proteins are the key molecules for the interaction between cells and biomaterials and are required for cell maturation, differentiation, and proliferation. Due to their large size, these are often used as a small sequence, having only active sites. These small sequences are called peptides and are easy to incorporate into the biomaterials [103]. The major problem associated with the usage of large protein molecules is the folding pattern, which cannot be manipulated easily, thereby making the active binding sites unavailable for the cells to bind and leading to adverse effects.

Osteogenic progenitor/stem cells form osteoblast (the most abundant cell type in bone tissue), and this differentiation is guided by specific growth factors.

Transforming growth factor-β (TGF-β) is a superfamily of pleiotropic proteins formed by interrelated dimeric polypeptide chains mainly involved in maintaining metabolic homeostasis of bone tissue [104]. Bone morphogenetic protein (BMP) is a growth factor family that plays a critical role in heart, neural, bone, and cartilage development in embryos. ***BMP-2*** and ***BMP-7*** are the most extensively researched for developing biomaterial-based therapies for clinical bone regeneration. BMPs bind with type-I and type-II serine/threonine kinase on target cells and activate *RUNX2* and *Osx*, the transcription factors for osteogenic genes such as *p21, RANKL, MMP2, MMP9, MMP13, VEGF, OP, and BSP* [105]. Osteoinductive properties of BMPs have been demonstrated using animal models with critical size-bone defects [106] and are also being researched in conjunction with other biomaterials for the treatment of fractures and other bone defects [107]. Recombinant human bone morphogenetic protein-2 (***rhBMP-2***)/dibotermin alfa and ***rhBMP-7*** are being used commercially as part of bone graft systems in Infuse®, Inductos®, Osigraft®, and OP-1 stryker. They have also been used with Col 1 Powder or gelatin capsules for treating bone defects [108].

Arginine–glycine–aspartic acid (RGD) is one the most common and specific sites for integrin receptors, which are key regulators of cell–cell interaction and cell–extracellular matrix communication. Hydrogels modified with RGD motifs have been researched for bone and cartilage regeneration [109]. BMP along with RGD sequence is another type of modification that can be used with scaffolds made from bioactive glasses, polyethylene terephthalate (PET), or self-assembled monolayers (SAMs), where RGD sequence facilitates stem cell binding and BMP promotes differentiation toward osteogenic lineage [103]. Along with RGD, Tyr-Ile-Gly-Ser-Arg (YIGSR)-bound PLGA films have shown enhanced attachment, proliferation, differentiation of immortalized murine myoblasts cell line, and topographic similarity to that of myotubes of human skeletal muscle [110]. Proteins of the TGF-β family are involved in the neo-cartilage formation and repair, specifically TGF-β3. It was found that TGF-β3 along with ghrelin promotes chondrogenic differentiation of mesenchymal stem cells, through a hydrogel-based delivery system [111].

2.4 Stem Cell Therapy

Stem cells are noncommitted, have a high proliferation rate, and can be differentiated into various cell types under specific conditions. Stem cells are the most promising type of therapeutics when it comes to regeneration. The most convenient part of using stem cells as a therapy is that they can be incorporated into a porous 3D scaffold; these biocompatible scaffolds provide mechanical support that promotes cell proliferation, migration, and ECM remodeling into specific tissue types.

Bone marrow therapy has been used clinically for bone regeneration by the introduction of mesenchymal stem cells directly at the site of injury. This therapy requires a long time for healing, as cells have to first differentiate into the osteogenic lineage and then into specific cell types for healing. An additional step before injecting cells can be in vitro pre-differentiation of mesenchymal stem cells into the osteogenic lineage. Incorporation of growth factors such as platelet lysate, with the differentiated stem cells has been shown to result in enhanced bone volume. But this additional step in the procedure has certain complications such as infection [112]. Bone regeneration can also be affected by the use of magnetic field. This therapy is called magneto therapy, which is a noninvasive, safe, and easy-to-use technology for treating bone fracture, osteoarthritis, spinal fusion, and wound healing [113].

The articular cartilage tissue is affected during joint inflammation in case of arthritis and other chronic diseases. These can be treated with the help of mesenchymal stem cells delivered with the help of different scaffolding materials. Cartilage tissue regeneration is one of the most studied topics, and various material platforms have been used for delivering stem cells. Macrophages have been found to be a key factor in muscular regeneration and along with monocytes are the first cells to be accumulated at the site of injury-induced inflammation [114]. Tissue regeneration is dependent on different stem cells recruited at the site of injury, and macrophages influence the stem cells during the regeneration process, particularly their polarization into inflammatory phenotype promotes myogenesis [115].

3 Prerequisites for Musculoskeletal Drug Delivery Systems

There are local drug delivery methods, such as intra-articular (IA) injection for synovial joint tissues, epidural injection for targeting IVD, and subacromial injection for shoulder bursa, which are mostly ineffective. For example, the half-lives of corticosteroid injections, which are used to treat localized pain and inflammation, only work for a few hours, thus requiring multiple injections of high medication doses [116]. Therefore, to restore the anatomy and physiology of damaged tissue, advanced drug delivery systems are required with prolonged retention and target localization. In the case of musculoskeletal tissues, such as cartilage, meniscus, and ligaments of the knee, ankle, shoulder, hip joints, and intervertebral discs (IVDs) in the spine, the delivery of drugs is still a challenging task due to the characteristic dense extracellular matrix and complex mechanical loading environment, which makes the diffusion and penetration of drugs difficult [117]. In addition to this, most of the existing drug delivery systems have limited loading capacity and lack sustained as well as the on-demand release of drugs. Meanwhile, constructing an effective delivery system for different musculoskeletal disorders may bring specific difficulties. In this perspective, mechanical and microenvironment-based requirements should be considered while developing a carrier system for the repair of musculoskeletal tissues.

3.1 Microenvironment-Based Requirements

Musculoskeletal tissues are characterized by a complex extracellular matrix composed of a dense meshwork of collagen fibrils, proteoglycans (PGs), containing highly negatively charged glycosaminoglycan chains, and cells that create barriers for drugs or drug carriers. For example, a dense network of type II collagen fibers and aggrecan make cartilage a barrier for drug penetration. However, for the availability of nutrients, growth factors, and other compounds, matrix permeability at various length scales is required. Studies have shown that the diffusion rate also depends on the pore size and connectivity of the matrix, which in turn determines the function of tissue structure [118]. Transport of nutrients, metabolites, and waste products is also restricted by pathophysiologic tissue changes, such as calcification of diseased cartilage. Therefore, the size of the delivery system should be designed according to the disease site. The very first requirement while designing the drug delivery system for musculoskeletal tissue is the size of the delivery system [119]. While treating musculoskeletal disorders, micro-/nano-based drug carriers are the most widely used ones as they incorporate drugs on the surface or inside the matrix and thus prevent enzymatic degradation of drugs and enhance penetration inside the extracellular matrix. The articular cartilage is 3–4 mm thick and mainly divided into five distinct zones (each zone has distinct matrix porosity), where the size of the particles and the pores of the cartilage matrix determine the depth at which micro/nanoparticles can diffuse. Due to the tight connective tissue structures in tendons and ligaments, which include a lot of thick collagen fibers that could prevent drug diffusion and penetration, local drug delivery is also constrained in these tissues [37].

While treating various musculoskeletal conditions such as osteoarthritis and rheumatoid arthritis, direct intra-articular injection delivers a high concentration of therapeutics in the joint spaces. However, the efficacy of the therapeutics reduces drastically due to the rapid clearance of the drug [117]. Furthermore, various small molecule-based drugs have limited action duration. Therefore, there is a need to design sustained and controlled drug delivery systems that in turn extend the action duration without causing side effects [120]. Numerous synthetic and natural biomaterials-based drug carriers have been employed to encapsulate a wide range of therapeutics that prevent the drug from in vivo degradation and rapid clearance as well as promote controlled release. There are mainly four categories of materials that are used as drug vehicles:

- *Natural materials* include chitosan, silk, collagen, alginate, and hyaluronan. These materials are biocompatible and bioresorbable but can be immunogenic [121, 122].
- *Inorganic materials* include calcium phosphate and bioglass. These materials mimic the bony structural composition [123–125].
- *Synthetic materials* include PLA, PLGA, and PCL. This class is used as biodegradable implants and drug delivery systems in clinics. They now hold a

prominent place in the field of degradable materials for orthopedic applications [126, 127].

- *Composite materials* include a combination of the above-mentioned categories in which the advantage of one is utilized to optimize other materials. For example, ceramic–polymer-based drug delivery systems have been reported for bone tissue regeneration where ceramic-based scaffold provides mechanical strength while the coating of diclofenac functionalized biopolymer poly(3-hydroxyoctanoate)-P(3HO) is used for controlled drug release [128, 129].

Therefore, materials used in drug delivery systems should fulfill essential requirements such as high biocompatibility and reduced immunological reactivity with no side effects. Among all, biodegradable polymers are the most interesting and commonly used as they do not need to be removed after treatment and degrade into nontoxic products. Even for a regulated drug release, the process of polymer degradation is crucial. There are various ways of degrading polymers such as thermal, mechanical, chemical, or enzymatic. In mechanical loading environments such as bone and cartilage, mechanical stress affects the degradation of the polymer [130]. However, biodegradable polymers are more prone to chemical degradation through either hydrolysis or enzyme-catalyzed hydrolysis [131]. The degradation rate is higher in the vascularized environment in comparison with the avascular region. Therefore, understanding the behavior of polymer degradation is a prerequisite so that drug-releasing kinetics can be regulated to obtain a therapeutic concentration and avoid toxic concentration.

The efficacy of drug delivery systems can be significantly strengthened by the regulated release of therapeutic compounds in response to pathologically relevant biochemical signals including low pH, elevated ROS, and overexpressed enzymes [132, 133]. In arthritis patients, local inflammatory reaction in joint tissues increases metabolic activity, which results in oxygen deprivation and shifts toward anaerobic glycolysis and lactate production. As a result, the microenvironment becomes more acidic; for example, pH values as low as 6.0 have been seen in the synovial tissue of rheumatoid arthritis patients [134]. Hence, for arthritis-affected sites, a smart pH-stimuli-responsive drug delivery system is required in order to administer medication without unwanted early drug leakage.

Overproduction of degradative enzymes is induced by inflammatory arthritis, which is partially due to the response to downstream pathways initiated by a variety of stimuli, including ROS production and an acid microenvironment [135]. The typical degradative enzymes in inflammatory arthritis include matrix metalloproteinases (MMPs), disintegrins, and metalloproteinase with thrombospondin motifs (ADAMTS). In synovial tissue at the edge of hyaline cartilage in the diseased joint, MMP-3 production is noticeably elevated [136]. Similar to this, it has been demonstrated that elevated levels of MMP-1, 2, 9, and 13 in the blood and synovial fluid of OA patients coordinate the breakdown of ECM proteins in articular cartilage [135, 137]. In order to create a carrier material responsive to MMP-12 in the inflammatory microenvironment of joints loaded with corticosteroid triamcinolone acetonide, Joshi et al. suggested a nanoparticle-hydrogel hybrid system (NP-gel) integrated with TG-18, a tiny hydrophilic and lipophilic chemical (TA) [138]. This system showed exemplary control over distant drug release kinetics and "on-demand" drug release. The specific release mechanism involved interactions between the MMPs' Zn (II) ions and external water molecules, which resulted in the formation of Zn (II)-bound hydroxides. These hydroxides served as nucleophiles to attack the carboxyl esters in TG-18, which led to the cleavage of the ester bond, the breakdown of the hydrogel, and the release of TA. Moreover, the systemic administration of enzyme-responsive growth factor delivery systems was studied by Qi et al. using nano-capsules that were sensitive to MMPs and caused selective breakdown for bone regeneration [139].

Targeted drug delivery continues to face significant challenges due to the abundance of dense, avascular negatively charged tissues found in the musculoskeletal system. Examples of such tissue include cartilage, meniscus, and ligament of the knee, shoulder, ankle, hip joints, etc., and the intervertebral disc of the spine. Degradation of these tissues is associated with several diseases such as osteoarthritis (OA), gout (synovial joints), spondylitis, and cervical pain (spine). The dense network of collagen fibrils, proteoglycans (PGs), which include highly negatively charged glycosaminoglycan (GAG) chains, and cells that make up the complex ECM of such avascular tissues acts as a barrier to entry for most drug or drug carriers. Particularly, the complex meshwork of type II collagen in cartilage, which is tightly packed with aggrecans, significantly resists the penetration of therapeutic compounds or carriers. Like cartilage, the meniscus also has a similar ECM structure, but it has lower fixed charge density (FCD) than cartilage (dry tissue weight aggrecan content of 20% against 35% in cartilage). The majority of the negatively charged PG gel that makes up the nucleus pulposus (NP) of IVDs is held loosely together by a sparse network of type II collagen fibrils. It is challenging for medications injected epidurally to permeate the NP via passive diffusion because the NP is covered by a thick fibrous annulus fibrosus (AF) layer with circumferentially aligned collagen. A high negative charge of tissues can be used to modulate therapeutics or their carriers by including optimally charged cationic domains so that electrostatic interaction facilitates their transportation. There are various studies showing cartilage penetrating cationic peptide-based carriers for drug delivery [140].

Vedadghavami et al. demonstrated, using short-length cationic peptide carrier motifs, that there exists an optimal net positive charge to deliver a drug or drug carrier of a given size to tissue of known fixed charge density. This will cause rapid penetration through the full thickness of cartilage before most of the drug is cleared from the joint space and can enable the highest intra-cartilage uptake and long-term retention. They demonstrated that intra-cartilage absorption increased with an increasing net charge of peptides to +14 but decreased as the charge went further because tighter binding interactions hampered peptide penetrability and uptake [141]. Thus, understanding the microenvironment of the site of delivery for the drug is of utmost importance. The drug delivery systems need to be designed according to the microenvironment and how it changes over time in case of diseases. These material systems must be stable and functional in the prevailing microenvironments of the musculoskeletal tissues.

3.2 Mechanical Requirements

The viscoelastic musculoskeletal system, which includes cartilaginous tissues, tendons, ligaments, and bone, is known for its mechanical properties. Each of these tissues can respond to the mechanical loads, therefore, triggering many cellular processes such as differentiation, matrix remodeling, and metabolism [142]. There are mainly two mechanisms by which mechanical stimuli are delivered to cells namely external forces (macroscopic forces) and cell contractility interactions with the matrix substratum (microscopic forces). External forces mainly include compression, tensile strain, shear stress, and hydrostatic pressure that are required for the formation of specific musculoskeletal tissues [143]. During growth and development, low to moderate tensile strains and low hydrostatic stresses support bone development; high hydrostatic pressure promotes cartilage formation; large tensile strains cause fibrous tissue formation; and hydrostatic pressure and high tensile strain together stimulate fibrocartilage formation. Therefore, tissue physical characteristics and biological functions are related to each other. The stiffness and mechanical strength of biological specimens can be determined by measuring both the force applied to a tissue and its deformation simultaneously. In general, the ultimate strengths of the bone produced are typically 200 MPa in compression, 125 MPa in tension, and 65 MPa in shear [144]; tendon is usually about 100 MPa [36].

Diseased tissues have compromised mechanical properties, which can be restored by tissue regeneration. For the development of an ideal carrier system for musculoskeletal tissue regeneration, vast research has been performed on material type and their processing conditions. In order to enhance the efficacy of treatment, researchers are utilizing tissue engineering and drug delivery strategies together [145]. A variety of drug delivery matrices and scaffolds have been used, such as porous scaffolds, nanofibrous membranes, microparticles, and hydrogels. These all have been synthesized using natural and synthetic polymers, inorganic materials, and their composites in which different kinds of proteins or growth factors are incorporated through covalent or noncovalent binding. In musculoskeletal degeneration conditions, the delivery of growth factors contributes to the synthesis of the extracellular matrix that can bear extreme loading in the body. Cells are also introduced into biomaterial carriers, or they are attracted by signaling molecules or growth factors. In addition to growth factors, mechanical forces are also known to affect tissue repair by stimulating extracellular protein production, which in turn impacts the structure of the tissue.

Elastic modulus is one parameter for the measurement of tissue stiffness. It is an intrinsic resistance to deformation under mechanical force. For example, bone is strong to protect organs and resist significant mechanical pressures while cartilage is softer and acts as a shock absorber. Therefore, tissue dysfunction has been linked to changes in tissue stiffness as a result of trauma (scarring), illness, or age. The stiffness of the substrate is strongly regulated by physiological levels in stem cell function [146]. Mesenchymal stem cells generated from human bone marrow (hBMSCs) have been shown to preserve their multipotency and remain dormant on flexible substrates with marrow-like flexibility. Surprisingly, on substrates with elasticity simulating the tissue modulus of the brain, muscle, and precalcified bone, respectively, hBMSCs preferentially develop toward neurogenic, myogenic, and osteogenic lineages [147].

Therefore, while developing drug delivery systems, smart biomaterials or composite biomaterials can be synthesized by modifying the bulk or surface properties that can resemble native tissue in terms of physical properties, e.g., stiffness. Tunable hydrogels for musculoskeletal tissue engineering and drug delivery have been introduced as a desirable alternative to invasive therapies. In a single composition, the density or length of polymer chains can be modulated in order to adjust the elastic property. This can also be done by altering the molecular weight of the polymers, the degree of cross-linking, or the water content in the hydrogels. However, because of the limited innate physical properties of a single composition, the level of modulation cannot be broad. Therefore, composites either between polymers or between polymers and inorganics—is a more attractive strategy. Recently, biomaterials with tunable physical properties have been developed, including collagen, glycosaminoglycans (GAGs), poly(acryl amide), and poly(ethylene glycol). Ziadlou et al. optimized hyaluronic acid-tyramine/silk-fibroin composite hydrogels as anti-inflammatory drug carriers and chondrocyte encapsulating matrices for cartilage tissue regeneration [148]. Furthermore, they have demonstrated that all hydrogel formulations were cytocompatible and could stimulate the expression of cartilage matrix proteins, allowing chondrocytes to form ECM. Hydrogels with HA20/SF80 polymeric ratios showed the most prominent chondrogenic effects. After 28 days of development in chondrogenic media, unrestricted mechanical testing revealed that the compressive modulus for HA20/SF80 chondrocyte-laden constructs increased about ten-fold. The increased polymer content in HA20/SF80 hydrogels led to more crosslinking sites being available, and the higher crosslink density (di-tyramine, di-tyrosine, and tyramine-tyrosine) produced a higher storage modulus (4 KPa) in comparison to other hydrogels in this study. The highest polymer content of any of the hydrogels examined was found in HA20/SF80 hydrogels, which also helped explain their higher storage modulus. The strain sweep revealed that HA0/SF100 and HA5/SF95 composite hydrogels could withstand more than 100% of strain before failing mechanically and that the strain at failure could be gradually decreased by raising the concentration of HA-Tyr in the composite hydrogels. Comparing the composite hydrogels to HA100/SF0 hydrogels, which began to break at less than 10% strain, a considerable improvement was seen due to their mechanical qualities. Composite hydrogels may be appropriate for tissue engineering of joint cartilage, which typically experiences physiological stresses of about 20% strain. In addition to this, Wu et al. designed and developed poly(p-dioxanone) (PPDO)/silk fibroin (SF) wavy nanofibrous scaffold (WNS) to promote tenogenesis for tendon tissue engineering as fibrous structure mimics collagen-based fibril architecture of native tendon extracellular matrix [149]. The ultimate tensile strength of WNSs, which was shown to be in the same range as the natural tendons (31 ± 1 MPa for PPDO WNSs, 25 ± 1 MPa for 4/1 PPDO/SF WNSs, and 17 ± 1 MPa for 2/1 PPDO/SF WNSs), is quite significant. For instance, for human Achilles tendons, human patellar tendons, and human rotator tendons, the ultimate strength was shown to be 28–86 MPa, 5–65 MPa, and 14–45 MPa, respectively. Unfortunately, the initial modulus of WNSs in the toe region was found to be lower than that of natural tendons (range: 40–400 MPa) (15 ± 4 MPa for pure PPDO WNSs, 28 ± 7 MPa for the 4/1 PPDO/SF WNSs, and 43 ± 6 for the 2/1 PPDO/SF WNSs). Therefore, modifying material selection is necessary to improve the structural and mechanical characteristics of WNSs.

As the basic function of the musculoskeletal tissues is to provide mechanical support and movement to the whole body, the drug delivery systems being developed for their treatment must comply with these prerequisites. The material systems need to be both stable and functional in these prevailing physiological and biomechanical regimens in order to provide successful drug delivery.

4 Drug Delivery Systems

A drug delivery system (DDS) can be defined as a formulation or device that aims to improve the specificity of therapeutics by sustaining drugs in vivo, modulating their release, and targeting their effects in a specific area. The efficacy of the therapeutics depends on numerous aspects mostly related to the delivery route [150]. A variety of natural polymers, such as silk fibroin, collagen, gelatin, cellulose, starch, etc., have been used for the fabrication of drug delivery systems [151–153]. Natural materials outperform artificial materials in terms of biocompatibility, accessibility, and convenience of modification. Various functional groups can potentially be introduced due to the presence of reactive groups on the natural materials, offering them varied functions or altering their physical and chemical properties [152, 153]. There are mainly two routes of administration of drugs—systemic and local. Systemic administration refers to the administration of drugs into the blood or circulatory system so that it reaches every part of the body and the whole body shows the effects of the drug, whereas local administration refers to the drug being applied or injected to the injured or local site. Depending upon the site of administration, they can also be classified as oral, subcutaneous, ocular, transdermal, nasal, vaginal, and urinary delivery systems [154]. The major limitations of systemic drug delivery is that most drugs are not delivered in a sustained or on-demand manner and are not released at the proper site. Local drug delivery, on the other hand, can precisely regulate and maintain relevant drug concentrations where they are demanded while effectively reducing toxicity and undesired side effects [155]. Local drug administration offers an appropriate dose of medication and accomplishes enhanced therapeutic efficacy by targeting diseased/injured musculoskeletal tissues in comparison to systemic drug delivery [156].

Musculoskeletal diseases like arthritis, metabolic disorders, etc., and injuries caused due to trauma are a big socioeconomic burden for the world. A variety of local and systemic therapeutic treatments are available for musculoskeletal disorders. The biological activity of the majority of therapeutic drugs can be influenced

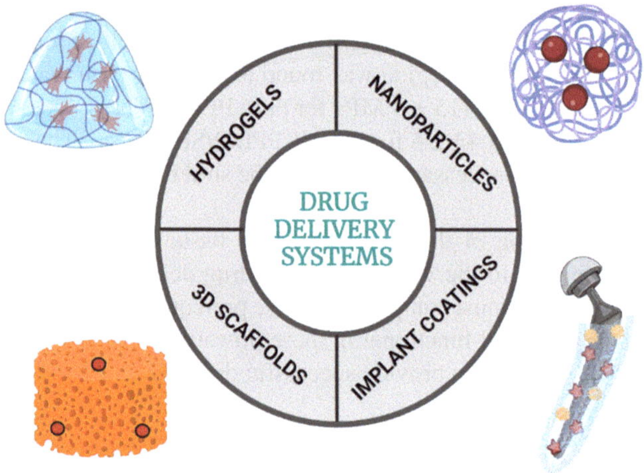

Fig. 1.5 Material systems for musculoskeletal drug delivery systems

by factors like pH, temperature, and other factors, which might limit their clinical efficacy. Local drug delivery systems are preferred for musculoskeletal tissues, particularly those that can offer drug concentrations within desired therapeutic windows for a specific period of time [156]. Various material systems such as dendrimers, nano/microparticles, degradable polymeric vehicles, swelling-controllable hydrogels, affinity-based delivery vehicles, implant coatings, and three-dimensional (3D) scaffolds have been explored and examined as local delivery vehicles of therapeutic drugs to treat musculoskeletal disorders (Fig. 1.5) [157–159].

4.1 Hydrogels

Hydrogels are 3D networks formed by the cross-linking of hydrophilic polymer chains with a high water content, which allows the efficient exchange of nutrients and oxygen, providing structural support for cell growth and tissue-like properties [160]. A suitable hydrogel should depict a brief gelation time, high water content, controlled and slow degradation, pH stability, and adequate mechanical properties to withstand the compressive and shear stresses of the musculoskeletal tissue [161]. Owing to the distinctive physical properties of hydrogels, they shield drugs from degradation, offer controlled dissolution, and regulate the release of cells, drugs, growth factors, etc. Therefore, hydrogels are suited for use in various drug delivery systems [162]. Hydrogels can be classified based on source, synthesis methods, cross-linking, the ionic charge on them, physical properties, etc.

Several polymeric materials used for the synthesis of hydrogels can be classified into natural, synthetic, and hybrid/semi-synthetic. Hydrogels derived from natural sources are excellent candidates for biomedical applications as they provide sticky surfaces for cell adhesion, have excellent swelling characteristics, non-toxic by-products, and are generally biocompatible and biodegradable. But poor stability, inadequate mechanical strength, and hastened degradation limit their usage [160]. Naturally derived hydrogels are further classified as protein-based hydrogels, polysaccharide-based hydrogels, and protein-polysaccharide-based hydrogels. Because of their excellent drug-eluting properties, hydrogels have been used widely for musculoskeletal drug delivery purposes (Table 1.2).

4.2 Nanotechnology-Based Drug Delivery Systems

There are various disadvantages of conventional methods of drug delivery for the treatment of musculoskeletal disorders, such as low targeting efficiency, difficulties in drug stability during transportation, prolonged drug release, and releasing drugs into individual cells, etc. [181]. The use of nanoparticles/microparticles overcome these limitations. Particles having a size in the range of 1–100 μm can be considered microparticles, whereas nanoparticles have a size in the range of 1–100 nm. There are various nano-biomaterial systems developed for the purpose of drug delivery. These nano-biomaterial-based structures include nanoparticles, dendrimers, exosomes, microparticles, nanotubes, and liposomes (Fig. 1.6). Nanoparticles (NPs) enhance drug release efficiency, target specificity, combined diagnosis and therapy and combat multi-drug resistance. NPs offer various advantages over conventional treatments in terms of better insoluble drug delivery, greater bioavailability, fewer side effects, and improved therapeutic efficacy. They enhance the medicines' half-life and preserve them from deteriorating. The persistent release of drugs can be facilitated by NPs functioning as local drug depots [182]. Dendrimers are multivalent macromolecules that resemble trees and have a well-defined, well-branched structure. They are adopted as nanocarriers for the delivery of a variety of drugs and bioactive substances [183]. One such example is cationic poly (amidoamine) PAMAM dendrimer that can be used for the delivery of drugs, growth factors, and small RNAs. Liposomes are one of the earliest and safest nanocarriers for administration. It consists of an exterior bilayer to carry lipophilic substances and a hydrophilic core to encapsulate hydrophilic drugs. Liposomes typically range in size from 80 to 200 nm and have the ability to transport a variety of drugs. Liposomal encapsulation improves drug effectiveness while limiting toxicity [184]. Despite being one of the most effective drug delivery systems, liposomes have disadvantages such as drug pre-leakage and rapid or insufficient drug release.

Micelles are self-assembling, amphiphilic nanocarriers that can transport drugs to the preferred location. Poorly water-soluble drugs can be incorporated into these nanocarriers to enable sustained or regulated drug release. Drugs in micelles are safeguarded from degradation and are cleared quickly [117]. Micelles are very attractive for both systemic and local drug delivery because of their modest size between 50 and 100 nm, sustained drug release kinetics, and controlled degradation under physiological conditions [185]. Exosomes are non-immunogenic

Table 1.2 List of hydrogels-based drug delivery systems in musculoskeletal disorders

Hydrogel system	Drug/cells/growth factors	Disease	References
Fibrin-HA composite hydrogel	Dexamethasone (DMX), galectin-3 inhibitor	Inflammatory joint diseases	[163]
Chitosan nanoparticles incorporated in silk fibroin hydrogel	Dexamethasone sodium phosphate	Arthritis	[164]
Silk fibroin hydrogel	*Sesbania sesban L.* extract	Rheumatoid arthritis	[164]
Alginate hydrogel	Transforming growth factor-β1 and Bone morphogenic protein 4	Osteochondral defects	[165]
Hybrid PEG-HA hydrogel	Iguratimod	Rheumatoid arthritis	[166]
Hyaluronic acid-hydrazide hydrogel	Infliximab	Rheumatoid arthritis	[167]
Dextran hydrogel	Adult chondrocytes or adipose-derived stem cells (ADSCs)	Cartilage regeneration	[168]
Tyramine-modified gellan gum (Ty-GG) hydrogels	Betamethasone	Rheumatoid arthritis	[169]
Pullulan-poly(ethylene glycol) hydrogels	Dexamethasone	Bone tissue regeneration	[170]
Poly(organophosphazene) – celecoxib nanoparticle hydrogel system	Celecoxib	Achilles tendon regeneration	[171]
Sericin hydrogel	Anti-RelA siRNA	Rheumatoid arthritis	[172]
Poloxamers hydrogel	Glucosamine	Osteoarthritis	[173]
GEL-MAN hydrogels	Triamcinolone acetonide	Osteoarthritis	[174]
mPEG-Polyalanine Hydrogel	Quercetin	Osteoarthritis	[175]
PLGA–PEG-PLGA Thermogel	Kartogenin	Osteoarthritis	[176]
Polyurethane Hydrogel	Diclofenac	Osteoarthritis	[177]
Carboxymethyl hexanoyl chitosan-hyaluronic acid Hydrogel	Berberine	Osteoarthritis	[178]
Hyaluronic acid-Methacrylate Hydrogel	Cordycepin	Osteoarthritis	[179]
Polyethylene glycol (PEG)-hydrogel	Agomir874	Intervertebral disc regeneration	[180]
Triglycerol monostearate hydrogel	Triamcinolone acetonide	Rheumatoid arthritis	[138]

membrane-based vesicles that can be an outstanding drug-delivering agent due to their high biocompatibility, good targeting, low toxicity, and inherent therapeutic potential. Exosomes offer a better delivery strategy to treat knee injuries and diseases because they have the same biological capabilities as their originating cells but lack immunogenicity and tumorigenicity [186]. It has been found that the size variation of drug carriers significantly affects blood circulation times and drug

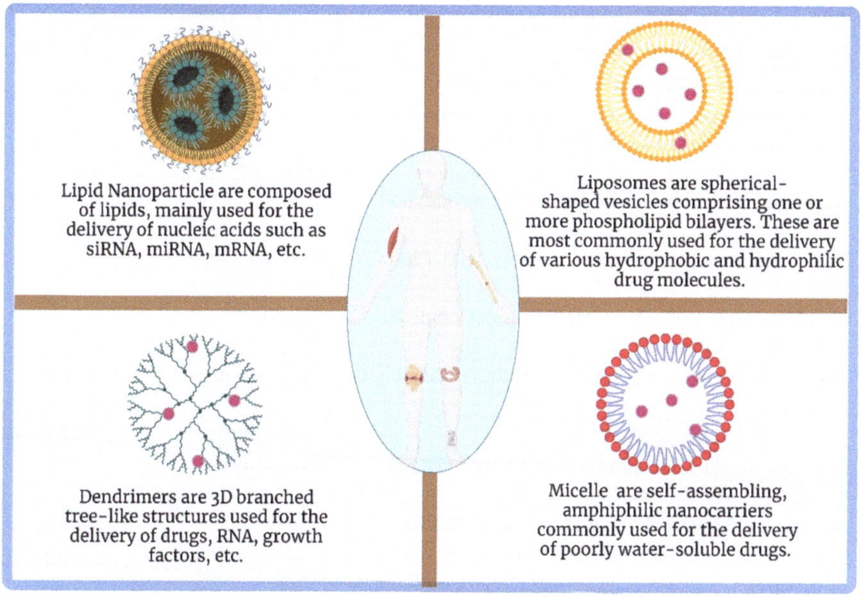

Fig. 1.6 Various nanotechnology-based drug delivery systems

absorption [187]. It has been recommended that drug carriers made up of nanomaterials should have a desirable size range of 10–100 nm [187]. Apart from particle size, the effect of nanocarrier shapes and architectures has also been researched. In particular, tubes, cages, shells, and other nanoscale hollow structures have shown excellent control over drug loading and release timing.

When it comes to targeted delivery of drugs for the musculoskeletal system, nanoparticles (NPs) can provide unique properties such as (i) improved drug-loading capacity, (ii) enhanced solubility of drugs, (iii) provided a protective barrier that increases medicinal stability and idrug retention time, (iv) targeted distribution, on-demand release, and minimal systemic detrimental effects on adjacent tissues or organs, and (v) enhanced capacity to transport through cell membranes, allowing for the administration of drugs into cells or certain organelles [188]. Because of these properties, nanoparticles have been widely used and researched for the treatment of musculoskeletal pathologies (Table 1.3). The safety of nanoparticles, including material toxicity, dosage toxicity, and metabolism rate, still remains a challenge.

Table 1.3 List of nanomaterial/nanoparticles used for drug delivery in musculoskeletal disorders

Nanomaterial/drug combination	Pathology	References
Cisplatin prodrug -Zn @PEG-Alendronate NPs	Bone cancer	[189]
Calcium phosphate nanoparticles functionalized with alendronate-conjugated polyethylene glycol (PEG)	Bone metastasis	[190]
Bisphosphonate-Functionalized Hydroxyapatite Nanoparticles for the Delivery of the Bromodomain Inhibitor JQ1	Osteosarcoma	[191]
Doxorubicin-loaded liposomes, modified with YSA-peptide to target EphA2	Osteosarcoma	[192]
Nanoparticulate Hydroxyapatite/Chitosan Composites loaded with antibiotics	Osteomyelitis	[193]
PDA-coated Gd2(CO3)3 nanoparticle for the delivery of hesperetin	Osteoarthritis	[194]
p66shc-siRNA-loaded PLGA NPs	Osteoarthritis	[195]
Etoricoxib-loaded bio-adhesive hybridized polylactic acid-based nanoparticles	Osteoarthritis	[196]
Kartogenin-conjugated polyurethane nanoparticles	Osteoarthritis	[197]
Hollow mesoporous silica nanoparticles for the delivery of celastrol	Osteoarthritis	[198]
Silk fibroin nanoparticles for celecoxib and curcumin delivery	Osteoarthritis	[199]
Silver nanoparticles for the delivery of Bauhinia acuminate flower extract	Meniscus injury healing	[200]
PLGA nanoparticles-mediated co-delivery of bFGF and VEGFA genes	Flexor tendon healing	[201]
Folate-conjugated BSA nanoparticles for the delivery of Etoricoxib	Rheumatoid arthritis	[202]
Poly-L-lysine Nanoparticles covalently bound to beta-cyclodextrin with the drug Bisphosphonate (Risedronate)	Cancer bone metastasis	[203]
PLGA Nanoparticles for the drugs Bisphosphonate (zoledronic acid) and doxorubicin	Cancer bone metastasis	[204, 205]
Poly (ethylene glycol) nanoparticles bearing paclitaxel and alendronate for the drug Aminobisphosphonate	Cancer bone metastasis	[206]
Polymerized liposomal nanoparticles for the drug Doxorubicin	Osteosarcoma	[207]
Magnetic arsenic trioxide nanoparticles for the drug Arsenic trioxide	Osteosarcoma	[208]
Calcium phosphate nanoparticles for the drug Cisplatin	Osteosarcoma	[209]
Lipid-modified dextran-based polymer nanoparticles for the drug Doxorubicin	Osteosarcoma	[210]
Dextran-PEI (Polyethylenimine) nanoparticles for the drug Doxorubicin	Osteosarcoma	[211]
PLGA-calcium phosphate Nanoparticles for the drug Tigecycline	Infection and inflammatory bone diseases	[212]

4.3 Implant Coatings

Implant coatings can be used as a drug delivery system for muscle and bone defects along with hydrogels and nanoparticles. Intramuscular or subcutaneous injections are widely used to administer drugs locally and steadily for treating muscle ailments. Implants or 3D scaffolds coated with drugs are employed mainly to treat bone deficiency or injury. Implant coating has attracted a lot of attention compared to other methods of drug delivery because it combines the merits of simplicity and versatility [156]. One disadvantage is that the structural function of delivery vehicles may be hampered by their poor mechanical characteristics [213, 214]. An implant should be placed at the injury site to offer mechanical stability as well as the missing biologically active compounds like drugs, growth factors, etc., which are necessary to help in the healing process. These can be provided by coatings enclosing drugs, growth factors, etc. on the implant surface. Several biologically active molecules are evaluated as promising options for coating implant surfaces with drugs and growth factors. Various coatings are inspired by the organic (extracellular matrix and proteins) and inorganic (hydroxyapatite) elements of bone or by a mixture of elements that closely resemble bone [215]. Coatings can be mainly divided into inorganic coatings and organic coatings [216]. They are also useful for enhancing the performance of metallic implants for bone fractures/dental restoration, which are the most commonly used implants worldwide.

4.4 3D Scaffolds

In several in vitro and in vivo tissue repair and growth scenarios, 3D scaffolds function as temporary substrates for supporting and directing tissue growth [217]. In addition to providing structural support, they can serve as a drug delivery system to deliver conventional drugs, growth factors, or cytokines. Scaffolds with the potential to physically/chemically incorporate drugs and then to locally deliver drugs at the injured site in a controlled fashion are highly desirable for treating musculoskeletal injuries and diseases [156]. Biodegradable polymers are one of the many materials that could be utilized for the manufacturing of 3D scaffolds. The main advantages of using these polymers for the development of 3D scaffolds are good biodegradability and biocompatibility, excellent plasticity, easy moldability, easy surface modifications, less toxicity, etc. [218]. Both natural and synthetic polymers have been widely used for drug delivery purposes. The most common strategies for incorporating growth factors/drugs in scaffolds are noncovalent and covalent immobilization, the choice of which depends on the biophysical and biochemical characteristics of the growth factors, their interactions with the polymers which are being used for drug delivery, and the kind of injury [219].

5 Outlook

The present chapter introduced musculoskeletal tissues and their developmental pathways, which are essential to understand when trying to devise a system that is meant to heal/regenerate the tissue back to its normal state. The pathophysiology of the diseases affecting the tissue for which a drug delivery system is being designed also needs to be understood, as the system needs to be stable and functional in the physicochemical conditions of the diseased tissue, which, in a lot of cases, do change with the stage of the disease. Specifically, for musculoskeletal tissues, the drug delivery system also needs to either match or be stable in the physiological loading regimes in these tissues, without which there will be high chances of failure. The subsequent chapters will focus on a particular musculoskeletal tissue, the associated pathologies, and different drug delivery systems that are being developed for treating these pathologies.

References

1. F.F. Safadi, M.F. Barbe, S.M. Abdelmagid, M.C. Rico, R.A. Aswad, J. Litvin, S.N. Popoff, Bone Structure, Development and Bone Biology, 2009. https://doi.org/10.1007/978-1-59745-347-9.
2. X. Lin, S. Patil, Y.G. Gao, A. Qian, The Bone Extracellular Matrix in Bone Formation and Regeneration, Front. Pharmacol. 11 (2020) 1–15. https://doi.org/10.3389/fphar.2020.00757.
3. M.S. Carvalho, J.M.S. Cabral, C.L. da Silva, D. Vashishth, Synergistic effect of extracellularly supplemented osteopontin and osteocalcin on stem cell proliferation, osteogenic differentiation, and angiogenic properties, J. Cell. Biochem. 120 (2019) 6555–6569. https://doi.org/10.1002/jcb.27948.
4. G.L. Koons, M. Diba, A.G. Mikos, Materials design for bone-tissue engineering, Nat. Rev. Mater. 5 (2020) 584–603. https://doi.org/10.1038/s41578-020-0204-2.
5. R.K. Fuchs, W.R. Thompson, S.J. Warden, 2 – Bone biology, Second Edi, Elsevier Ltd, 2019. https://doi.org/10.1016/B978-0-08-102451-5.00002-0.
6. A.C. Karaplis, Embryonic Development of Bone and Regulation of Intramembranous and Endochondral Bone Formation, Princ. Bone Biol. Two-Volume Set. 1 (2008) 53–84. https://doi.org/10.1016/B978-0-12-373884-4.00025-2.
7. A.D. Berendsen, B.R. Olsen, Bone development, Bone. 80 (2015) 14–18. https://doi.org/10.1016/j.bone.2015.04.035.
8. E.J. Mackie, Y.A. Ahmed, L. Tatarczuch, K.S. Chen, M. Mirams, Endochondral ossification: How cartilage is converted into bone in the developing skeleton, Int. J. Biochem. Cell Biol. 40 (2008) 46–62. https://doi.org/10.1016/j.biocel.2007.06.009.
9. Y. Krishnan, A.J. Grodzinsky, Cartilage diseases, Matrix Biol. 71–72 (2018) 51–69. https://doi.org/10.1016/j.matbio.2018.05.005.
10. R.F. Ker, THE DESIGN OF SOFT COLLAGENOUS LOAD-BEARING TISSUES, 3324 (1999) 3315–3324.
11. Abousleiman, R. I., V.I. Sikavitsas, BIOREACTORS FOR TISSUES OF THE MUSCULOSKELETAL SYSTEM, in: In Advances in Experimental Medicine and Biology, 2000: pp. 243–259. https://doi.org/10.1007/978-0-387-34133-0_17.
12. S. Bliss, Musculoskeletal Structure and Physiology, (2018) 43–71.
13. H. Alizadeh Sardroud, T. Wanlin, X. Chen, B.F. Eames, Cartilage Tissue Engineering Approaches Need to Assess Fibrocartilage When Hydrogel Constructs Are Mechanically

Loaded, Front. Bioeng. Biotechnol. 9 (2022) 1–19. https://doi.org/10.3389/fbioe.2021.787538.
14. X. Wang, L. Rackwitz, N. Ulrich, R.S. Tuan, Cartilage Development, Physiology, Pathologies, and Regeneration, (2009) 367–393. https://doi.org/10.1007/978-0-387-74660-9.
15. P.J. Yang, J.S. Temenoff, D. Ph, Engineering Orthopedic Tissue Interfaces, 15 (2009).
16. N. Nischal, K.P. Iyengar, D. Herlekar, R. Botchu, Imaging of Cartilage and Chondral Defects : An Overview, (2023).
17. W.R. Frontera, J. Ochala, Skeletal Muscle: A Brief Review of Structure and Function, Behav. Genet. 45 (2015) 183–195. https://doi.org/10.1007/s00223-014-9915-y.
18. S.M. Greising, H.M. Gransee, C.B. Mantilla, G.C. Sieck, Systems biology of skeletal muscle : fiber type as an organizing principle, (2012). https://doi.org/10.1002/wsbm.1184.
19. Y. Sakamoto, Histological Features of Endomysium, Perimysium and Epimysium in Rat Lateral Pterygoid Muscle, 119 (1996) 113–119.
20. G.H. Cardinet, Skeletal Muscle Function, Clin. Biochem. Domest. Anim. (1997) 407–440. https://doi.org/10.1016/b978-012396305-5/50017-8.
21. S. V. Brooks, Current topics for teaching skeletal muscle physiology, Am. J. Physiol. – Adv. Physiol. Educ. 27 (2003) 171–182. https://doi.org/10.1152/advan.2003.27.4.171.
22. A. Amelot, C. Mazel, The Intervertebral Disc: Physiology and Pathology of a Brittle Joint, World Neurosurg. 120 (2018) 265–273. https://doi.org/10.1016/j.wneu.2018.09.032.
23. Y. Choi, Pathophysiology of Degenerative Disc Disease, 3 (2009) 39–44.
24. I.L. Mohd Isa, S.A. Mokhtar, S.A. Abbah, M.B. Fauzi, A. Devitt, A. Pandit, Intervertebral Disc Degeneration: Biomaterials and Tissue Engineering Strategies toward Precision Medicine, Adv. Healthc. Mater. 11 (2022). https://doi.org/10.1002/adhm.202102530.
25. S. Sara, E. Wachtel, P. Roughley, Biochimica et Biophysica Acta Structure, function, aging and turnover of aggrecan in the intervertebral disc, BBA – Gen. Subj. 1840 (2014) 3181–3189. https://doi.org/10.1016/j.bbagen.2014.07.013.
26. S.Z. Wang, Y.F. Rui, J. Lu, C. Wang, Cell and molecular biology of intervertebral disc degeneration : current understanding and implications for potential therapeutic strategies, (2014) 381–390. https://doi.org/10.1111/cpr.12121.
27. P.P. Raj, Intervertebral Disc : Pathophysiology-Treatment, 8 (2008) 18–44.
28. Y. Dou, X. Sun, X. Ma, X. Zhao, Q. Yang, Intervertebral Disk Degeneration: The Microenvironment and Tissue Engineering Strategies, Front. Bioeng. Biotechnol. 9 (2021) 1–18. https://doi.org/10.3389/fbioe.2021.592118.
29. S. Williams, B. Alkhatib, R. Serra, Development of the axial skeleton and intervertebral disc, 1st ed., Elsevier Inc., 2019. https://doi.org/10.1016/bs.ctdb.2018.11.018.
30. B.K. Connizzo, S.M. Yannascoli, L.J. Soslowsky, Structure – function relationships of postnatal tendon development : A parallel to healing, Matrix Biol. 32 (2013) 106–116. https://doi.org/10.1016/j.matbio.2013.01.007.
31. M. Benjamin, E. Kaiser, S. Milz, Structure-function relationships in tendons: A review, J. Anat. 212 (2008) 211–228. https://doi.org/10.1111/j.1469-7580.2008.00864.x.
32. J.L. Zitnay, J.A. Weiss, Load Transfer, Damage, and Failure in Ligaments and Tendons, (2018). https://doi.org/10.1002/jor.24134.
33. M. Benjamin, J.R. Ralphs, Fibrocartilage in tendons and ligaments — an adaptation to compressive load, (1998) 481–494.
34. K.J. Karunaseelan, O. Dandridge, R.J. Van Arkel, J.R.T. Jeffers, Capsular ligaments provide a passive stabilizing force to protect the hip against edge loading, 10 (2021) 594–601. https://doi.org/10.1302/2046-3758.109.BJR-2020-0536.R1.
35. G. Schulze-tanzil, Intraarticular Ligament Degeneration Is Interrelated with Cartilage and Bone Destruction in Osteoarthritis, (2019).
36. D.T. Kirkendall, W.E. Garrett, Function and biomechanics of tendons, Scand. J. Med. Sci. Sport. 7 (1997) 62–66. https://doi.org/10.1111/j.1600-0838.1997.tb00120.x.
37. S. Zhang, M. Xing, B. Li, Acta Biomaterialia Recent advances in musculoskeletal local drug delivery q, Acta Biomater. (2019). https://doi.org/10.1016/j.actbio.2019.01.043.

38. A.D. Woolf, Global burden of osteoarthritis and musculoskeletal diseases, BMC Musculoskelet. Disord. 16 (2015) 2474. https://doi.org/10.1186/1471-2474-16-s1-s3.
39. G.A. Rodan, T.J. Martin, Therapeutic Approaches to Bone Diseases, 289 (2000) 1508–1515.
40. G. Chindamo, S. Sapino, E. Peira, D. Chirio, M.C. Gonzalez, M. Gallarate, Bone diseases: Current approach and future perspectives in drug delivery systems for bone targeted therapeutics, Nanomaterials. 10 (2020). https://doi.org/10.3390/nano10050875.
41. L. Tong, H. Yu, L. Chen, H. Wang, L. Xing, Current understanding of osteoarthritis pathogenesis and relevant new approaches, (2022) 1–17. https://doi.org/10.1038/s41413-022-00226-9.
42. H.U. Scherer, T. Häupl, G.R. Burmester, The etiology of rheumatoid arthritis, J. Autoimmun. 110 (2020) 102400. https://doi.org/10.1016/j.jaut.2019.102400.
43. A. Ghouri, P.G. Conaghan, Update on novel pharmacological therapies for osteoarthritis, Ther. Adv. Musculoskelet. Dis. 11 (2019) 1–11. https://doi.org/10.1177/1759720X19864492.
44. J. Bullock, S.A.A. Rizvi, A.M. Saleh, S.S. Ahmed, D.P. Do, R.A. Ansari, J. Ahmed, Rheumatoid arthritis: A brief overview of the treatment, Med. Princ. Pract. 27 (2019) 501–507. https://doi.org/10.1159/000493390.
45. D. Kravchenko, C. Behning, R. Bergner, How to Differentiate Gout, Calcium Pyrophosphate Deposition Disease, and Osteoarthritis Using Just Four Clinical Parameters, (2021) 1–13.
46. R. Lozano, N. Fullman, D. Abate, S.M. Abay, C. Abbafati, N. Abbasi, E. Beghi, Measuring progress from 1990 to 2017 and projecting attainment to 2030 of the health-related Sustainable Development Goals for 195 countries and territories : a systematic analysis for the Global Burden of Disease Study 2017, Lancet. 10159 (2018) 2091–2138. https://doi.org/10.1016/S0140-6736(18)32281-5.
47. T. Oichi, T. Saito, Y. Taniguchi, Y. Oshima, Pathomechanism of intervertebral disc degeneration, (2020) 1–9. https://doi.org/10.1002/jsp2.1076.
48. A.J. Freemont, The cellular pathobiology of the degenerate intervertebral disc and discogenic back pain, (2009) 5–10. https://doi.org/10.1093/rheumatology/ken396.
49. M. V. Risbud, I.M. Shapiro, Role of cytokines in intervertebral disc degeneration: pain and disc content., Nat Rev Rheumatol. 10 (2015) 44–56. https://doi.org/10.1038/nrrheum.2013.160.Role.
50. A. Hiyama, K. Suyama, M. Watanabe, Correlational analysis of chemokine and inflammatory cytokine expression in the intervertebral disc and blood in patients with lumbar disc disease, (2021) 1–10. https://doi.org/10.1002/jor.25136.
51. T. Benzakour, V. Igoumenou, A.F. Mavrogenis, A. Benzakour, Current concepts for lumbar disc herniation, Int. Orthop. 43 (2019) 841–851. https://doi.org/10.1007/s00264-018-4247-6.
52. N.E. Foster, J.R. Anema, D. Cherkin, R. Chou, S.P. Cohen, D.P. Gross, P.H. Ferreira, J.M. Fritz, B.W. Koes, W. Peul, J.A. Turner, C.G. Maher, R. Buchbinder, J. Hartvigsen, M. Underwood, M. van Tulder, L. Menezes Costa, P. Croft, M. Ferreira, S. Genevay, M.J. Hancock, D. Hoy, J. Karppinen, A. Kongsted, Q. Louw, B. Öberg, G. Pransky, M. Schoene, J. Sieper, R.J. Smeets, A. Woolf, Prevention and treatment of low back pain: evidence, challenges, and promising directions, Lancet. 391 (2018) 2368–2383. https://doi.org/10.1016/S0140-6736(18)30489-6.
53. D.S. Kreiner, S.W. Hwang, J.E. Easa, D.K. Resnick, J.L. Baisden, S. Bess, C.H. Cho, M.J. Depalma, P. Dougherty, R. Fernand, G. Ghiselli, A.S. Hanna, T. Lamer, A.J. Lisi, D.J. Mazanec, R.J. Meagher, R.C. Nucci, R.D. Patel, J.N. Sembrano, A.K. Sharma, J.T. Summers, C.K. Taleghani, W.L. Tontz, J.F. Toton, An evidence-based clinical guideline for the diagnosis and treatment of lumbar disc herniation with radiculopathy, Spine J. 14 (2014) 180–191. https://doi.org/10.1016/j.spinee.2013.08.003.
54. R.W. Jackman, S.C. Kandarian, The molecular basis of skeletal muscle atrophy, (2004) 834–843.
55. S. Schiaffino, C. Mammucari, Regulation of skeletal muscle growth by the IGF1-Akt / PKB pathway : insights from genetic models, (2011) 1–14.
56. K. Mukund, S. Subramaniam, Skeletal muscle : A review of molecular structure and function, in health and disease, (2020) 1–46. https://doi.org/10.1002/wsbm.1462.

57. V.C. Foletta, L.J. White, A.E. Larsen, B. Léger, A.P. Russell, The role and regulation of MAFbx / atrogin-1 and MuRF1 in skeletal muscle atrophy, (2011) 325–335. https://doi.org/10.1007/s00424-010-0919-9.
58. Y. Li, Y. Chen, J. John, J. Moylan, B. Jin, L. Douglas, M.B. Reid, TNF-α acts via p38 MAPK to stimulate expression of the ubiquitin ligase atrogin1/MAFbx in skeletal muscle, 19 (2011) 362–370. https://doi.org/10.1096/fj.04-2364com.TNF-.
59. R.W. Jackman, E.W. Cornwell, C.-L. Wu, S.C. Kandarian, NF-kappaB signaling and transcriptional regulation in skeletal muscle, 98 (2014) 19–24. https://doi.org/10.1113/expphysiol.2011.063321.NF-kappaB.
60. Y. Shi, Z. Liu, Y. Yang, X. Xu, Y. Li, T. Li, Design of poly(mPEGMA-co-MAA) hydrogel-based mPEG-b-PCL nanoparticles for oral meloxicam delivery, Mater. Sci. Eng. C. 76 (2017) 975–984. https://doi.org/10.1016/j.msec.2017.03.163.
61. Z. Karami, S. Sadighian, K. Rostamizadeh, M. Parsa, S. Rezaee, Naproxen conjugated mPEG-PCL micelles for dual triggered drug delivery, Mater. Sci. Eng. C. 61 (2016) 665–673. https://doi.org/10.1016/j.msec.2015.12.067.
62. M. Singh, G.S. Gonegandla, Bisphosphonate-Induced Osteonecrosis of the Jaws (BIONJ), J. Maxillofac. Oral Surg. 19 (2020) 162–167. https://doi.org/10.1007/s12663-019-01211-2.
63. K. Sumedh, V. Pegah, M. Markus, P. Théophile, R. Ramin, Ac ce ed us cr t, Pharm. Dev. Technol. 0 (2020) 000. https://doi.org/10.1080/10837450.2020.1798458.
64. C.F. Thorn, C. Oshiro, S. Marsh, T. Hernandez-Boussard, H. McLeod, T.E. Klein, R.B. Altman, Doxorubicin pathways, Pharmacogenet. Genomics. 21 (2011) 440–446. https://doi.org/10.1097/fpc.0b013e32833ffb56.
65. T.T. Pham, H.T. Nguyen, C.D. Phung, S. Pathak, S. Regmi, D. Ha, J.O. Kim, C.S. Yong, S.K. Kim, J. Choi, S. Yook, J. Park, J. Jeong, Targeted delivery of doxorubicin for the treatment of bone metastasis from breast cancer using alendronate–functionalized graphene oxide nanosheets, J. Ind. Eng. Chem. (2019). https://doi.org/10.1016/j.jiec.2019.03.055.
66. A.Y. Dna, E. Bagheri, M. Alibolandi, K. Abnous, Targeted delivery and controlled release of doxorubicin to cancer cells by smart, (2021) 1351–1363. https://doi.org/10.1039/d0tb01960g.
67. H. Ma, C. He, Y. Cheng, Z. Yang, J. Zang, J. Liu, X. Chen, Localized Co-delivery of Doxorubicin, Cisplatin, and Methotrexate by Thermosensitive Hydrogels for Enhanced Osteosarcoma Treatment, ACS Appl. Mater. Interfaces. 7 (2015) 27040–27048. https://doi.org/10.1021/acsami.5b09112.
68. S. Ghosh, Bioorganic Chemistry Cisplatin : The first metal based anticancer drug, Bioorg. Chem. 88 (2019) 102925. https://doi.org/10.1016/j.bioorg.2019.102925.
69. A.M. Florea, D. Büsselberg, Cisplatin as an anti-tumor drug: Cellular mechanisms of activity, drug resistance and induced side effects, Cancers (Basel). 3 (2011) 1351–1371. https://doi.org/10.3390/cancers3011351.
70. U. Hess, S. Shahabi, L. Treccani, P. Streckbein, C. Heiss, K. Rezwan, Co-delivery of cisplatin and doxorubicin from calcium phosphate beads/matrix scaffolds for osteosarcoma therapy, Mater. Sci. Eng. C. 77 (2017) 427–435. https://doi.org/10.1016/j.msec.2017.03.164.
71. K. Sridharan, N.J. Gogtay, Therapeutic nucleic acids: current clinical status, Br. J. Clin. Pharmacol. (2016) 659–672. https://doi.org/10.1111/bcp.12987.
72. C.A. Stein, D. Castanotto, FDA-Approved Oligonucleotide Therapies in 2017, Mol. Ther. 25 (2017) 1069–1075. https://doi.org/10.1016/j.ymthe.2017.03.023.
73. Y. Zhang, W. Ma, Y. Zhan, C. Mao, X. Shao, X. Xie, X. Wei, Y. Lin, Nucleic acids and analogs for bone regeneration, Bone Res. 6 (2018). https://doi.org/10.1038/s41413-018-0042-7.
74. Y. Cai, E. López-Ruiz, J. Wengel, L.B. Creemers, K.A. Howard, A hyaluronic acid-based hydrogel enabling CD44-mediated chondrocyte binding and gapmer oligonucleotide release for modulation of gene expression in osteoarthritis, J. Control. Release. 253 (2017) 153–159. https://doi.org/10.1016/j.jconrel.2017.03.004.
75. H. Pan, Q. Zheng, S. Yang, X. Guo, B. Wu, Z. Zou, Z. Duan, A novel peptide-modified and gene-activated biomimetic bone matrix accelerating bone regeneration, J. Biomed. Mater. Res. – Part A. 102 (2014) 2864–2874. https://doi.org/10.1002/jbm.a.34961.

76. H. Hu, L. Dong, Z. Bu, Y. Shen, J. Luo, H. Zhang, S. Zhao, F. Lv, Z. Liu, miR-23a-3p-abundant small extracellular vesicles released from Gelma/nanoclay hydrogel for cartilage regeneration, J. Extracell. Vesicles. 9 (2020). https://doi.org/10.1080/20013078.2020.1778883.
77. H. Li, J. Fan, L. Fan, T. Li, Y. Yang, H. Xu, J. Li, T. Li, X. Weng, S. Wang, R.C. Zhao, MiRNA-10b Reciprocally Stimulates Osteogenesis and Inhibits Adipogenesis Partly through the TGF- β / SMAD2 Signaling Pathway, 9 (2018) 1058–1073.
78. G. He, J. Chen, D. Huang, miR – 877 – 3p promotes TGF – β 1 – induced osteoblast differentiation of MC3T3 – E1 cells by targeting Smad7, (2019) 312–319. https://doi.org/10.3892/etm.2019.7570.
79. A. Akkouch, S. Eliason, M.E. Sweat, M. Romero-bustillos, M. Zhu, F. Qian, B.A. Amendt, L. Hong, Enhancement of MicroRNA-200c on Osteogenic Differentiation and Bone Regeneration by Targeting Sox2-Mediated Wnt Signaling and Klf4, 30 (2019) 1–3. https://doi.org/10.1089/hum.2019.019.
80. X. Su, L. Liao, Y. Shuai, H. Jing, S. Liu, H. Zhou, Y. Liu, Y. Jin, MiR-26a functions oppositely in osteogenic differentiation of BMSCs and ADSCs depending on distinct activation and roles of Wnt and BMP signaling pathway, (2015) 1–12. https://doi.org/10.1038/cddis.2015.221.
81. Y. Huang, Y. Xu, miR-19b enhances osteogenic differentiation of mesenchymal stem cells and promotes fracture healing through the WWP1 / Smurf2-mediated KLF5 / β -catenin signaling pathway, Exp. Mol. Med. (2021) 973–985. https://doi.org/10.1038/s12276-021-00631-w.
82. R. Gu, N. Liu, S. Luo, W. Huang, Z. Zha, J. Yang, MicroRNA-9 regulates the development of knee osteoarthritis through the NF-kappaB1 pathway in chondrocytes, Med. (United States). 95 (2016) 1–7. https://doi.org/10.1097/MD.0000000000004315.
83. O. Ham, B. Song, S. Lee, E. Choi, M. Cha, C. Youn, J. Park, I. Kim, W. Chang, S. Lim, C. Hyun, S. Kim, Biomaterials The role of microRNA-23b in the differentiation of MSC into chondrocyte by targeting protein kinase A signaling, Biomaterials. 33 (2012) 4500–4507. https://doi.org/10.1016/j.biomaterials.2012.03.025.
84. G. Mao, Z. Zhang, S. Hu, Z. Zhang, Z. Chang, Z. Huang, W. Liao, Y. Kang, Exosomes derived from miR-92a-3p- overexpressing human mesenchymal stem cells enhance chondrogenesis and suppress cartilage degradation via targeting WNT5A, (2018) 1–13.
85. X. Chen, Y. Shi, P. Xue, X. Ma, J. Li, J. Zhang, Mesenchymal stem cell-derived exosomal microRNA-136-5p inhibits chondrocyte degeneration in traumatic osteoarthritis by targeting ELF3, (2020) 1–13.
86. T.A. Karlsen, G.A. De Souza, B. Ødegaard, L. Engebretsen, J.E. Brinchmann, microRNA-140 Inhibits Inflammation and Stimulates Chondrogenesis in a Model of Interleukin 1 β -induced Osteoarthritis, Mol. Ther. – Nucleic Acids. 5 (2016) e373. https://doi.org/10.1038/mtna.2016.64.
87. R.P. Goodman, A.J. Turberfield, The single-step synthesis of a DNA tetrahedron, Chem. Commun. 4 (2004) 1372–1373. https://doi.org/10.1039/b402293a.
88. Y. Lin, Q. Li, L. Wang, Q. Guo, S. Liu, S. Zhu, Y. Sun, Y. Fan, Y. Sun, H. Li, X. Tian, D. Luo, S. Shi, Advances in regenerative medicine applications of tetrahedral framework nucleic acid-based nanomaterials: an expert consensus recommendation, Int. J. Oral Sci. 14 (2022). https://doi.org/10.1038/s41368-022-00199-9.
89. Q. Peng, X.R. Shao, J. Xie, S.R. Shi, X.Q. Wei, T. Zhang, X.X. Cai, Y.F. Lin, Understanding the Biomedical Effects of the Self-Assembled Tetrahedral DNA Nanostructure on Living Cells, ACS Appl. Mater. Interfaces. 8 (2016) 12733–12739. https://doi.org/10.1021/acsami.6b03786.
90. M. Zhou, N. Liu, Q. Zhang, T. Tian, Q. Ma, T. Zhang, X. Cai, Effect of tetrahedral DNA nanostructures on proliferation and osteogenic differentiation of human periodontal ligament stem cells, Cell Prolif. 52 (2019) 1–10. https://doi.org/10.1111/cpr.12566.
91. S. Shi, S. Lin, X. Shao, Q. Li, Z. Tao, Y. Lin, Modulation of chondrocyte motility by tetrahedral DNA nanostructures, Cell Prolif. 50 (2017) 1–9. https://doi.org/10.1111/cpr.12368.

92. S. Sirong, C. Yang, T. Taoran, L. Songhang, L. Shiyu, Z. Yuxin, S. Xiaoru, Z. Tao, L. Yunfeng, C. Xiaoxiao, Effects of tetrahedral framework nucleic acid/wogonin complexes on osteoarthritis, Bone Res. 8 (2020). https://doi.org/10.1038/s41413-019-0077-4.
93. X.R. Shao, S.Y. Lin, Q. Peng, S.R. Shi, X.L. Li, T. Zhang, Y.F. Lin, Effect of tetrahedral DNA nanostructures on osteogenic differentiation of mesenchymal stem cells via activation of the Wnt/β-catenin signaling pathway, Nanomedicine Nanotechnology, Biol. Med. 13 (2017) 1809–1819. https://doi.org/10.1016/j.nano.2017.02.011.
94. X. Shao, S. Lin, Q. Peng, S. Shi, X. Wei, T. Zhang, Y. Lin, Tetrahedral DNA Nanostructure: A Potential Promoter for Cartilage Tissue Regeneration via Regulating Chondrocyte Phenotype and Proliferation, Small. 13 (2017). https://doi.org/10.1002/smll.201602770.
95. M. Zhou, S. Gao, X. Zhang, T. Zhang, T. Zhang, T. Tian, S. Li, Y. Lin, X. Cai, The protective effect of tetrahedral framework nucleic acids on periodontium under inflammatory conditions, Bioact. Mater. 6 (2021) 1676–1688. https://doi.org/10.1016/j.bioactmat.2020.11.018.
96. S. Shi, T. Tian, Y. Li, D. Xiao, T. Zhang, P. Gong, Y. Lin, Tetrahedral Framework Nucleic Acid Inhibits Chondrocyte Apoptosis and Oxidative Stress through Activation of Autophagy, ACS Appl. Mater. Interfaces. 12 (2020) 56782–56791. https://doi.org/10.1021/acsami.0c17307.
97. J.S. Park, H.J. Lee, D.Y. Lee, H.S. Jo, J.H. Jeong, D.H. Kim, D.C. Nam, C.J. Lee, S.C. Hwang, Chondroprotective effects of wogonin in experimental models of osteoarthritis in vitro and in vivo, Biomol. Ther. 23 (2015) 442–448. https://doi.org/10.4062/biomolther.2015.045.
98. Y. Gao, T. Zhang, J. Zhu, D. Xiao, M. Zhang, Y. Sun, Y. Li, Y. Lin, X. Cai, Effects of the tetrahedral framework nucleic acids on the skeletal muscle regeneration: In vitro and in vivo, Mater. Chem. Front. 4 (2020) 2731–2743. https://doi.org/10.1039/d0qm00329h.
99. W. Cui, X. Chen, J. Zhu, M. Zhang, D. Xiao, X. Qin, T. Zhang, Y. Lin, Preventive effect of tetrahedral framework nucleic acids on bisphosphonate-related osteonecrosis of the jaw, Nanoscale. 12 (2020) 17196–17202. https://doi.org/10.1039/d0nr03731a.
100. D. Zhao, W. Cui, M. Liu, J. Li, Y. Sun, S. Shi, S. Lin, Y. Lin, Tetrahedral Framework Nucleic Acid Promotes the Treatment of Bisphosphonate-Related Osteonecrosis of the Jaws by Promoting Angiogenesis and M2 Polarization, ACS Appl. Mater. Interfaces. 12 (2020) 44508–44522. https://doi.org/10.1021/acsami.0c13839.
101. M. Zhang, X. Zhang, T. Tian, Q. Zhang, Y. Wen, J. Zhu, D. Xiao, W. Cui, Y. Lin, Anti-inflammatory activity of curcumin-loaded tetrahedral framework nucleic acids on acute gouty arthritis, Bioact. Mater. 8 (2022) 368–380. https://doi.org/10.1016/j.bioactmat.2021.06.003.
102. Q. Li, D. Zhao, X. Shao, S. Lin, X. Xie, M. Liu, W. Ma, S. Shi, Y. Lin, Aptamer-Modified Tetrahedral DNA Nanostructure for Tumor-Targeted Drug Delivery, ACS Appl. Mater. Interfaces. 9 (2017) 36695–36701. https://doi.org/10.1021/acsami.7b13328.
103. G. Bullock, J. Atkinson, P. Gentile, P. Hatton, C. Miller, Osteogenic peptides and attachment methods determine tissue regeneration in modified bone graft substitutes, J. Funct. Biomater. 12 (2021). https://doi.org/10.3390/jfb12020022.
104. L.A. Poniatowski, P. Wojdasiewicz, R. Gasik, D. Szukiewicz, Transforming growth factor beta family: Insight into the role of growth factors in regulation of fracture healing biology and potential clinical applications, Mediators Inflamm. 2015 (2015). https://doi.org/10.1155/2015/137823.
105. T. Komori, Department of Cell Biology, Unit of Basic Medical Sciences, Nagasaki University Graduate School of Biomedical Sciences, Nagasaki 852-8588, Japan, Bone. (2008) 898–903.
106. D. Chen, M. Zhao, G.R. Mundy, Bone morphogenetic proteins, Growth Factors. 22 (2004) 233–241. https://doi.org/10.1080/08977190412331279890.
107. D. Halloran, H.W. Durbano, A. Nohe, Developmental review bone morphogenetic protein-2 in development and bone homeostasis, J. Dev. Biol. 8 (2020) 28–30. https://doi.org/10.3390/JDB8030019.
108. G.S. Krishnakumar, A. Roffi, D. Reale, E. Kon, G. Filardo, Clinical application of bone morphogenetic proteins for bone healing: a systematic review, Int. Orthop. 41 (2017) 1073–1083. https://doi.org/10.1007/s00264-017-3471-9.

109. M. Yang, Z.C. Zhang, Y. Liu, Y.R. Chen, R.H. Deng, Z.N. Zhang, J.K. Yu, F.Z. Yuan, Function and Mechanism of RGD in Bone and Cartilage Tissue Engineering, Front. Bioeng. Biotechnol. 9 (2021) 1–10. https://doi.org/10.3389/fbioe.2021.773636.
110. P.Y. Wang, T.H. Wu, W.B. Tsai, W.H. Kuo, M.J. Wang, Grooved PLGA films incorporated with RGD/YIGSR peptides for potential application on skeletal muscle tissue engineering, Colloids Surfaces B Biointerfaces. 110 (2013) 88–95. https://doi.org/10.1016/j.colsurfb.2013.04.016.
111. J. Lin, L. Wang, J. Lin, Q. Liu, Systems for Cartilage Regeneration, (2021) 1–11.
112. J. Bolte, C. Vater, A.C. Culla, T. Ahlfeld, J. Nowotny, P. Kasten, A.C. Disch, S.B. Goodman, M. Gelinsky, M. Stiehler, S. Zwingenberger, Two-step stem cell therapy improves bone regeneration compared to concentrated bone marrow therapy, J. Orthop. Res. 37 (2019) 1318–1328. https://doi.org/10.1002/jor.24215.
113. Y. Xia, J. Sun, L. Zhao, F. Zhang, X.J. Liang, Y. Guo, M.D. Weir, M.A. Reynolds, N. Gu, H.H.K. Xu, Magnetic field and nano-scaffolds with stem cells to enhance bone regeneration, Biomaterials. 183 (2018) 151–170. https://doi.org/10.1016/j.biomaterials.2018.08.040.
114. J.G. Tidball, Mechanisms of muscle injury, repair, and regeneration, Compr. Physiol. 1 (2011) 2029–2062. https://doi.org/10.1002/cphy.c100092.
115. J. Dort, P. Fabre, T. Molina, N.A. Dumont, Macrophages Are Key Regulators of Stem Cells during Skeletal Muscle Regeneration and Diseases, Stem Cells Int. 2019 (2019). https://doi.org/10.1155/2019/4761427.
116. C.H. Evans, V.B. Kraus, L.A. Setton, Progress in intra-articular therapy, Nat Rev Rheumatol. 10 (2015) 11–22. https://doi.org/10.1038/nrrheum.2013.159.Progress.
117. M.F. Rai, C.T. Pham, Intra-articular drug delivery systems for joint diseases, Curr. Opin. Pharmacol. 40 (2018) 67–73. https://doi.org/10.1016/j.coph.2018.03.013.
118. L. Ngo, M.L. Knothe Tate, Osteoarthritis: New Strategies for Transport and Drug Delivery across Length Scales, ACS Biomater. Sci. Eng. 6 (2020) 6009–6020. https://doi.org/10.1021/acsbiomaterials.0c01081.
119. A.G. Bajpayee, A.J. Grodzinsky, Cartilage-targeting drug delivery: can electrostatic interactions help?, Nat. Publ. Gr. (2017). https://doi.org/10.1038/nrrheum.2016.210.
120. W. Deng, Y. Yan, P. Zhuang, X. Liu, K. Tian, W. Huang, C. Li, Synthesis of nanocapsules blended polymeric hydrogel loaded with bupivacaine drug delivery system for local anesthetics and pain management, Drug Deliv. 29 (2022) 399–412. https://doi.org/10.1080/10717544.2021.2023702.
121. H. Idrees, S.Z.J. Zaidi, A. Sabir, R.U. Khan, X. Zhang, S.U. Hassan, A review of biodegradable natural polymer-based nanoparticles for drug delivery applications, Nanomaterials. 10 (2020) 1–22. https://doi.org/10.3390/nano10101970.
122. X. Tong, W. Pan, T. Su, M. Zhang, W. Dong, X. Qi, Recent advances in natural polymer-based drug delivery systems, React. Funct. Polym. 148 (2020) 104501. https://doi.org/10.1016/j.reactfunctpolym.2020.104501.
123. Z. Shi, Y. Zhou, T. Fan, Y. Lin, H. Zhang, L. Mei, Inorganic nano-carriers based smart drug delivery systems for tumor therapy, Smart Mater. Med. 1 (2020) 32–47. https://doi.org/10.1016/j.smaim.2020.05.002.
124. A.Y. Cai, Y.J. Zhu, C. Qi, Biodegradable Inorganic Nanostructured Biomaterials for Drug Delivery, Adv. Mater. Interfaces. 7 (2020) 1–30. https://doi.org/10.1002/admi.202000819.
125. Y. Zou, B. Huang, L. Cao, Y. Deng, J. Su, Tailored Mesoporous Inorganic Biomaterials: Assembly, Functionalization, and Drug Delivery Engineering, Adv. Mater. 33 (2021) 1–27. https://doi.org/10.1002/adma.202005215.
126. Y. Arun, R. Ghosh, A.J. Domb, Biodegradable Hydrophobic Injectable Polymers for Drug Delivery and Regenerative Medicine, Adv. Funct. Mater. 31 (2021) 1–22. https://doi.org/10.1002/adfm.202010284.
127. V. Ogay, E.A. Mun, G. Kudaibergen, M. Baidarbekov, K. Kassymbek, Z. Zharkinbekov, A. Saparov, Progress and prospects of polymer-based drug delivery systems for bone tissue regeneration, Polymers (Basel). 12 (2020) 1–25. https://doi.org/10.3390/polym12122881.

128. S. Sharma, P. Sudhakara, J. Singh, R.A. Ilyas, M.R.M. Asyraf, M.R. Razman, Critical review of biodegradable and bioactive polymer composites for bone tissue engineering and drug delivery applications, Polymers (Basel). 13 (2021). https://doi.org/10.3390/polym13162623.
129. S. Skibiński, E. Cichoń, K. Haraźna, E. Marcello, I. Roy, M. Witko, A. Ślósarczyk, J. Czechowska, M. Guzik, A. Zima, Functionalized tricalcium phosphate and poly(3-hydroxyoctanoate) derived composite scaffolds as platforms for the controlled release of diclofenac, Ceram. Int. 47 (2021) 3876–3883. https://doi.org/10.1016/j.ceramint.2020.09.248.
130. P. Li, X. Feng, X. Jia, Y. Fan, Influences of tensile load on in vitro degradation of an electrospun poly(L-lactide-co-glycolide) scaffold, Acta Biomater. 6 (2010) 2991–2996. https://doi.org/10.1016/j.actbio.2010.02.023.
131. C. Engineer, J. Parikh, A. Raval, Review on hydrolytic degradation behavior of biodegradable polymers from controlled drug delivery system, Trends Biomater. Artif. Organs. 25 (2011) 79–85.
132. Y. Dou, C. Li, L. Li, J. Guo, J. Zhang, Bioresponsive drug delivery systems for the treatment of inflammatory diseases, J. Control. Release. 327 (2020) 641–666. https://doi.org/10.1016/j.jconrel.2020.09.008.
133. F. Xiong, Z. Qin, H. Chen, Q. Lan, Z. Wang, N. Lan, Y. Yang, L. Zheng, J. Zhao, D. Kai, PH-responsive and hyaluronic acid-functionalized metal-organic frameworks for therapy of osteoarthritis, J. Nanobiotechnology. 18 (2020) 1–14. https://doi.org/10.1186/s12951-020-00694-3.
134. C. Li, H. Li, Q. Wang, M. Zhou, M. Li, T. Gong, Z. Zhang, X. Sun, pH-sensitive polymeric micelles for targeted delivery to inflamed joints, J. Control. Release. 246 (2017) 133–141. https://doi.org/10.1016/j.jconrel.2016.12.027.
135. C.J. Malemud, Matrix Metalloproteinases and Synovial Joint Pathology, 1st ed., Elsevier Inc., 2017. https://doi.org/10.1016/bs.pmbts.2017.03.003.
136. Y. Rengel, C. Ospelt, S. Gay, Proteinases in the joint: Clinical relevance of proteinases in joint destruction, Arthritis Res. Ther. 9 (2007). https://doi.org/10.1186/ar2304.
137. X. Feng, Y. Chen, Drug delivery targets and systems for targeted treatment of rheumatoid arthritis, J. Drug Target. 26 (2018) 845–857. https://doi.org/10.1080/1061186X.2018.1433680.
138. N. Joshi, J. Yan, S. Levy, S. Bhagchandani, K. V. Slaughter, N.E. Sherman, J. Amirault, Y. Wang, L. Riegel, X. He, T.S. Rui, M. Valic, P.K. Vemula, O.R. Miranda, O. Levy, E.M. Gravallese, A.O. Aliprantis, J. Ermann, J.M. Karp, Towards an arthritis flare-responsive drug delivery system, Nat. Commun. 9 (2018) 1–11. https://doi.org/10.1038/s41467-018-03691-1.
139. H. Qi, L. Yang, X. Li, X. Sun, J. Zhao, X. Hou, Z. Li, X. Yuan, Z. Cui, X. Yang, Systemic administration of enzyme-responsive growth factor nanocapsules for promoting bone repair, Biomater. Sci. 7 (2019) 1675–1685. https://doi.org/10.1039/c8bm01632a.
140. A. Vedadghavami, E.K. Wagner, S. Mehta, T. He, C. Zhang, A.G. Bajpayee, Cartilage penetrating cationic peptide carriers for applications in drug delivery to avascular negatively charged tissues, 2019. https://doi.org/10.1016/j.actbio.2018.12.004.
141. A. Vedadghavami, C. Zhang, A.G. Bajpayee, Overcoming negatively charged tissue barriers: Drug delivery using cationic peptides and proteins, Nano Today. 34 (2020) 100898. https://doi.org/10.1016/j.nantod.2020.100898.
142. D.E. Jaalouk, J. Lammerding, Mechanotransduction gone awry, Nat. Rev. Mol. Cell Biol. 10 (2009) 63–73. https://doi.org/10.1038/nrm2597.
143. C.N. McCutchen, D.L. Zignego, R.K. June, Metabolic responses induced by compression of chondrocytes in variable-stiffness microenvironments, J. Biomech. 64 (2017) 49–58. https://doi.org/10.1016/j.jbiomech.2017.08.032.
144. W.C. Hayes, Bone Mechanics: From Tissue Mechanical Properties to an Assessment of Structural Behavior, Front. Biomech. (1986) 196–209. https://doi.org/10.1007/978-1-4612-4866-8_15.

145. W.L.K. Chen, C.A. Simmons, Lessons from (patho)physiological tissue stiffness and their implications for drug screening, drug delivery and regenerative medicine, Adv. Drug Deliv. Rev. 63 (2011) 269–276. https://doi.org/10.1016/j.addr.2011.01.004.
146. A.J. Engler, S. Sen, H.L. Sweeney, D.E. Discher, Matrix Elasticity Directs Stem Cell Lineage Specification, Cell. 126 (2006) 677–689. https://doi.org/10.1016/j.cell.2006.06.044.
147. J.P. Winer, P.A. Janmey, M.E. McCormick, M. Funaki, Bone marrow-derived human mesenchymal stem cells become quiescent on soft substrates but remain responsive to chemical or mechanical stimuli, Tissue Eng. – Part A. 15 (2009) 147–154. https://doi.org/10.1089/ten.tea.2007.0388.
148. R. Ziadlou, S. Rotman, A. Teuschl, E. Salzer, A. Barbero, I. Martin, M. Alini, D. Eglin, S. Grad, Optimization of hyaluronic acid-tyramine/silk-fibroin composite hydrogels for cartilage tissue engineering and delivery of anti-inflammatory and anabolic drugs, Mater. Sci. Eng. C. 120 (2021) 111701. https://doi.org/10.1016/j.msec.2020.111701.
149. S. Wu, J. Liu, Y. Qi, J. Cai, J. Zhao, B. Duan, S. Chen, Tendon-bioinspired wavy nanofibrous scaffolds provide tunable anisotropy and promote tenogenesis for tendon tissue engineering, Mater. Sci. Eng. C. 126 (2021). https://doi.org/10.1016/j.msec.2021.112181.
150. M.W. Tibbitt, J.E. Dahlman, R. Langer, Emerging Frontiers in Drug Delivery, J. Am. Chem. Soc. 138 (2016) 704–717. https://doi.org/10.1021/jacs.5b09974.
151. J. Jacob, J.T. Haponiuk, S. Thomas, S. Gopi, Biopolymer based nanomaterials in drug delivery systems: A review, Mater. Today Chem. 9 (2018) 43–55. https://doi.org/10.1016/j.mtchem.2018.05.002.
152. X. Qi, W. Wei, J. Li, Y. Liu, X. Hu, J. Zhang, L. Bi, W. Dong, Fabrication and Characterization of a Novel Anticancer Drug Delivery System: Salecan/Poly(methacrylic acid) Semi-interpenetrating Polymer Network Hydrogel, ACS Biomater. Sci. Eng. 1 (2015) 1287–1299. https://doi.org/10.1021/acsbiomaterials.5b00346.
153. X. Qi, L. Lin, L. Shen, Z. Li, T. Qin, Y. Qian, X. Wu, X. Wei, Q. Gong, J. Shen, Efficient Decontamination of Lead Ions from Wastewater by Salecan Polysaccharide-Based Hydrogels, ACS Sustain. Chem. Eng. 7 (2019) 11014–11023. https://doi.org/10.1021/acssuschemeng.9b02139.
154. P. Trucillo, Drug carriers: Classification, administration, release profiles, and industrial approach, Processes. 9 (2021). https://doi.org/10.3390/pr9030470.
155. Local and Targeted Drug Delivery for Bone Regeneration Introduction and Motivation, 2019.
156. S. Zhang, M. Xing, B. Li, Recent advances in musculoskeletal local drug delivery, Acta Biomater. 93 (2019) 135–151. https://doi.org/10.1016/j.actbio.2019.01.043.
157. Biocompatible nanocomposite scaffolds based on copolymer-grafted chitosan for bone tissue engineering with drug delivery capability _ Elsevier Enhanced Reader, (n.d.).
158. A.K. Gaharwar, A. Arpanaei, T.L. Andresen, A. Dolatshahi-Pirouz, 3D Biomaterial Microarrays for Regenerative Medicine: Current State-of-the-Art, Emerging Directions and Future Trends, Adv. Mater. 28 (2016) 771–781. https://doi.org/10.1002/adma.201503918.
159. https://doi.org/10.1016/S1748-0132(07)70084-1 _ Elsevier Enhanced Reader, (n.d.).
160. A. Gupta, J. Lee, T. Ghosh, V.Q. Nguyen, A. Dey, B. Yoon, W. Um, J.H. Park, Polymeric Hydrogels for Controlled Drug Delivery to Treat Arthritis, Pharmaceutics. 14 (2022). https://doi.org/10.3390/pharmaceutics14030540.
161. M. Hafezi, S.N. Khorasani, M. Zare, R.E. Neisiany, P. Davoodi, Advanced hydrogels for cartilage tissue engineering: Recent progress and future directions, Polymers (Basel). 13 (2021). https://doi.org/10.3390/polym13234199.
162. S. Jacob, A.B. Nair, J. Shah, N. Sreeharsha, S. Gupta, P. Shinu, Emerging role of hydrogels in drug delivery systems, tissue engineering and wound management, Pharmaceutics. 13 (2021). https://doi.org/10.3390/pharmaceutics13030357.
163. N. Storozhylova, J. Crecente-Campo, D. Cabaleiro, L. Lugo, C. Dussouy, S. Simões, M. Monteiro, C. Grandjean, M.J. Alonso, An In Situ Hyaluronic Acid-Fibrin Hydrogel Containing Drug-Loaded Nanocapsules for Intra-Articular Treatment of Inflammatory

Joint Diseases, Regen. Eng. Transl. Med. 6 (2020) 201–216. https://doi.org/10.1007/s40883-020-00154-2.
164. M. Akrami-Hasan-Kohal, M. Eskandari, A. Solouk, Silk fibroin hydrogel/dexamethasone sodium phosphate loaded chitosan nanoparticles as a potential drug delivery system, Colloids Surfaces B Biointerfaces. 205 (2021) 111892. https://doi.org/10.1016/j.colsurfb.2021.111892.
165. E. Ruvinov, T. Tavor Re'em, F. Witte, S. Cohen, Articular cartilage regeneration using acellular bioactive affinity-binding alginate hydrogel: A 6-month study in a mini-pig model of osteochondral defects, J. Orthop. Transl. 16 (2019) 40–52. https://doi.org/10.1016/j.jot.2018.08.003.
166. Z. Ma, C. Tao, L. Sun, S. Qi, Y. Le, J. Wang, C. Li, X. Liu, J. Zhang, J. Zhao, In situ forming injectable hydrogel for encapsulation of nanoiguratimod and sustained release of therapeutics, Int. J. Nanomedicine. 14 (2019) 8725–8738. https://doi.org/10.2147/IJN.S214507.
167. Y. Zhao, C. Gao, H. Liu, H. Liu, Y. Feng, Z. Li, H. Liu, stem cell survival, engraftment, and function in rheumatoid arthritis, Acta Biomater. (2020). https://doi.org/10.1016/j.actbio.2020.12.005.
168. L. Fan, C. Lin, P. Zhao, X. Wen, G. Li, An Injectable Bioorthogonal Dextran hydrogel for Enhanced Chondrogenesis of Primary Stem Cells, 2018. https://doi.org/10.1089/ten.TEC.2018.0085.
169. I.M. Oliveira, C. Gonçalves, M.E. Shin, S. Lee, R.L. Reis, G. Khang, Enzymatically crosslinked tyramine-gellan gum hydrogels as drug delivery system for rheumatoid arthritis treatment, (2021) 1288–1300.
170. N. Chauhan, P. Gupta, L. Arora, D. Pal, Y. Singh, Dexamethasone-loaded, injectable pullulan-poly(ethylene glycol) hydrogels for bone tissue regeneration in chronic inflammatory conditions, Mater. Sci. Eng. C. 130 (2021) 112463. https://doi.org/10.1016/j.msec.2021.112463.
171. J. Kim, B.B. Seo, K.H. Hong, S.E. Kim, Y.M. Kim, S.C. Song, Long-term anti-inflammatory effects of injectable celecoxib nanoparticle hydrogels for Achilles tendon regeneration, Acta Biomater. 144 (2022) 183–194. https://doi.org/10.1016/j.actbio.2022.03.033.
172. T. Kanazawa, K. Tamano, K. Sogabe, T. Endo, H. Ibaraki, Y. Takashima, Y. Seta, Intra-articular retention and anti-arthritic effects in collagen-induced arthritis model mice by injectable small interfering RNA containing hydrogel, Biol. Pharm. Bull. 40 (2017) 1929–1933. https://doi.org/10.1248/bpb.b17-00481.
173. T. Zhang, S. Chen, H. Dou, Q. Liu, G. Shu, J. Lin, W. Zhang, G. Peng, Z. Zhong, H. Fu, Novel glucosamine-loaded thermosensitive hydrogels based on poloxamers for osteoarthritis therapy by intra-articular injection, Mater. Sci. Eng. C. 118 (2021) 111352. https://doi.org/10.1016/j.msec.2020.111352.
174. K. Chen, S. Li, F. Yuan, P. Sun, Y. Zhang, GEL-MAN Hydrogel Loaded With Triamcinolone Acetonide for the Treatment of Osteoarthritis, Front. Bioeng. Biotechnol. 8 (2020) 1–8. https://doi.org/10.3389/fbioe.2020.00872.
175. S.W. Mok, S.C. Fu, Y.C. Cheuk, I.M. Chu, K.M. Chan, L. Qin, S.H. Yung, K.W. Kevin Ho, Intra-Articular Delivery of Quercetin Using Thermosensitive Hydrogel Attenuate Cartilage Degradation in an Osteoarthritis Rat Model, Cartilage. 11 (2020) 490–499. https://doi.org/10.1177/1947603518796550.
176. S.J. Wang, J.Z. Qin, T.E. Zhang, C. Xia, Intra-articular Injection of Kartogenin-Incorporated Thermogel Enhancing Osteoarthritis Treatment, Front. Chem. 7 (2019) 1–8. https://doi.org/10.3389/fchem.2019.00677.
177. A. Sulistio, F.M. Mansfeld, F. Reyes-Ortega, A.M. D'Souza, S.M.Y. Ng, S. Birkett, A. Blencowe, G.G. Qiao, C.B. Little, C.C. Shu, A.M. Bendele, D. Valade, A.C. Donohue, J.F. Quinn, M.R. Whittaker, T.P. Davis, R.J. Tait, Intra-articular Treatment of Osteoarthritis with Diclofenac-Conjugated Polymer Reduces Inflammation and Pain, ACS Appl. Bio Mater. 2 (2019) 2822–2832. https://doi.org/10.1021/acsabm.9b00232.
178. K.Y. Lu, Y.C. Lin, H.T. Lu, Y.C. Ho, S.C. Weng, M.L. Tsai, F.L. Mi, A novel injectable in situ forming gel based on carboxymethyl hexanoyl chitosan/hyaluronic acid polymer blending

for sustained release of berberine, Carbohydr. Polym. 206 (2019) 664–673. https://doi.org/10.1016/j.carbpol.2018.11.050.
179. C. Xia, P. Chen, S. Mei, L. Ning, C. Lei, J. Wang, J. Zhang, J. Ma, S. Fan, Photo-crosslinked HAMA hydrogel with cordycepin encapsulated chitosan microspheres for osteoarthritis treatment, Oncotarget. 8 (2017) 2835–2849. https://doi.org/10.18632/oncotarget.13748.
180. Advanced Science – 2019 – Chen – Gene-Hydrogel Microenvironment Regulates Extracellular Matrix Metabolism Balance in.pdf, (n.d.).
181. L. Yang, T.J. Webster, Nanotechnology controlled drug delivery for treating bone diseases, Expert Opin. Drug Deliv. 6 (2009) 851–864. https://doi.org/10.1517/17425240903044935.
182. A. Younas, H. Gu, Y. Zhao, N. Zhang, Novel approaches of the nanotechnology-based drug delivery systems for knee joint injuries: A review, Int. J. Pharm. 608 (2021). https://doi.org/10.1016/j.ijpharm.2021.121051.
183. M. Nasr, E. Elmowafy, M.E. Soliman, The evolution of dendrimers to composite dendrimers: a review of the state of the art, in: Nanoparticles Pharmacother., Elsevier, 2019: pp. 217–249. https://doi.org/10.1016/B978-0-12-816504-1.00004-1.
184. I.H. Tarner, U. Müller-Ladner, † Justus, Drug delivery systems for the treatment of rheumatoid arthritis, (1742). https://doi.org/10.1517/17425240802358220.
185. M.L. Kang, S.Y. Jeong, G. Il Im, Hyaluronic Acid Hydrogel Functionalized with Self-Assembled Micelles of Amphiphilic PEGylated Kartogenin for the Treatment of Osteoarthritis, Tissue Eng. – Part A. 23 (2017) 630–639. https://doi.org/10.1089/ten.tea.2016.0524.
186. Q. Zhou, Y. Cai, Y. Jiang, X. Lin, Exosomes in osteoarthritis and cartilage injury: Advanced development and potential therapeutic strategies, Int. J. Biol. Sci. 16 (2020) 1811–1820. https://doi.org/10.7150/ijbs.41637.
187. M. Goldberg, R. Langer, X. Jia, Nanostructured materials for applications in drug delivery and tissue engineering * NIH Public Access NANOSTRUCTURED MATERIALS AS DRUG-DELIVERY SYSTEMS, 2006.
188. H. Cheng, A. Chawla, Y. Yang, Y. Li, J. Zhang, H.L. Jang, A. Khademhosseini, Development of nanomaterials for bone-targeted drug delivery, Drug Discov. Today. 22 (2017) 1336–1350. https://doi.org/10.1016/j.drudis.2017.04.021.
189. Y. He, Y. Huang, Z. Huang, Y. Jiang, X. Sun, Y. Shen, W. Chu, C. Zhao, Bisphosphonate-functionalized coordination polymer nanoparticles for the treatment of bone metastatic breast cancer, J. Control. Release. 264 (2017) 76–88. https://doi.org/10.1016/j.jconrel.2017.08.024.
190. W. Chu, Y. Huang, C. Yang, Y. Liao, X. Zhang, M. Yan, S. Cui, C. Zhao, Calcium phosphate nanoparticles functionalized with alendronate-conjugated polyethylene glycol (PEG) for the treatment of bone metastasis, Int. J. Pharm. 516 (2017) 352–363. https://doi.org/10.1016/j.ijpharm.2016.11.051.
191. V.M. Wu, J. Mickens, V. Uskoković, Bisphosphonate-Functionalized Hydroxyapatite Nanoparticles for the Delivery of the Bromodomain Inhibitor JQ1 in the Treatment of Osteosarcoma, ACS Appl. Mater. Interfaces. 9 (2017) 25887–25904. https://doi.org/10.1021/acsami.7b08108.
192. F. Haghiralsadat, G. Amoabediny, S. Naderinezhad, K. Nazmi, J.P. De Boer, B. Zandieh-Doulabi, T. Forouzanfar, M.N. Helder, EphA2 Targeted Doxorubicin-Nanoliposomes for Osteosarcoma Treatment, Pharm. Res. 34 (2017) 2891–2900. https://doi.org/10.1007/s11095-017-2272-6.
193. V. Uskoković, T.A. Desai, In vitro analysis of nanoparticulate hydroxyapatite/chitosan composites as potential drug delivery platforms for the sustained release of antibiotics in the treatment of osteomyelitis, J. Pharm. Sci. 103 (2014) 567–579. https://doi.org/10.1002/jps.23824.
194. Z. Ouyang, T. Tan, C. Liu, J. Duan, W. Wang, X. Guo, Q. Zhang, Z. Li, Q. Huang, P. Dou, T. Liu, Targeted delivery of hesperetin to cartilage attenuates osteoarthritis by bimodal imaging with Gd 2 (CO 3) 3 @PDA nanoparticles via TLR-2/NF-κB/Akt signaling, Biomaterials. 205 (2019) 50–63. https://doi.org/10.1016/j.biomaterials.2019.03.018.
195. H.J. Shin, H. Park, N. Shin, J. Shin, D.H. Gwon, H.H. Kwon, Y. Yin, J.A. Hwang, J. Hong, J.Y. Heo, C.S. Kim, Y. Joo, Y. Kim, J. Kim, J. Beom, D.W. Kim, p66shc siRNA nanoparticles

ameliorate chondrocytic mitochondrial dysfunction in osteoarthritis, Int. J. Nanomedicine. 15 (2020) 2379–2390. https://doi.org/10.2147/IJN.S234198.
196. A.H. Salama, A.A. Abdelkhalek, N.A. Elkasabgy, Etoricoxib-loaded bio-adhesive hybridized polylactic acid-based nanoparticles as an intra-articular injection for the treatment of osteoarthritis, Int. J. Pharm. 578 (2020) 119081. https://doi.org/10.1016/j.ijpharm.2020.119081.
197. A.L. Boechat, C.P. De Oliveira, A.M. Tarrago, A.G. Da Costa, A. Malheiro, S.S. Guterres, A.R. Pohlmann, Methotrexate-loaded lipid-core nanocapsules are highly effective in the control of inflammation in synovial cells and a chronic arthritis model, Int. J. Nanomedicine. 10 (2015) 6603–6614. https://doi.org/10.2147/IJN.S85369.
198. T. Jin, D. Wu, X.M. Liu, J.T. Xu, B.J. Ma, Y. Ji, Y.Y. Jin, S.Y. Wu, T. Wu, K. Ma, Intra-articular delivery of celastrol by hollow mesoporous silica nanoparticles for pH-sensitive anti-inflammatory therapy against knee osteoarthritis, J. Nanobiotechnology. 18 (2020) 1–15. https://doi.org/10.1186/s12951-020-00651-0.
199. B. Crivelli, E. Bari, S. Perteghella, L. Catenacci, M. Sorrenti, M. Mocchi, S. Faragò, G. Tripodo, A. Prina-Mello, M.L. Torre, Silk fibroin nanoparticles for celecoxib and curcumin delivery: ROS-scavenging and anti-inflammatory activities in an in vitro model of osteoarthritis, Eur. J. Pharm. Biopharm. 137 (2019) 37–45. https://doi.org/10.1016/j.ejpb.2019.02.008.
200. D. Hu, X. Gu, W. Si, W. Qin, J. Jiao, Y.F. Hao, Biosynthesis of Silver nanoparticles using Bauhinia acuminate flower extract and their effect to promote osteogenesis of MSCs and improve meniscus injury healing, J. Photochem. Photobiol. B Biol. 197 (2019) 111536. https://doi.org/10.1016/j.jphotobiol.2019.111536.
201. Q.Q. Yang, Y.X. Shao, L.Z. Zhang, Y.L. Zhou, Therapeutic strategies for flexor tendon healing by nanoparticle-mediated co-delivery of bFGF and VEGFA genes, Colloids Surfaces B Biointerfaces. 164 (2018) 165–176. https://doi.org/10.1016/j.colsurfb.2018.01.031.
202. U. Bilthariya, N. Jain, V. Rajoriya, A.K. Jain, Folate-conjugated albumin nanoparticles for rheumatoid arthritis-targeted delivery of etoricoxib, Drug Dev. Ind. Pharm. 41 (2015) 95–104. https://doi.org/10.3109/03639045.2013.850705.
203. F. Daubiné, D. Cortial, G. Ladam, H. Atmani, Y. Haïkel, J.C. Voegel, P. Clézardin, N. Benkirane-Jessel, Nanostructured polyelectrolyte multilayer drug delivery systems for bone metastasis prevention, Biomaterials. 30 (2009) 6367–6373. https://doi.org/10.1016/j.biomaterials.2009.08.002.
204. M. Salerno, E. Cenni, C. Fotia, S. Avnet, D. Granchi, F. Castelli, D. Micieli, R. Pignatello, M. Capulli, N. Rucci, A. Angelucci, A. Del Fattore, A. Teti, N. Zini, A. Giunti, N. Baldini, Bone-Targeted Doxorubicin-Loaded Nanoparticles as a Tool for the Treatment of Skeletal Metastases, 2010.
205. K. Ramanlal Chaudhari, A. Kumar, V.K. Megraj Khandelwal, M. Ukawala, A.S. Manjappa, A.K. Mishra, J. Monkkonen, R.S. Ramachandra Murthy, Bone metastasis targeting: A novel approach to reach bone using Zoledronate anchored PLGA nanoparticle as carrier system loaded with Docetaxel, J. Control. Release. 158 (2012) 470–478. https://doi.org/10.1016/j.jconrel.2011.11.020.
206. C. Clementi, K. Miller, A. Mero, R. Satchi-Fainaro, G. Pasut, Dendritic poly(ethylene glycol) bearing paclitaxel and alendronate for targeting bone neoplasms, Mol. Pharm. 8 (2011) 1063–1072. https://doi.org/10.1021/mp2001445.
207. M. Susa, A.K. Iyer, K. Ryu, F.J. Hornicek, H. Mankin, M.M. Amiji, Z. Duan, Doxorubicin loaded polymeric nanoparticulate delivery system to overcome drug resistance in osteosarcoma, BMC Cancer. 9 (2009). https://doi.org/10.1186/1471-2407-9-399.
208. A. Barroug, L.T. Kuhn, L.C. Gerstenfeld, M.J. Glimcher, Interactions of cisplatin with calcium phosphate nanoparticles: In vitro controlled adsorption and release, J. Orthop. Res. 22 (2004) 703–708. https://doi.org/10.1016/j.orthres.2003.10.016.
209. N. Federman, J. Chan, J.O. Nagy, E.M. Landaw, K. McCabe, A.M. Wu, T. Triche, H. Kang, B. Liu, J.D. Marks, C.T. Denny, Enhanced growth inhibition of osteosarcoma by cytotoxic polymerized liposomal nanoparticles targeting the alcam cell surface receptor, Sarcoma. 2012 (2012). https://doi.org/10.1155/2012/126906.

210. X.S. Li, W.Q. Li, W.B. Wang, Using targeted magnetic arsenic trioxide nanoparticles for osteosarcoma treatment, Cancer Biother. Radiopharm. 22 (2007) 772–778. https://doi.org/10.1089/cbr.2007.352.
211. K. Sun, J. Wang, J. Zhang, M. Hua, C. Liu, T. Chen, Dextran-g-PEI nanoparticles as a carrier for co-delivery of adriamycin and plasmid into osteosarcoma cells, Int. J. Biol. Macromol. 49 (2011) 173–180. https://doi.org/10.1016/j.ijbiomac.2011.04.007.
212. W. Gu, C. Wu, J. Chen, Y. Xiao, Nanotechnology in the targeted drug delivery for bone diseases and bone regeneration, Int. J. Nanomedicine. 8 (2013) 2305–2317. https://doi.org/10.2147/IJN.S44393.
213. J.R. Woodard, A.J. Hilldore, S.K. Lan, C.J. Park, A.W. Morgan, J.A.C. Eurell, S.G. Clark, M.B. Wheeler, R.D. Jamison, A.J. Wagoner Johnson, The mechanical properties and osteoconductivity of hydroxyapatite bone scaffolds with multi-scale porosity, Biomaterials. 28 (2007) 45–54. https://doi.org/10.1016/j.biomaterials.2006.08.021.
214. S. Bose, S. Tarafder, Calcium phosphate ceramic systems in growth factor and drug delivery for bone tissue engineering: A review, Acta Biomater. 8 (2012) 1401–1421. https://doi.org/10.1016/j.actbio.2011.11.017.
215. L.T. De Jonge, S.C.G. Leeuwenburgh, J.G.C. Wolke, J.A. Jansen, Organic-inorganic surface modifications for titanium implant surfaces, Pharm. Res. 25 (2008) 2357–2369. https://doi.org/10.1007/s11095-008-9617-0.
216. B. Trajkovski, A. Petersen, P. Strube, M. Mehta, G.N. Duda, Intra-operatively customized implant coating strategies for local and controlled drug delivery to bone, Adv. Drug Deliv. Rev. 64 (2012) 1142–1151. https://doi.org/10.1016/j.addr.2012.05.016.
217. Y. Luo, G. Engelmayr, D.T. Auguste, L. da Silva Ferreira, J.M. Karp, R. Saigal, R. Langer, 3D Scaffolds, in: Princ. Tissue Eng. Fourth Ed., Elsevier Inc., 2013: pp. 475–494. https://doi.org/10.1016/B978-0-12-398358-9.00024-0.
218. A. Iovene, Y. Zhao, S. Wang, K. Amoako, Bioactive polymeric materials for the advancement of regenerative medicine, J. Funct. Biomater. 12 (2021). https://doi.org/10.3390/jfb12010014.
219. H. Li, Z. Yang, L. Fu, Z. Yuan, C. Gao, X. Sui, S. Liu, J. Peng, Y. Dai, Q. Guo, Advanced Polymer-Based Drug Delivery Strategies for Meniscal Regeneration, Tissue Eng. – Part B Rev. 27 (2021) 266–293. https://doi.org/10.1089/ten.teb.2020.0156.

Chapter 2
Drug Delivery Systems for Bone

Jacob Miszuk and Hongli Sun

Contents

J. Miszuk · H. Sun (✉)
Department of Oral and Maxillofacial Surgery, University of Iowa College of Dentistry, Iowa City, IA, USA

Iowa Institute for Oral Health Research, University of Iowa College of Dentistry, Iowa City, IA, USA
e-mail: hongli-sun@uiowa.edu

S. Murab (ed.), *Drug Delivery Systems for Musculoskeletal Tissues*,
https://doi.org/10.1007/978-3-031-55653-1_2

1 Bone Defects and Existing Surgical Treatment Strategies

Bone tissue is the second-most transplanted tissue worldwide, second only to blood. As a highly functionalized tissue type that plays a significant number of roles in the mechanical support of the body, blood cell production, and stem cell reservoir, bone is an incredibly vital tissue to the human body. Injuries and defects to bone tissue can be repaired naturally by the body under normal circumstances, but many factors can complicate defect repair, including but not limited to the size and location of the injury, the age of the patient, and other factors. Defect repair is characterized through a multistep process, encompassing numerous different cell types and signaling factors that work cohesively to generate new, functional bone to regenerate the defect site. For defects too large or complicated for the body to repair by itself, outside intervention is required to assist in complete and fast repair. This can take the form of mechanical support by biomaterial scaffolds, treatment with drugs to strengthen the body's natural repair processes, or a combination of material and scaffold, among other things. This chapter will discuss an introduction to bone defect repair and describe traditional and new drug delivery strategies aimed at aiding the repair of bone defects.

1.1 General Pathophysiology for Bone Defect Repair

Bone fracture healing is typically characterized by three major phases: (1) inflammatory phase, (2) Reparative phase, and (3) Remodeling phase. Each major phase is transiently marked by the presence of differing cell types, a varying signaling environment, with each stage overlapping to some degree. Bone typically regenerates through the endochondral ossification process—damaged tissue undergoes hypertrophy and recruitment of mesenchymal stem cells, which form a cartilaginous mid-step allowing the invasion of blood vessels, hematopoietic cells, and eventually osteogenic precursors leading to the production of mature bone and hematopoietic tissue.

The inflammatory phase generally occurs immediately after an injury or bone resection—this phase is characterized by local vasodilation, nonspecific acute inflammatory exudate production, and hyperemia in surrounding tissues. The body will begin the removal of necrotic tissue as monocytes and macrophages migrate to the damaged tissue area, osteocyte degradation occurs, and a hematoma forms at the injury or defect site. This phase is also characterized by a high volume of proinflammatory cytokines and has a high risk of infection, where many drugs or pharmaceutical treatments attempt to blunt the potentially damaging effects of an extended inflammatory phase or infection.

The reparative phase is marked by the formation of a fracture callus, a naturally formed internal fixation of the fracture site, usually within 24–48 hours after the presence of injury or defect. A hypoxic environment and lack of control in the defect zone encourage the start of the formation of a soft callus. Proliferation of periosteal multipotential cells and vasculature begins to penetrate the defect site, and a transient cartilaginous region is formed via collagen production by fibroblasts. An initial osteoid is formed by day 4 on average at the fracture ends, and complete callus formation is denoted by the presence of early-stage vasculature, type II collagen, and cartilage tissue. By the end of the first month, the callus begins to progressively mineralize and harden as it transitions to a hard callus in preparation for remodeling—this is characterized by a shift from type II to type I collagen, an increase in stiffness, and new osteoblastic differentiation to begin replacement of the cartilage tissue.

Remodeling is the final, longest phase—similar to the constant ongoing remodeling that bone tissue naturally undergoes under normal conditions. Osteoclast and osteoblast activities resorb and re-lay bone tissue and slowly change the shape of the hard callus in case of deformation, and return to an equilibrium of bone mineral absorption/resorption activity.

In the case of injury too severe or too large for full repair of bone it is termed a "nonunion." Nonunions typically happen when there is a severe lack of stability or blood flow to the affected region or some combination of both. Many additional factors can increase the risk of nonunions, which include but are not limited to:

- Use of tobacco
- Older age
- Anemia
- Diabetes
- Vitamin D levels
- Hypothyroidism
- Poor nutrition
- Complicated or compound breaks

In most cases, there is a poor supply of oxygen/blood or regenerative source of stem cells or lack of osteoinductive signals that are nonconducive to regeneration. In cases such as these, outside intervention is necessitated in order to bridge defect gaps and provide adequate healing to fully regenerate functional bone tissue and restore quality of life to the affected individual.

1.2 Existing Strategies for Critically Sized Bone Defect Repair

Critical-sized bone defects unable to self-heal without outside intervention result in nonunions, leading to a severe loss of function significantly reducing the quality of life and increasing the mortality risk of the affected patients. Multiple clinical strategies to repair critical-sized defects currently exist, each of which presents positives and drawbacks. Of these, autogenic bone grafting is described as the "gold standard" technique, providing the highest success rate for implants clinically, yet still comes with multiple undesirable drawbacks, leading us to continue searching for alternative strategies for critical-sized defect repair. The most common techniques to repair these defects and their notable advantages/disadvantages are as follows:

- Autogenic bone grafting: Nonimmunogenic, low supply, donor site morbidity, and additional surgery required
- Allogenic/Xenogenic bone grafting: Higher potential supply, potential immunogenic concerns
- Decellularized matrix grafting: Biocompatible, low osseointegration, and harsh sterilization procedure required
- Metallic/Ceramic or composite implant: Inexpensive, non-functional repair, additional surgery required, and potential stress-shielding effects
- Biodegradable polymeric scaffold: Inexpensive, highly customizable, natural and synthetic polymers available for use, and mechanical strength is relatively low compared to metals and ceramics

Of the grafting techniques, supply is limited, and nonautogenous graft sources have a high potential for immune response concerns. For man-made implants, metals and ceramics are nonbiodegradable, leading to further procedures and blocking the formation of new functional bone tissue. Due to these significant drawbacks, polymeric scaffolds are one promising alternative material that is highly customizable and biodegradable, allowing time for the slow degradation of scaffolding to give way to new functional bone tissue.

As the biomaterial scaffold is only one-third of the components of the classically described tissue engineering triangle, there is often a complementary drug-delivery strategy paired with the biomaterial in order to synergistically enhance the new bone formation generated by the healing response. Tissue healing cascades typically have a very robust growth factor signaling environment, but can be damaged or intensely disrupted due to injury or other compounding factors. Bone healing especially occupies a long period of time and involves a large number of signaling factors that drive a complex environment of cell types to migrate, proliferate, and differentiate to develop entirely new vasculature and functional bone tissue in the injury site. There are many well-researched drugs and bioactive factors that have shown demonstrable improvements in bone repair, but many hurdles to their applications exist—as sustained high dosages are often necessary to induce complete repair. Systemic drug administration faces challenges with the metabolism of drugs prior to achieving their therapeutic effects, further increasing dosage requirements, and many of the potent drugs carry high price tags, leading to costs spiraling out of control for clinical and even research usage. Due to these concerns, alternative strategies to develop different approaches to drug development and drug release are incredibly important to overcome the barriers facing functional repair of bone tissue in humans going forward.

2 Bioactive Factors for Healing of Bone Defects

2.1 Drug Delivery for Improved Osteogenesis

The primary driving force behind bone repair and increased success of defect repair strategies is the differentiation of stem cells into the osteoblast fate—cells directly responsible for the laying down of new bone tissue. Bone morphogenic proteins (BMPs) are a key player in driving new bone formation and osteogenesis as part of the Transforming Growth Factor-Beta (TGF-β) superfamily of signaling proteins. Of these, BMP2 is one of the most well-known and widely studied growth factors in the field of bone tissue engineering. BMP2 plays a central role in the development of bone and cartilage tissue, and its use is approved by the United States Food and Drug Administration (FDA) for biomedical and orthopedic applications. BMP2 signaling activates a series of SMAD proteins that eventually form a complex and translocate to the nucleus, leading to downstream gene expression regulation. Recombinant BMP2 is a potent inducer of osteogenic differentiation; however, it carries a number of significant drawbacks that impede clinical usage, namely, a very high price tag and low half-life in solution. Because bone defect healing regeneration spans a multi-month duration, repeated doses are essential to keep up the activity of the drug during healing, requiring additional upkeep and significantly skyrocketing costs. Additionally, high, repeated doses can cause unintended side effects such as ectopic bone formation. As such, strategies to find alternate drugs or protect against the degradation of BMP2 are essential to improve the efficacy of drug delivery strategies and reduce the impact on health-care costs.

Parallel agonists of BMP2 can potentially synergistically enhance BMP2-induced bone formation, strengthening new bone repair and significantly reducing the dosage of BMP2 required to generate new bone tissue in critical-sized defects. One such small molecule drug, phenamil-methanesulfonate (Phe), is a known parallel BMP2 agonist via its ability to downregulate Smurf1, a known SMAD inhibitor. In our previously published research [7], a mild dose of 20 μM of Phe is shown to significantly increase the early-stage osteogenic marker alkaline phosphatase (ALP) activity of C2C12 (mouse myoblast cell line) cells after 7 days of treatment (Fig. 2.1). However, when combined with BMP2 in solution, Phe is able to synergistically enhance the activity of even a lower dose BMP2, generating far higher ALP activity of C2C12 cells after one week in culture, as shown in Fig. 2.2. This demonstrates the capability of a small drug to enhance the activity of BMP2 at a relatively low cost to BMP2, reducing the costs and the potential for serious unwanted side effects of BMP2 treatment.

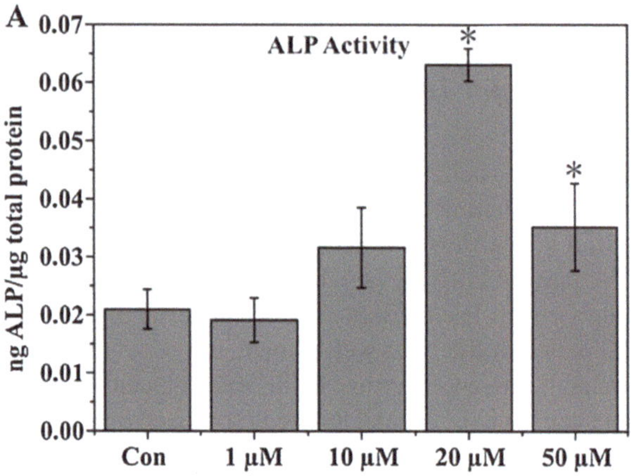

Fig. 2.1 ALP Activity of C2C12 cells incubated with various doses of Phe after 7 days. (Adapted from App. Mater. Today [7], with permission from Elsevier)

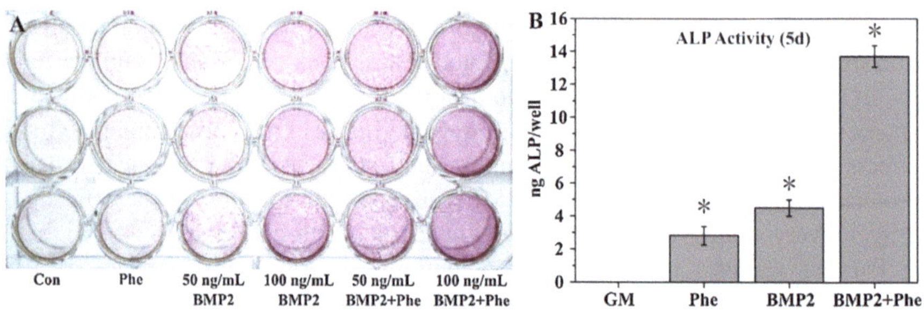

Fig. 2.2 ALP activity of C2C12 cells incubated with various doses of BMP2 and/or Phe after 7 days. ALP stain (**a**) on the left and ALP quantification (**b**) on the right. *$p < 0.05$. (Adapted from App. Mater. Today [7], with permission from Elsevier)

2.2 *Drug Delivery for Managing the Inflammatory Response*

The foreign body response to biomaterial implants and severe injury/trauma routinely have a significant inflammatory response, characterized by early-stage pro-inflammatory macrophages and later on pro-healing macrophages. A prolonged proinflammatory environment can lead to chronic inflammation, causing significant tissue damage and hampers the body's ability to regenerate lost tissue. In bone, macrophages and differentiating stem cells cross-communicate during the repair process, making it critical to manage the response of macrophages in order to effectively guide the regeneration of new bone.

Cytokines produced by immune cells such as TGF-β and interleukin (IL)-4 have shown demonstrable effects on inducing osteoblast migration, proliferation, and the production of extracellular matrix (ECM) [8]. Antagonistic production of tumor necrosis factor-alpha (TNF-α) or IL-1β from macrophages shows contrasting effects on osteoblasts, inhibiting their proliferation and differentiation in addition to promoting the activity of bone-resorbing osteoclasts. Strategies to mitigate and influence these interactions are becoming more prevalent, as biomaterial design iterates and new drugs are being developed to mediate these inflammatory responses, as they play a central role in the signaling environment during bone regeneration.

Heparin and heparin sulfates are examples of widely studied anticoagulant drugs that have been used in the treatment of embolisms, and the recent literature describes heparin's ability to modify bone substitute biomaterials to bind and neutralize pro-inflammatory proteins and guard against sepsis-associated inflammation [5]. Heparin's ability to bind growth factors via several avenues including electrostatic charge, hydrogen bonding, and hydrophobic interactions makes it an incredibly robust molecule in terms of its ability to capture many different types of drugs in addition to its inherent anticoagulant activity. Controlling/influencing the activity of macrophages and other immune cells can be an extremely potent tool to leverage the signaling environment during repair, as natural sources of growth factors can perform a far greater variety of tasks than most synthetically produced factors currently are capable of.

2.3 Drug Delivery for Improving Neovascularization During Repair

Bone is among the most vascularized tissues in the body, and the regeneration of complex three-dimensional shapes demands adequate nutrient perfusion. As such, rapid vascularization is essential for successful bone regeneration. Osteogenic and angiogenic processes in bone regeneration are closely coupled, as many growth factors play direct and indirect roles in regenerating both bone and new vasculature. Vascular endothelial growth factor (VEGF) is one critical component that plays a significant role in angiogenesis, among other proangiogenic factors such as basic fibroblast growth factor (bFGF), erythropoietin (EPO), thrombin, fibrinogen fragments, and thymosin-β4. Additionally, the cell population can be influenced by scaffold morphology and growth factors to naturally produce proangiogenic factors like VEGF through M2-polarized macrophages.

Hypoxia is a major concern during bone regeneration, as defects cause disruption in blood supply and the reduced oxygen availability results in a hypoxic environment. Hypoxia-inducible factor-1 alpha (HIF-1α) is a key upstream transcription factor in response to hypoxia stress and plays a central role in the regeneration of bone tissue by the production of VEGF, stromal-cell-derived factor 1 (SDF-1), and other signals driving recruitment of progenitor cells and angiogenesis. Drug delivery strategies aimed at activating HIF-1α could help mediate angiogenesis and downstream lead to enhanced new bone formation.

One hypoxia-mimicking agent, desferoxamine (DFO), is an FDA-approved iron chelator for several iron overload diseases. DFO acts as a downstream agonist of

HIF-1α by inhibiting the activity of the prolyl hydroxylase enzyme (PHD), a significant enzyme responsible for degrading HIF-1α. DFO has been shown in the literature to promote bone regeneration; however, it must contend with the drawbacks of a very short plasma half-life (5-and-a-half minutes on average), and high cytotoxic effects at higher dosages. Thus, local delivery of DFO at sustained low dosage is an alternative strategy to blunt the significant cytotoxic effects of higher dose administration and reduce the cost of application.

In the described work, DFO is used to treat multiple cell types to observe VEGF expression response [13]. HUVEC (human umbilical vein endothelial cells) cells and hMSC (human mesenchymal stem cells) were treated with DFO for 24 h in culture, where HUVECs showed a moderate increase in VEGF expression, and the hMSCs showed a much more significant increase in a dose-dependent manner. Even with a relatively mild dose of 20 μM in hMSCs, a significant increase in VEGF was noted after DFO administration. Downstream gene expression of BMP2 was also assayed in both cell types after 24 and 48 h, where significant increases were noted in both cell types in a dose-dependent manner (Fig. 2.3). This treatment with DFO suggests that a moderate dose would be able to both enhance new neovascularization to later lead to enhanced new bone formation.

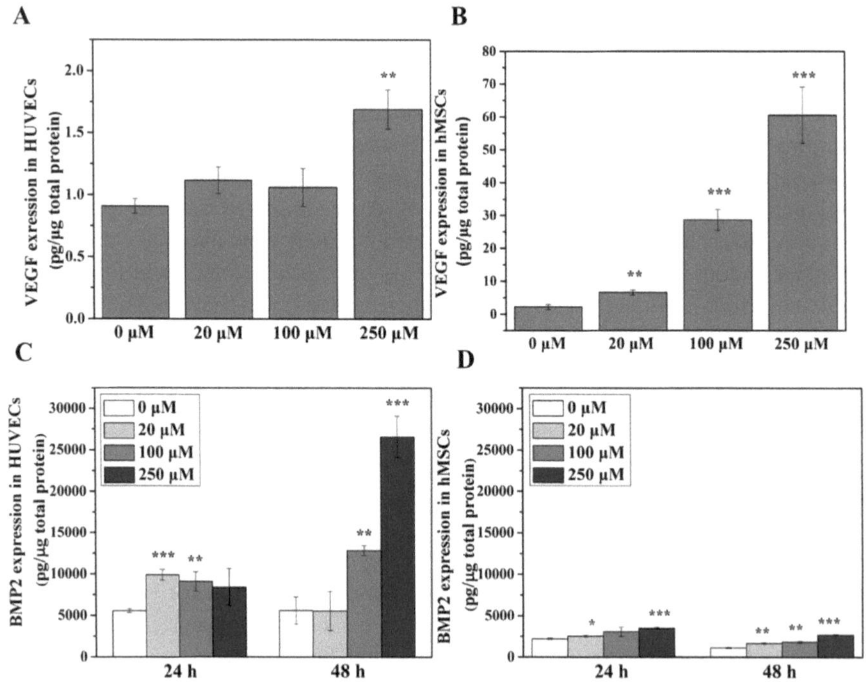

Fig. 2.3 DFO-induced expression of VEGF (**a**, **b**) and BMP2 (**c**, **d**) of HUVECs and hMSCS. (Adapted with permission from Applied Materials [13]. Copyright 2016 American Chemical Society)

Another hormone of interest in blood supply is erythropoietin (EPO). EPO is a well-studied primary erythropoietic hormone that stimulates red blood cell production in the marrow and has recently been shown to modulate BMP2-induced cranial bone formation partly by its function in improving angiogenesis [9], although its other function in bone repair is still largely elusive. In this research, EPO was introduced into gelatin scaffolds with or without BMP2 via subcutaneous injection in a mouse craniofacial critical-sized defect model after implantation. Several improvements in angiogenesis were noted after treatment in combination with BMP, as shown in Fig. 2.4, by first noting more blood vessels were observed in regenerated

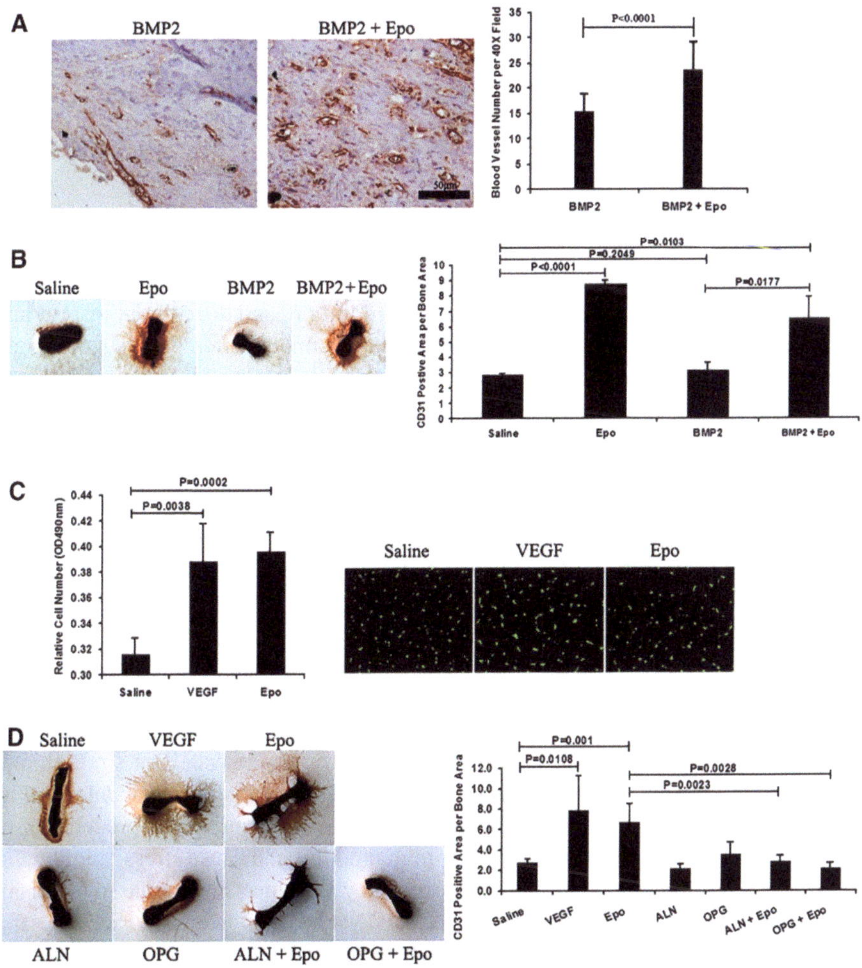

Fig. 2.4 CD31+ blood vessels in regenerated tissues at 1 week (**a**). CD31+ area of fetal mouse metatarsal assay normalized to the total bone area (**b**). Proliferation of HUVECs ($n = 4$). Data are expressed as mean ± SD (*$p < 0.05$; ***$p < 0.001$). (Adapted from Tissue Eng. Part A [9] Copyright 2012 Mary Ann Liebert, Inc)

tissues after 1 week of implantation. Fetal mouse metatarsal assay demonstrated the ability of EPO to increase angiogenesis compared to BMP2 alone or with saline. The mechanism for this improvement was not entirely clear, so the proliferation and endothelial tube formation of HUVECs were subsequently investigated, where it was found that EPO induced similar proliferation rates to VEGF and improved tube formation by VEGF in vitro. It was also noted that EPO-induced angiogenesis was largely reduced when osteoclastogenesis was blocked by ALN or OPN in metatarsal organ culture assay—leading to a suggestion that osteoclastogenesis plays an essential role in EPO-mediated angiogenesis. This is supported by the more potent improvements noted in the metatarsal bone model with the inclusion of osteoclastogenesis versus the endothelial tube formation model.

3 Biomaterials-Based Drug Delivery Platforms for Bone Defect Repair

3.1 Scaffold-Based Release Systems

Complementary to many drug delivery platforms is a biomaterial scaffold—initially developed as a matrix for cell attachment, proliferation, and differentiation, but has evolved over time to act as a depot and carrier for many different types of drugs, small molecules, genes, proteins, and other growth factors to assist with tissue regeneration. Biomaterial scaffolds can be seeded with external growth factors or be synthesized to effectively capture and act as a reservoir for endogenously produced factors, increasing their effectiveness in generating new functional bone tissue in critical-sized defects. Numerous techniques to enhance scaffolds with drug are described in the literature, including but not limited to soluble diffusible coatings, encapsulated drug released by material degradation, and physically or chemically adsorbed/embedded drugs on material surfaces. Each strategy must be carefully chosen to complement the type of scaffold with which it is implemented—as differing types of biomaterial scaffolds have their sets of advantages and disadvantages for bone defect repair.

Broadly, biomaterial scaffolds for bone defect repair can be split into three major categories: metals, ceramics, and polymers. Metals provide excellent strength and physical properties to support defects; however, they are nonbiodegradable, and as a result, they fail to restore bone function and require subsequent procedures to maintain and remove. Additionally, their high modulus can cause stress-shielding at defect sites resulting in reduced healing efficiency. Ceramics are similar to metals in providing good mechanical support to defects, yet are also less biodegradable and are very brittle, increasing the risk of fracture and causing damage to surrounding tissue. Polymeric biomaterials, while not as mechanically strong as metals, offer a biodegradable route to defect repair, opening up the door for the restoration of function and laying down of new endogenous tissue. Additionally, polymer-based scaffolds have a high degree of customizability (especially synthetic polymers), allowing them to be tailored to degrade at specific speeds, be functionalized with bioactive moieties, and have biomimetic nanofibrous morphologies, among other advantages.

Polymer-based scaffolds are one type of biomaterial scaffold that has seen frequent usage as base materials for drug-delivery platforms due to their aforementioned customizability. Tailorable degradation rates can readily be paired with the release rate of drugs, and functional groups on the surfaces of polymers can link several different types of proteins or small molecules or be supplemented with other physical attachments such as micro- or nano-particle carriers (to be discussed later in the chapter). Biomimetic scaffolds seeking to emulate the endogenous extracellular matrix (ECM) are a recently popular type of scaffold as they are favorable for cell attachment and the high surface area of the small nanofibers allows for a high degree of drug or functional conjugations. An example of 3D nanofibrous scaffolds is shown in Fig. 2.5, where the polymer polycaprolactone (PCL) is synthesized via thermally induced self-agglomeration (TISA) of electrospun fibers (A, B) and thermally induced phase separation with porogen (TIPS&P) (C, D) [10]. In this research, scaffolds synthesized via the TISA technique and treated with BMP2 could form some new functional bone in an in vivo ectopic model (Fig. 2.6); however, the new bone area was limited to the outer edges of scaffolds. Due to this, significant

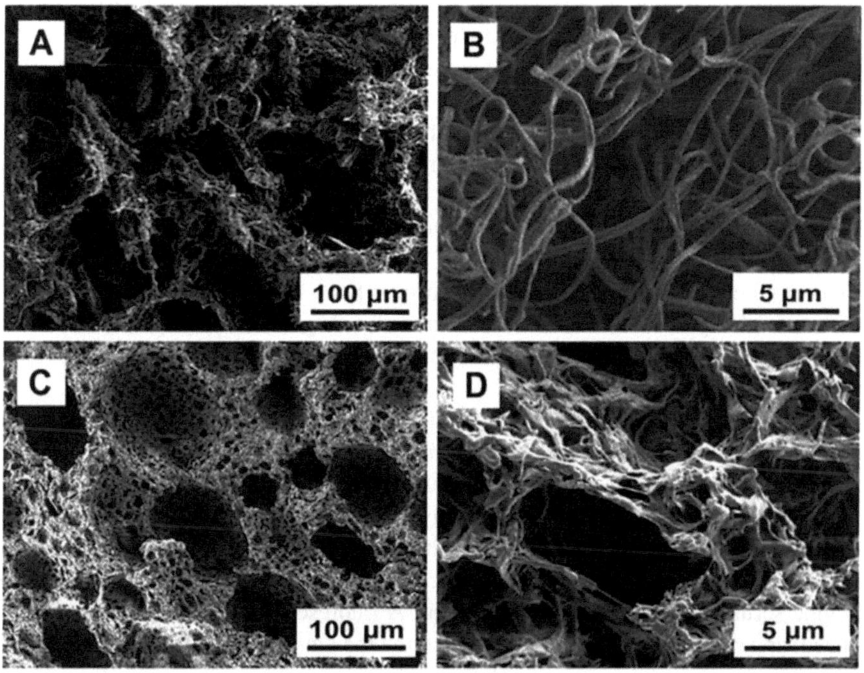

Fig. 2.5 Scanning electron microscope image showing morphology of 3D PCL scaffolds. (Adapted from Adv. Healthcare Mats. [10]. Copyright John Wiley and Sons)

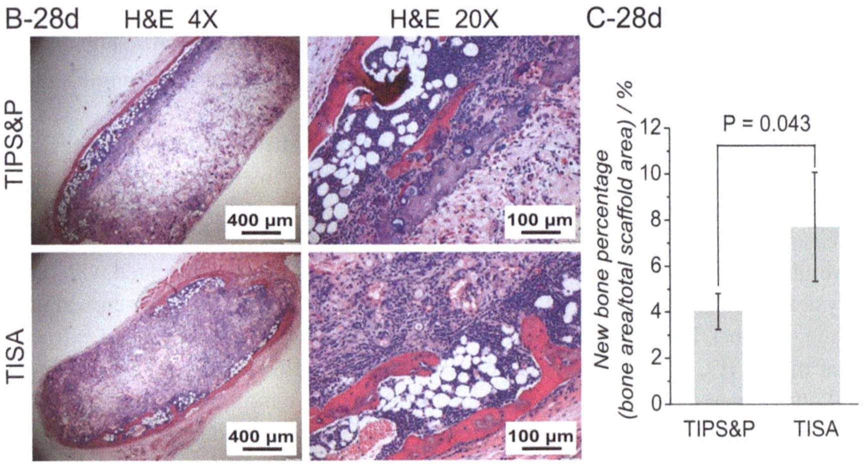

Fig. 2.6 BMP2-induced new ectopic bone formation in PCL scaffolds after 28-day implantation. (Adapted from Adv. Healthcare Mats. [10]. Copyright John Wiley and Sons)

advances could be made to polymer chemistry and improvements to add drugs or other soluble factors to enhance the new bone formation in vivo.

One method for capturing drug and growth factors to scaffolds is physical adsorption/attachment. The small molecule Phenamil, as mentioned earlier in the chapter, is a logical treatment to help improve new bone formation generated in vivo. Choosing an ideal method of conjugation for attachment and release is critical in order to effectively capture and release the drug while limiting any potential side effects. For this modification, a synthetic hydroxyapatite (HA) crystal coating is applied to the TISA (Fig. 2.7)—a natural bone-like mimicking functionalization that emulates the chemistry of the natural bone niche and is also able to improve the function of scaffolds when in combination with Phe [7]. In vivo mouse ectopic bone formation showed that a combination of HA and Phe administered to TISA scaffolds resulted in a significant increase in new bone formed in harvested ossicles with a single application of both BMP2 and Phe (Fig. 2.8). Moving further in this modification, a sustained release of drug could potentially further improve the function of the biomaterial scaffold.

In the follow-up work, the TISA scaffolds are coated completely with an even coating of HA (Fig. 2.9), and the Phe is co-precipitated into the HA layer [6]. Dissolution of the HA layer while in solution would lead to a sustained release of Phe over time, as compared to Phe simply physically dropping onto the surface of the same type of scaffolds (Fig. 2.10c). The HA coating technique was also demonstrated to work with not only small molecules such as Phe but also a model protein bovine serum albumin (BSA) (Fig. 2.10a). A fluorescent label on BSA was introduced, and confocal microscopy demonstrated the amount of BSA remaining in the scaffolds after up to 3 weeks in solution, where the group with physically surface adsorbed BSA (Fig. 2.10b, top row) was significantly released after 2 weeks, but still much remained in the HA layer after up to 3 weeks (Fig. 2.10b, bottom row).

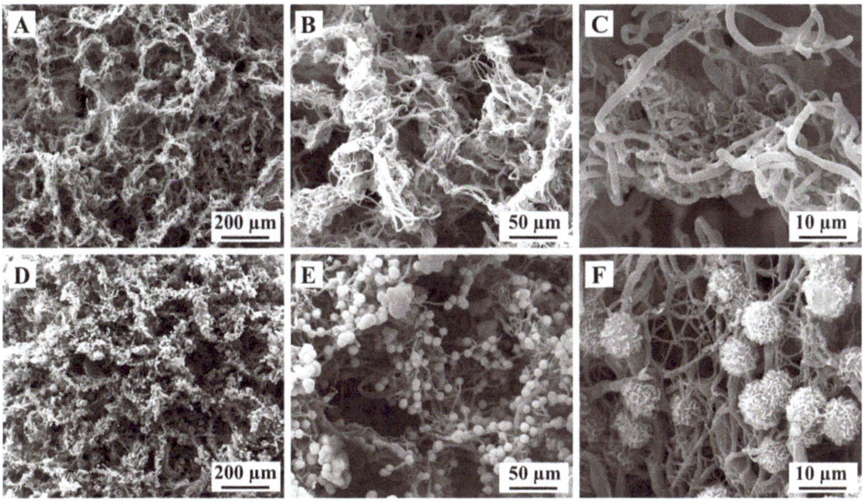

Fig. 2.7 SEM morphology of TISA scaffolds (top row) and TISA-HA scaffolds with HA crystals (bottom row). (Adapted from App. Mater. Today [7], with permission from Elsevier)

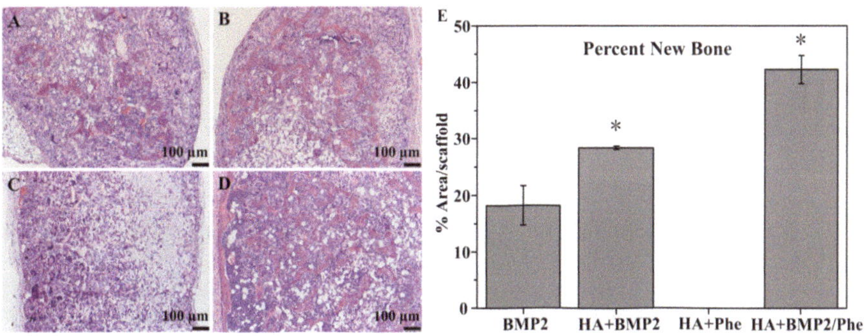

Fig. 2.8 BMP2-induced new ectopic bone formation in PCL scaffolds after 28-day implantation. BMP2 only (**a**), BMP2+HA (**b**), Phe+HA (**c**), BMP2/Phe+HA (**d**). (Adapted from App. Mater. Today [7], with permission from Elsevier)

Follow-up analysis of the osteogenic effect induced by the Phe-incorporated TISA+HA scaffolds noted a few significant changes, notably that the ALP activity of C2C12 cells was significantly improved in the incorporated Phe group as compared to the physically adsorbed group (Fig. 2.11a).

Gene expression of master osteogenic transcription factor Runx2 was also significantly enhanced after 10 days of C2C12 cells cultured on the scaffolds with the incorporated Phe as well (Fig. 2.11b). Through this method, physical incorporation within a mineral apatite layer, sustained release of Phe demonstrated a strong capability of improving multiple early- and late-stage osteogenic markers that strongly suggest success in the formation of new bone in vivo.

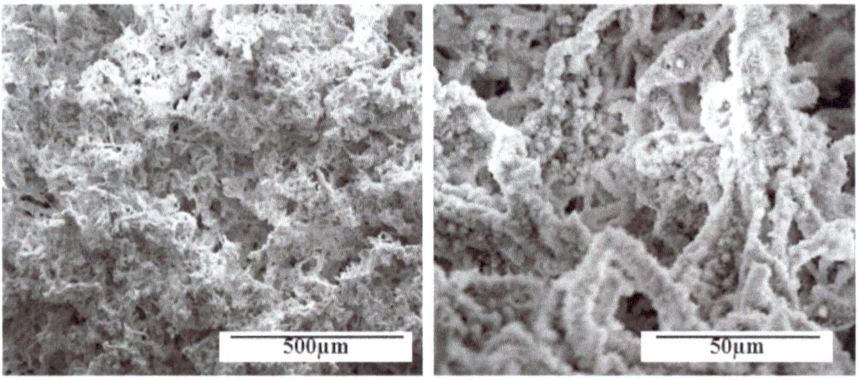

Fig. 2.9 SEM morphology of TISA-HA coated scaffolds. (Adapted with permission from ACS App. Bio. Mater. [6]. Copyright 2021 American Chemical Society)

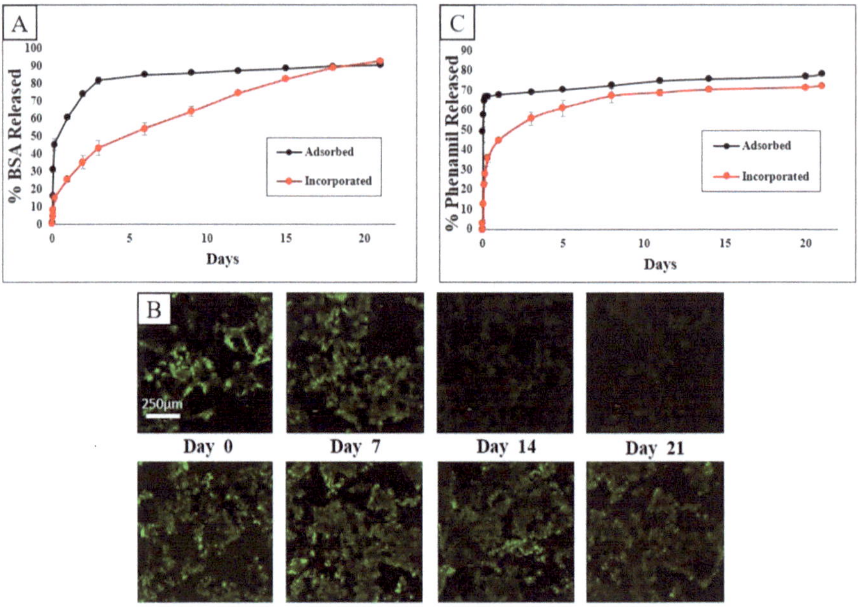

Fig. 2.10 Release profile of BSA from scaffolds visualized as percentage (**a**) and fluorescence time-lapse of adsorbed BSA (**b**, top row) and incorporated BSA (**b**, bottom row). Release profile of Phenamil (**c**). (Adapted with permission from ACS App. Bio. Mater. [6]. Copyright 2021 American Chemical Society)

Some types of drugs or small molecules are unable to physically adhere/attach or are difficult to do so due to various reasons—in this case, a chemical attachment or conjugation may be necessary to provide sustained release of drug. Described earlier in the chapter was DFO's contribution to angiogenesis—how moderate doses can generate significant improvements to angiogenic markers. However, higher

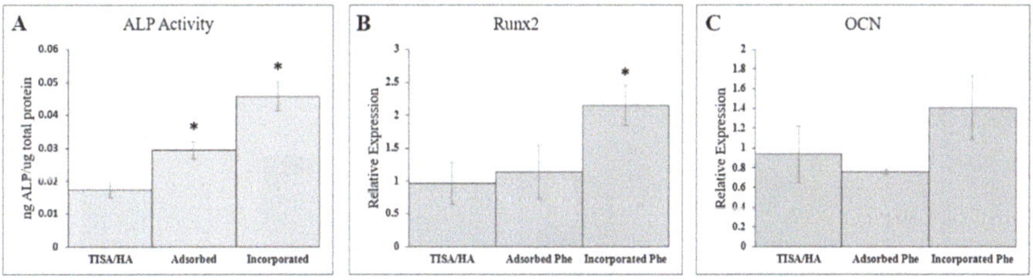

Fig. 2.11 ALP activity of C2C12 cells cultured on dishes in the presence of Phe (**a**). Osteogenic gene marker expression of Runx2 (**b**) and OCN (**c**), as measured after culturing for 10 days within TISA/HA scaffolds. Data are expressed as mean ± SD ($n = 3$, *$p < 0.01$). (Adapted with permission from ACS App. Bio. Mater. [6]. Copyright 2021 American Chemical Society)

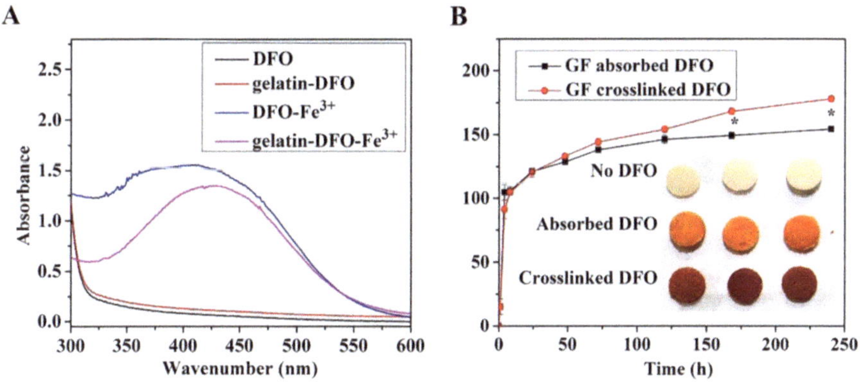

Fig. 2.12 Iron binding test and in vitro release of DFO from GF scaffolds. (Adapted with permission from ACS Appl. Mater. Interfaces [13] Copyright 2016 American Chemical Society)

doses showed more significant improvements but at the cost of cytotoxicity. This is addressed in the following research by covalent immobilization of DFO on to gelatin scaffolds (GF-DFO) [13]. DFO released from GF-DFO scaffolds is visualized in Fig. 2.12, where a higher amount of DFO was released overall from the crosslinked GF scaffolds after 10 days in solution. Additionally, more DFO is still retained in the crosslinked scaffolds as noted by the captured Fe^{3+}, suggesting even further potential for longer sustained release, and that the crosslinking is able to capture significantly more DFO that by simple surface absorption. Following up, cells cultured on the GF-DFO scaffolds showed improved viability for low doses of DFO for the HUVEC cells and no significant change in viability for hMSCs (Fig. 2.13a). DFO also induced no significant change in the morphology of either cell type after 16 h of culture, indicating cells could still easily attach and spread to GF-DFO scaffolds (Fig. 2.13b). VEGF expression of both cell types was notably increased on GF-DFO scaffolds after 24 h of culture (Fig. 2.14), indicating the DFO release was capable of improving angiogenesis by mimicking hypoxic conditions. Furthermore, a mouse cranial bone defect model was used to investigate the DFO release function

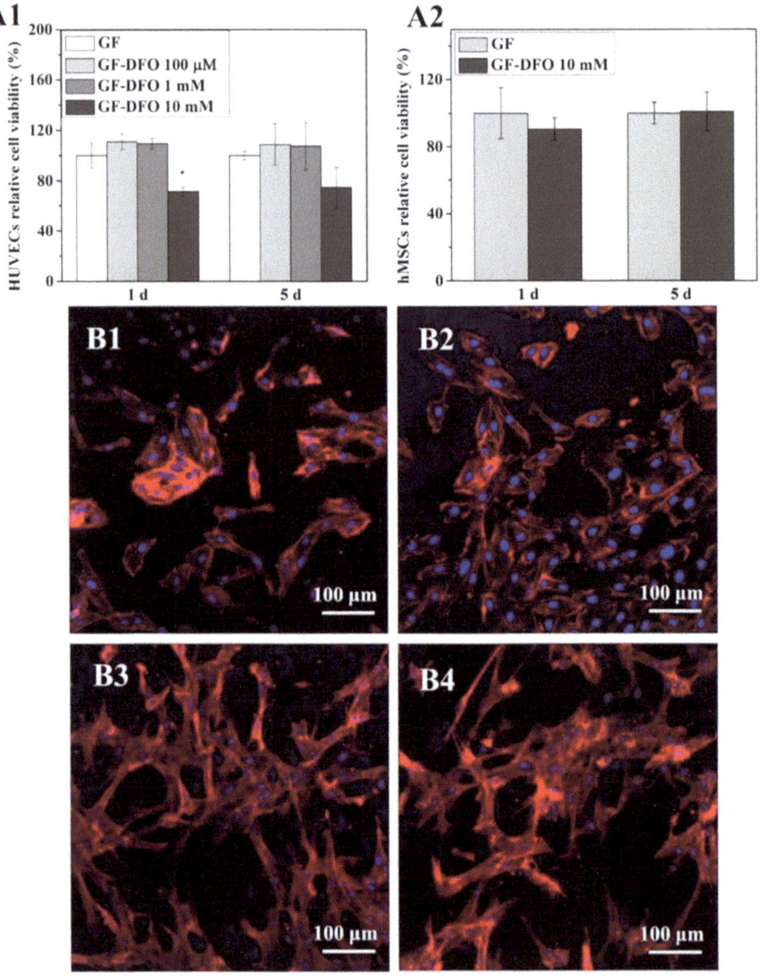

Fig. 2.13 Cell viability on GF vs GF-DFO scaffolds (A). Confocal microscopy detailing cell morphology of HUVECs on GF (B1) and GF-DFO (B2) and hMSCs on GF (B3) and GF-DFO (B4) scaffolds. (Adapted with permission from ACS Appl. Mater. Interfaces [13] Copyright 2016 American Chemical Society)

in vivo, where it was found that a significant amount of a new bone was formed in the GF-DFO group as compared to just GF alone. Only 1/5 samples in the GF group had any new bone tissue at all in the defect area, while 6/8 samples in the GF-DFO group had a new bone area (Fig. 2.15). The total area fraction of new bone tissue and number of scaffolds with a positive amount of new bone is given in Table 2.1. Overall, the elevated VEGF expression in both hMSCs and HUVEC cells suggests that the new bone formation is mediated through improved angiogenesis, although further experiments to analyze new blood vessel formation in vivo could confirm this in the future at different time points.

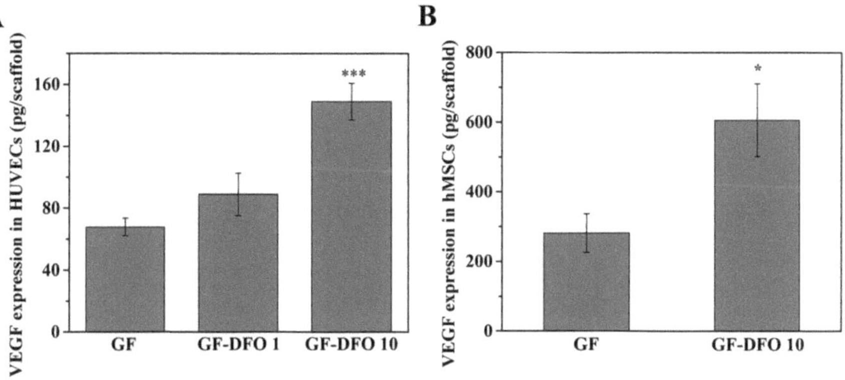

Fig. 2.14 VEGF expression of HUVECs and hMSCs cultured on GF and GF-DFO scaffolds after culturing for 24 h ($n = 3$). Data are expressed as mean ± SD ($*p < 0.05$; $***p < 0.001$). (Adapted with permission from ACS Appl. Mater. Interfaces [13] Copyright 2016 American Chemical Society)

3.2 Synthetic Micro- and Nano-particle Carrier Systems

Micro- and Nano-particle carriers represent a significant portion of drug delivery techniques, commonly utilizing a polymeric core to encapsulate drugs to aid in bone repair. Other types of nanoparticles can link and capture charged drugs or small molecules and bind them for sustained release or be utilized to more effectively capture and protect endogenously produced factors such as BMP2 during the repair process. Particulate carriers enjoy many advantages over biomaterial scaffolds, namely, their ability to be injected in a minimally invasive manner. Carriers can also be customized to carry numerous types of drugs, tailored release rates, and attach targeting modes to enable precision-release of payload, maximizing the effectiveness of the release system. Additionally, many types of nanoparticles can be conjugated or linked to 3D scaffolds to combine their benefits.

One example particle recently described for drug release is nanosilicate nanoclay (NS), a 2D nanomaterial for bone tissue engineering, which has demonstrated some capacity for improving mechanical properties of scaffolds, possesses intrinsic osteoinductivity, and capacity for drug release [1]. NS have a high specific surface area and charge anisotropy, making them effective particles for drug binding and sustained release. In the previously described article, NS alone is capable of improving several osteogenic markers of hMSCs, including ALP activity after 7 days, mineralization of calcium contents after 21 days, and osteogenic marker gene expression of ALP and bone sialoprotein (BSP) after 7 days of culture (Fig. 2.16) [11]. While noting these significant benefits of osteogenic markers, the mechanism and contribution of NS toward these increases were still unclear. It is noted that NS themselves could degrade into pro-osteblastic products (i.e., SiO_4^{4-}, Mg^{2+}, and Li^{+}), which is one potential mechanism for the improvement in several of the osteogenic markers. However, the functionalization of GF scaffolds with only NS alone failed

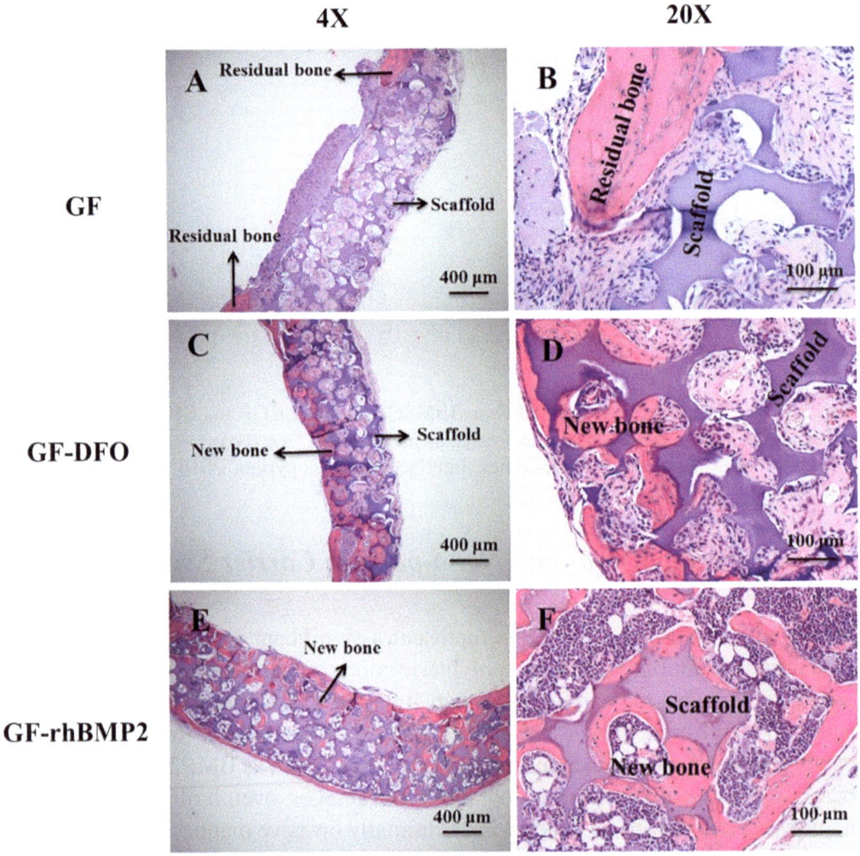

Fig. 2.15 Histological examination of repaired calvarias of the GF group (**a**, **b**), GF-DFO group (**c**, **d**), and GF-rhBMP2 group (**e**, **f**) after 6 weeks implantation in vivo. (Adapted with permission from ACS Appl. Mater. Interfaces [13] Copyright 2016 American Chemical Society)

Table 2.1 New bone formation in GF, GF-DFO, and GF-rhBMP2 groups after 6 weeks of implantation

	GF	GF-DFO	GF-rhBMP2
Scaffolds with new bone formation/total scaffolds	1/5	6/8	7/7
New bone area fraction	0.20%	3.87%	22.08%

Adapted with permission from ACS Appl. Mater. Interfaces [13] Copyright 2016 American Chemical Society

to translate to new bone formation in a mouse ectopic bone model, but a combination with BMP2 did improve significant improvements in new bone (Fig. 2.17).

The significant improvement in new BMP2-induced bone formation in GF scaffolds led to further questions on the potential mechanism of NS role in promoting new bone formation. One potential reason was NS ability to protect the activity of

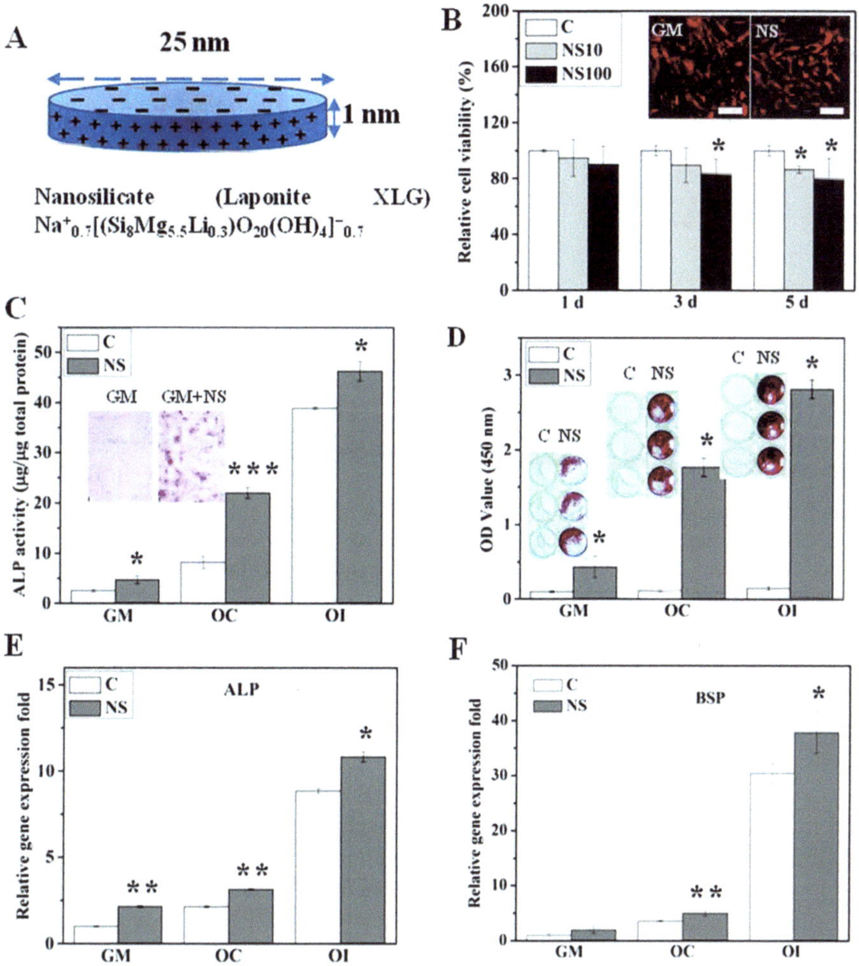

Fig. 2.16 NS representative morphology and formula (**a**). Cytotoxicity assay of hMSCs after 1 day of culture with NS (**b**). NS effect on ALP activity of hMSCs after 7 days culture (**c**). NS effect on the calcium content of hMSCs after 21 days culture (**d**). Osteogenic marker gene expressions after 7 days culture (**e**, **f**). (Reproduced from J. Mater. Chem. B. [11] with permission from the Royal Society of Chemistry)

BMP2 and prevent degradation in solution. This hypothesis was strongly supported by data from a follow-up experiment using calvarial-derived preosteoblast MC3T3-E1 cells in Fig. 2.18 [3]. A single dose of BMP2 combined with NS was compared to multiple dose application of BMP2 and early-stage osteoblast marker ALP and late-stage mineral content was evaluated. ALP activity of NS plus a single dose of BMP2 had a comparable level to that of the group treated with multiple doses of BMP2 without NS. After 3 weeks of treatment, a single dose of BMP2 combined with NS had a similar mineral content to the group treated with six doses

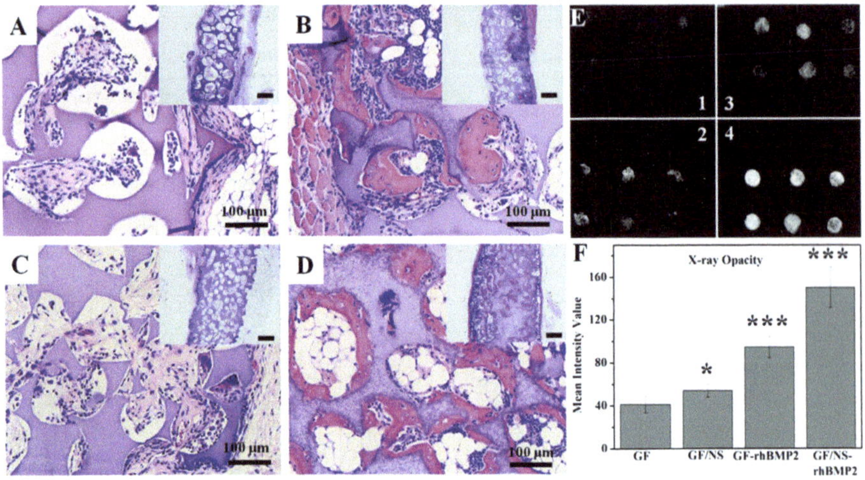

Fig. 2.17 Histological examination of repaired ossicles of the GF group (**a**), GF-rhBMP2 group (**b**), GF-NS group (**c**), and GF-NS/rhBMP2 group (**d**) after 6 weeks implantation in vivo. (Reproduced from J. Mater. Chem. B. [11] with permission from the Royal Society of Chemistry)

over time, suggesting that NS could bind and protect the activity of BMP2 in solution over a long period of time, and still maintain its potent pro-osteogenic capabilities. These data were further supported by real-time PCR analysis of osteogenic gene markers Runx2, BSP, and OCN (data not shown), displaying similar trends to the ALP and calcification data.

Evaluating the sustained release of BMP2 from NS, it became clearer the role of NS in protecting the activity of BMP2—physical adsorption of an equal amount of BMP2 dropped onto scaffolds without NS resulted in higher burst release and overall lower total amount released after four weeks in solution (Fig. 2.19). In contrast, significantly more BMP2 was released from the GF+BMP2/NS group, suggesting direct NS binding with BMP2 protected the activity of BMP2 over a longer period of time. As new bone regeneration is the ultimate goal, in order to better understand the role of NS in functional bone regeneration, new bone formation was evaluated with NS functionalized scaffolds in a mouse calvarial defect model. After 6 weeks of implantation, the GF+BMP2/NS group demonstrated a significantly higher volume of newly formed bone in defects compared to other groups (Fig. 2.20). Radiographic and histological examination showed very limited new bone formed in groups with lose dose (0.5 μg per scaffold) of BMP2, but the same dose complemented with NS completely covered the defect sites with new bone tissue. More new bone tissue was observed in H&E staining, and bone volume per tissue volume was significantly increased in the GF+BMP2/NS group compared to each other group. Overall, these data strongly suggest NS function is the ability to modulate the binding of BMP2 and provide sustained release, leading to significantly enhanced pro-osteogenic markers in stem cells and ultimately strong new bone formation in a cranial defect animal model.

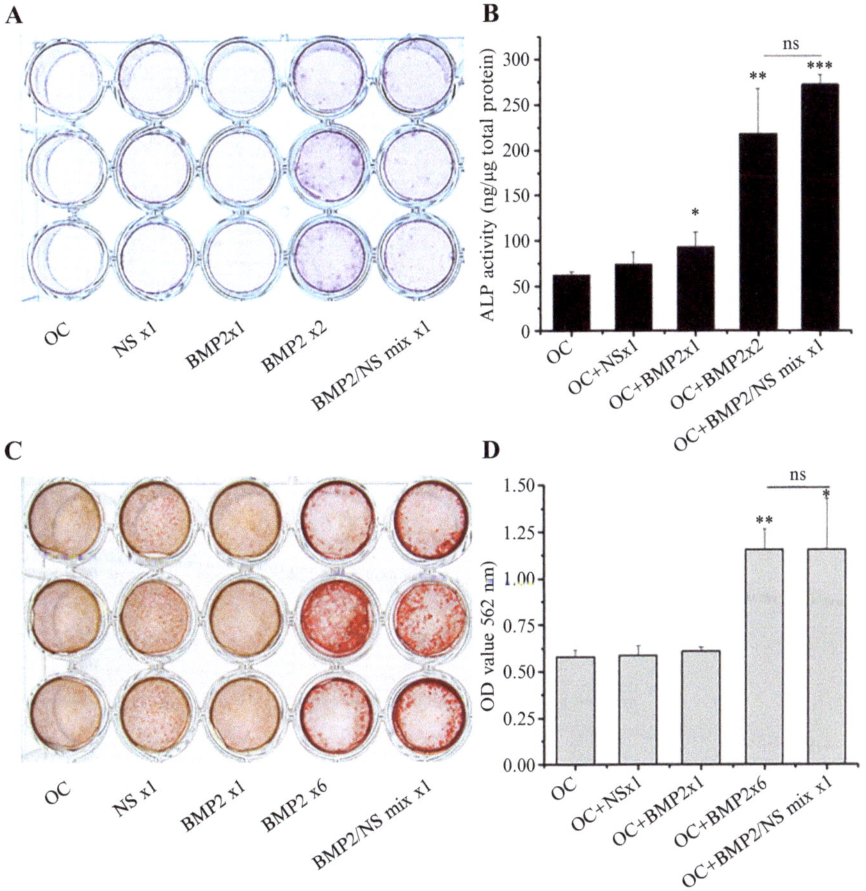

Fig. 2.18 ALP staining (**a**) and quantification (**b**) of MC3T3-E1 cells after 7 days. Alizarin Red-S staining (**c**) and CPC quantification (**d**) of MC3T3-E1 cells after 21 days. (Adapted from Appl. Mater. Today [3], with permission from Elsevier)

3.3 Multiphasic Drug Release

Single-drug strategies usually have limited effectiveness in guiding new bone repair, as natural bone repair is driven by numerous factors, encompasses many differing cell types, and lasts from weeks to several months. No currently known single-drug application strategies can completely and effectively encompass osteogenesis, inflammation control, and angiogenesis, among other facets of the long temporal process of bone repair. Emerging strategies to fit multiple drugs into release systems with staggered release profiles and timings are beginning to be described in the literature, aimed at further customizing the abilities of these systems to deliver timely sustained doses of drugs to improve the healing process.

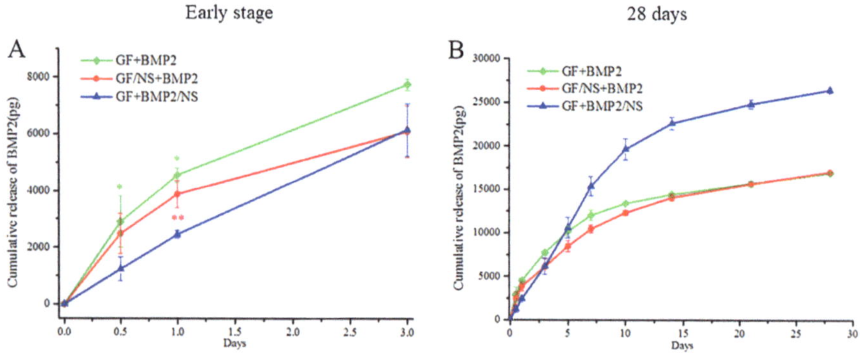

Fig. 2.19 In vitro release curves of BMP2 from scaffolds at early stage (**a**) and later stage (**b**). (Adapted from Appl. Mater. Today [3], with permission from Elsevier)

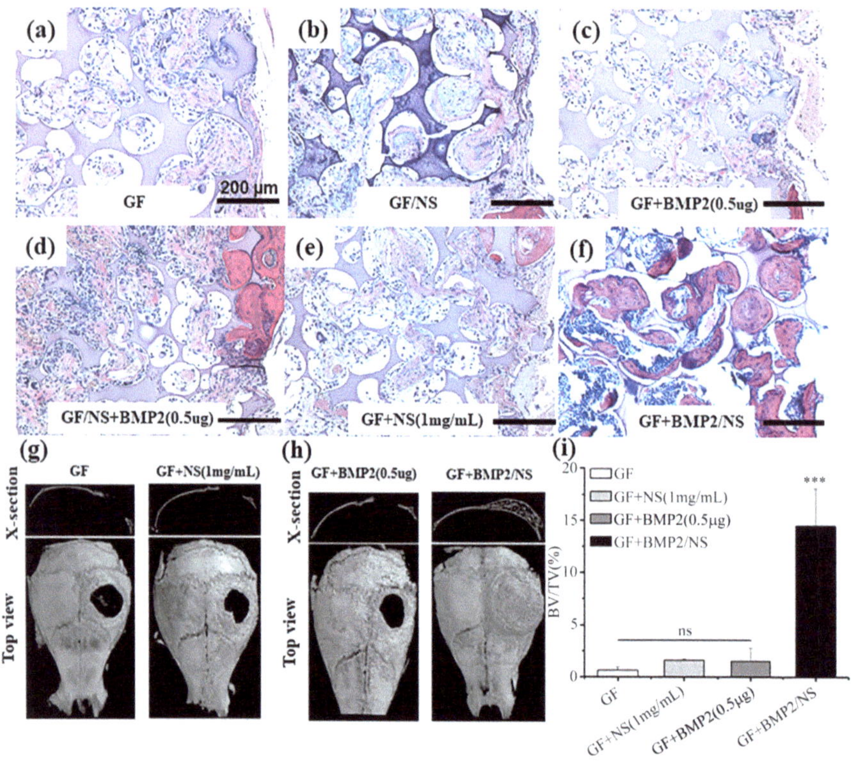

Fig. 2.20 H&E staining (**a–f**) and uCT analysis (**g–i**) of restored calvaries after 6 weeks implantation. Data are expressed as mean ± SD (n = 4~5, ***$p < 0.001$ versus GF group). (Adapted from Appl. Mater. Today [3], with permission from Elsevier)

Angiogenesis alone is driven by several factors during bone repair. As earlier described in the chapter, DFO is a hypoxia-mimicking drug that can activate HIF-1α. BMP2 is another growth factor that strongly encourages angiogenesis as well, and it has been hypothesized that a combination delivery of both factors could significantly enhance the new generation of vasculature in bone regeneration. Delivery of both drugs individually may require high dosage and incur significant side effects, and a combined delivery could synergistically enhance the angiogenesis at lower doses, saving on costs and reducing the risk of nonspecific effects or cytotoxicity.

In the described research, we designed a nanofibrous, biomimetic 3D scaffold loaded with both DFO and BMP2 to deliver each drug with distinct release kinetics to aid in bone regeneration in vivo ([12]. GF scaffolds were conjugated with BMP2 via mesoporous silica nanoparticles (MSN), and DFO was crosslinked with chitosan (CTS) and added to the GF/MSN-BMP2 scaffolds. Scaffolds were then evaluated for release of both drugs over time, and their effects on osteogenic differentiation of stem cells were investigated (Fig. 2.21).

Release curves of BMP2 from MSN and DFO from the CTS conjugation were first evaluated (Fig. 2.22). Data indicated that the mesoporous structure of the MSN

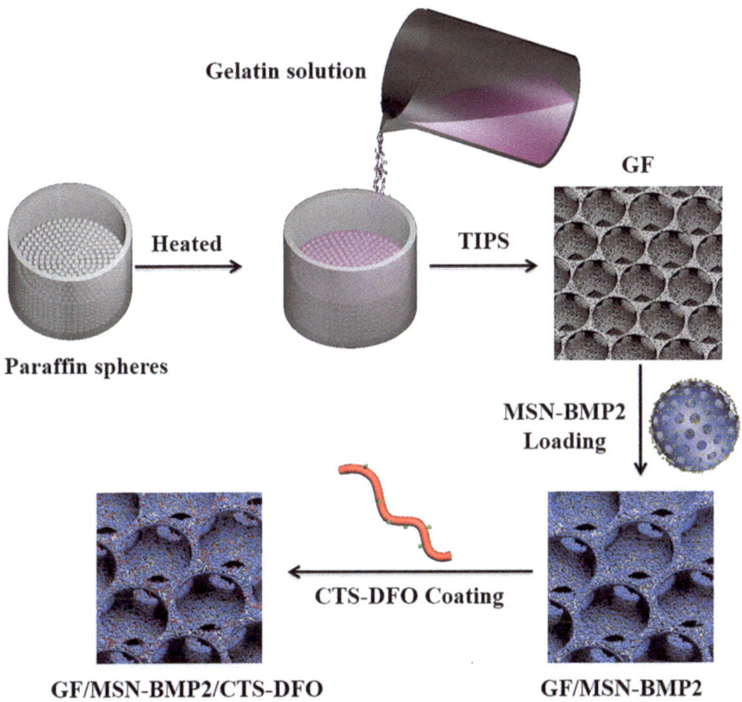

Fig. 2.21 Schematic for the synthesis of GF/MSN-BMP2/CTS-DFO scaffolds. (Adapted from J Control Release [12], with permission from Elsevier)

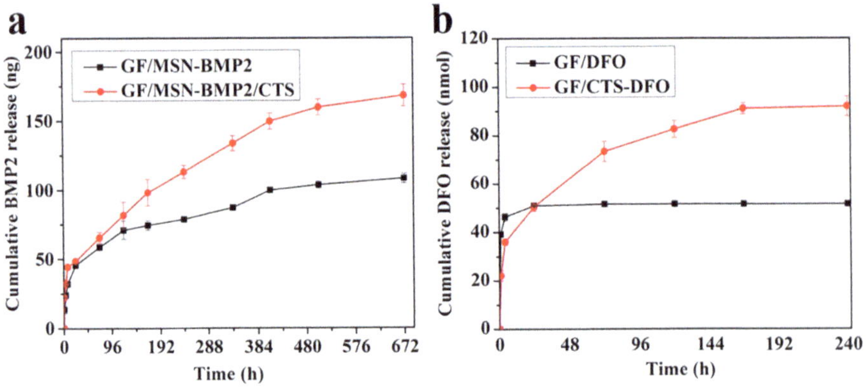

Fig. 2.22 Release curves of BMP2 and DFO from GF scaffolds. (Adapted from J Control Release [12], with permission from Elsevier)

could support long-term release of BMP2 from scaffolds, and the coating of CTS could further protect the activity over time to prolong its release. DFO incorporated via covalent conjugation with the CTS also showed a smaller burst release and much higher sustained release amount over 10 days, indicating the CTS could also protect the activity of the DFO as well. Also, interestingly to note, the BMP2 was continually detected up to 4 weeks of release while DFO was mostly completely released by 10 days. As DFO is intended to promote angiogenesis and the BMP2 is more focused on the generation of new bone, this difference in release rate could more effectively improve the early-stage angiogenesis during bone repair via DFO release and also improve the new bone formation in the later stages via the BMP2.

hMSCs showed no significant changes in cell viability with either coating or conjugation of drugs and CTS, indicating general safety (Fig. 2.23). VEGF expression was significantly increased in the DFO scaffold group, while even further significantly increased in the combined BMP2/DFO group, but interestingly the BMP2-alone group did not enhance the VEGF expression. This further lends credence to the hypothesis that angiogenesis is affected by multiple factors during bone repair. Further investigation into osteogenic factors would shed light on the combined effect of DFO+BMP2 release for bone regeneration, where real-time PCR analysis of early-stage marker Runx2 and late-stage marker OCN were observed (Fig. 2.24). Early-stage marker Runx2 was noticeably lower in the combined DFO+BMP2 group than the BMP2 alone, suggesting that DFO inhibits an early-stage osteogenic differentiation. However, later-stage marker OCN and the subsequent total calcium content from scaffolds after 21 days was the highest in the combined group, also suggesting the combined treatment still resulted in a significantly improved effect on osteogenesis.

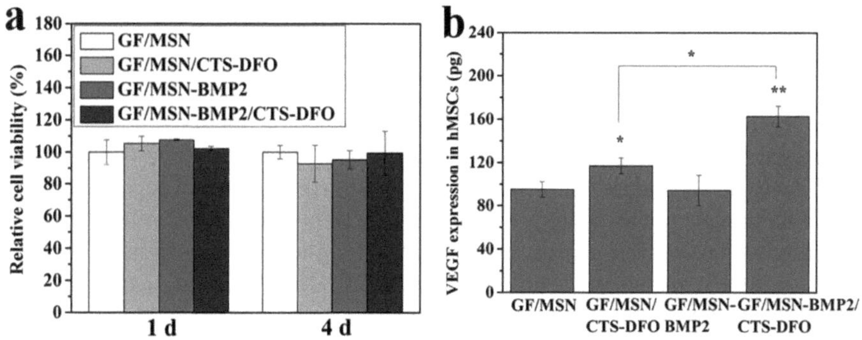

Fig. 2.23 C2C12 cell viability on scaffolds (a) and VEGF expression after 24 hours in hMSCs. Data are expressed as mean ± SD ($n = 3$,*$p < 0.05$, **$p < 0.01$). (Adapted from J Control Release [12], with permission from Elsevier)

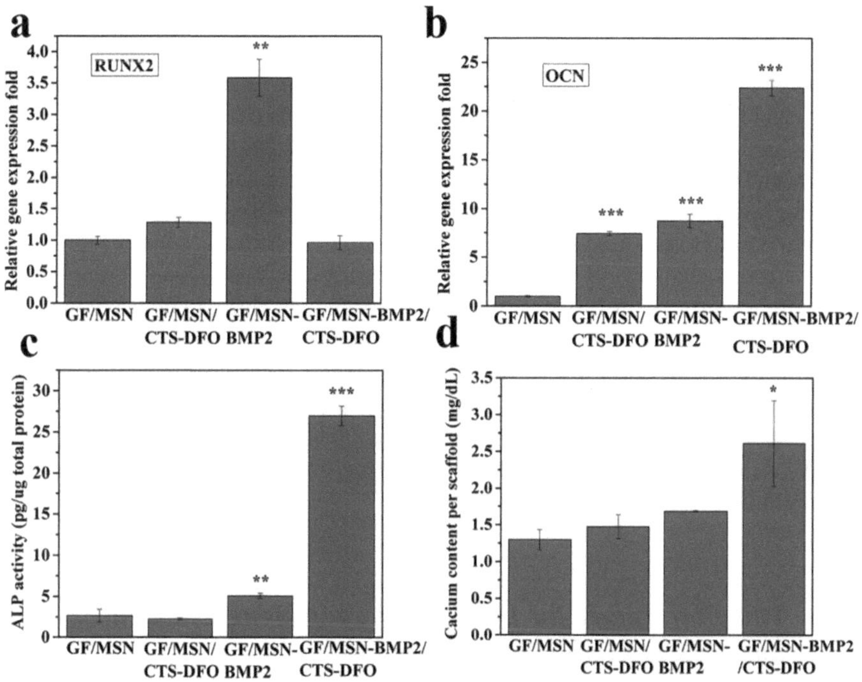

Fig. 2.24 C2C12 cells osteogenic gene marker expression of Runx2 (**a**), OCN (**b**) as studied by the real-time PCR assay. Quantitative ALP activity of C2C12 cells (**c**) and total calcium content of hMSCs (**d**). ($n = 3$,*$p < 0.05$, **$p < 0.01$, ***$p < 0.001$). (Adapted from J Control Release [12], with permission from Elsevier)

3.4 Triggering Controlled Release Systems (Smart Release)

Due to the high complexity and transient nature of the signaling environment during bone regeneration, a high degree of control in drug release offers significant benefits in improving the tissue regeneration process. While many material-based release systems typically release their payload as a function of biomaterial degradation rate, some limitations to this strategy include an overall poor control of release rate and kinetics, which can be especially critical during long-term implantation during bone healing. To this, there is a growing field of "smart release" systems that can be triggered to begin their release in response to an internal or external stimulus. These types of drug release platforms are able to enable precise, simultaneous, or sequential release of single or multiple drug payloads. Some examples of environmental triggers that can induce the release of drugs include pH, temperature, enzymes, or mechanical strain, while external triggers can be set off via elements such as magnetic fields, ultrasound, or infrared radiation. Local pH ranges can change wildly during bone regeneration, as osteoblast/clast activity can induce changes in local pH and be used as a trigger to release pH-sensitive linkages in functional groups or coatings to release drugs. Temperature increase in response to infection and/or inflammation can be used to trigger the release of antibacterial or anti-inflammatory drugs and blunt the acute damaging effects they can cause in regenerating tissue. Ultrasound-triggered release systems can be protected against most environmental effects and be used to release drugs at a much later time point locally without necessitating further procedures or systemic administration of drugs.

While smart-release systems demonstrate high promise in bone regeneration, there are still challenges facing their application in the harsh, long-duration bone regeneration process. It is difficult to engineer materials that can withstand a harsh biological environment for an extremely long period of time and retain the ability and function to precisely trigger the release of drugs. Investigations into long-term safety and toxicity in bone tissue and osteogenic potential are still unclear, especially with very limited clinical information available. Yet, despite these concerns, many of these technologies are rapidly evolving to meet and overcome the challenges of bone regeneration.

3.5 Potential Future for Drug Delivery in Bone: Natural and Nonimmunogenic Carriers

Two significant drawbacks of using synthetic carriers in drug delivery are the potential immune response and harmful degradation by-product(s). Rapid clearance by the mononuclear phagocyte system can significantly hamper attempts to deliver drugs via synthetic nanocarriers. Typically, these limitations can be mitigated through smart engineering and the use of materials that avoid toxic or acidic by-products, but the effects can rarely be eliminated completely. New emerging techniques in drug delivery that utilize natural or nonimmunogenic carriers such as exosomes or DNA/RNA-based aptamers can take advantage of their natural ability for cell–cell communications and specific targeting proteins on their surfaces to enable precision targeting for their payload release.

Exosomes are mediators of intercellular communication, where donor cells are able to transfer numerous types of materials, such as proteins, mRNAs, and microRNAs (miRNAs), among other types of material, to other cells. Possessing a wide variety of surface adhesion proteins and vector ligands, exosomes are able to uniquely target and deliver their payload to specific receptor cells—this delivery method can be utilized to even further enable precision delivery of drug and reduce the effects of off-target activity. Additionally, exosomes are allogenic, allowing them to evade capture and destruction by the host immune system, further bolstering the effectiveness of their payload. Exosomes derived from bone-marrow-derived mesenchymal stem cells have seen some preliminary studies in wound healing capacity, yet few to no studies utilizing exosomes for BMPs or other pro-osteogenic drugs have yet to be described [2]. The future of exosomal strategies looks bright; however, their current studies in bone regeneration are very limited partially because of the difficulty of production/purification of high quality and quantity of exosomes for clinical applications. Additionally, there are concerns in evaluating how altering surface targeting ligands affects stability, and mechanisms of cellular entry of payloads remain currently unclear.

Nucleic acid aptamers are another natural vehicle for potential use in drug delivery; they are single-stranded DNA or RNA molecules that have high specificity for cellular targets similar to the antibody–antigen interaction. Aptamers also enjoy other advantages over many other targeted vectors, such as a much smaller size and high stability, allowing them to maintain high bioavailability for targeted therapeutics. The ligand-receptor binding interaction also allows for the internalization of typically noncell-permeable drugs, leading to rapid re-expression of ligands on the cell surface, priming the target tissue for repeated future targeting of drugs. RNA aptamters have seen use as a fibroblast growth factor (FGF) signaling agonist in the treatment of bone disease, helping restore osteoblast differentiation in a bone cancer model; however, their role in new bone regeneration/healing so far is underexplored [4]. Further roadblocks in the RNA aptamer technology still remain, particularly in achieving large-scale production, as the current cost-effectiveness is suboptimal. Additionally, information on potential toxicity and long-term usage of repeated doses is currently limited. Despite these barriers, the aptamer-based techniques have a high ceiling in drug delivery for bone tissue due to their ability to potentially selectively target among the high amount of cell types present during bone healing.

References

1. Carrow, J. K., L. M. Cross, R. W. Reese, M. K. Jaiswal, C. A. Gregory, R. Kaunas, I. Singh and A. K. Gaharwar (2018). "Widespread changes in transcriptome profile of human mesenchymal stem cells induced by two-dimensional nanosilicates." Proc Natl Acad Sci U S A **115**(17): E3905–e3913.
2. Hoang, D. H., T. D. Nguyen, H. P. Nguyen, X. H. Nguyen, P. T. X. Do, V. D. Dang, P. T. M. Dam, H. T. H. Bui, M. Q. Trinh, D. M. Vu, N. T. M. Hoang, L. N. Thanh and U. T. T. Than (2020). "Differential Wound Healing Capacity of Mesenchymal Stem Cell-Derived Exosomes Originated From Bone Marrow, Adipose Tissue and Umbilical Cord Under Serum- and Xeno-Free Condition." Front Mol Biosci **7**: 119.
3. Hu, J., J. M. Miszuk, K. M. Stein and H. Sun (2020). "Nanoclay Promotes Mouse Cranial Bone Regeneration Mainly through Modulating Drug Binding and Sustained Release." Appl Mater Today **21**.
4. Jin, L., Y. Nonaka, S. Miyakawa, M. Fujiwara and Y. Nakamura (2016). "Dual Therapeutic Action of a Neutralizing Anti-FGF2 Aptamer in Bone Disease and Bone Cancer Pain." Mol Ther **24**(11): 1974–1986.
5. Li, X. and X. Ma (2017). "The role of heparin in sepsis: much more than just an anticoagulant." Br J Haematol **179**(3): 389–398.
6. Miszuk, J., Z. Liang, J. Hu, H. Sanyour, Z. Hong, H. Fong and H. Sun (2021). "Elastic Mineralized 3D Electrospun PCL Nanofibrous Scaffold for Drug Release and Bone Tissue Engineering." ACS Applied Bio Materials.
7. Miszuk, J. M., T. Xu, Q. Yao, F. Fang, J. D. Childs, Z. Hong, J. Tao, H. Fong and H. Sun (2018). "Functionalization of PCL-3D electrospun nanofibrous scaffolds for improved BMP2-induced bone formation." Applied Materials Today **10**: 194–202.
8. Mori, G., P. D'Amelio, R. Faccio and G. Brunetti (2013). "The Interplay between the Bone and the Immune System." Clinical and Developmental Immunology **2013**: 720504.
9. Sun, H., Y. Jung, Y. Shiozawa, R. S. Taichman and P. H. Krebsbach (2012). "Erythropoietin modulates the structure of bone morphogenetic protein 2-engineered cranial bone." Tissue Eng Part A **18**(19-20): 2095–2105.
10. Xu, T., J. M. Miszuk, Y. Zhao, H. Sun and H. Fong (2015). "Electrospun polycaprolactone 3D nanofibrous scaffold with interconnected and hierarchically structured pores for bone tissue engineering." Adv Healthc Mater **4**(15): 2238–2246.
11. Yao, Q., K. E. Fuglsby, X. Zheng and H. Sun (2020). "Nanoclay-functionalized 3D nanofibrous scaffolds promote bone regeneration." J Mater Chem B **8**(17): 3842–3851.
12. Yao, Q., Y. Liu, B. Selvaratnam, R. T. Koodali and H. Sun (2018). "Mesoporous silicate nanoparticles/3D nanofibrous scaffold-mediated dual-drug delivery for bone tissue engineering." J Control Release **279**: 69–78.
13. Yao, Q., Y. Liu, J. Tao, K. M. Baumgarten and H. Sun (2016). "Hypoxia-Mimicking Nanofibrous Scaffolds Promote Endogenous Bone Regeneration." ACS Appl Mater Interfaces **8**(47): 32450–32459.

Chapter 3
Drug Delivery Systems for Cartilage

Akash Yadav, Raghavendra Dhanenawar, and Akshay Srivastava

Contents

A. Yadav · R. Dhanenawar
Department of Biotechnology, National Institute of Pharmaceutical Education and Research (NIPER) – Ahmedabad, Gandhinagar, Gujarat, India

A. Srivastava (✉)
Department of Medical Devices, National Institute of Pharmaceutical Education and Research (NIPER) – Ahmedabad, Gandhinagar, Gujarat, India
e-mail: akshay.srivastava@niperahm.res.in

S. Murab (ed.), *Drug Delivery Systems for Musculoskeletal Tissues*,
https://doi.org/10.1007/978-3-031-55653-1_3

1 Introduction

Cartilage is a connective tissue and an important component of the bone and other avascular tissues. It is mainly found in the joints (synovial), spine, ribs, ears, bronchial tubes, and nose [1]. Humans have four diverse forms of cartilage, viz, hyaline, fibrous, articular, and elastic cartilage. The specific type of cartilage that is most often seen in the body is hyaline cartilage. It lines the joints and embraces the edges of the bones, exhibiting a glass-like appearance. Hyaline cartilage helps the bones move through the joints smoothly because it is slick and slippery. It is flexible and robust enough to prevent the joints' structural integrity from being lost. The following areas of the body contain hyaline cartilage: In the nasal passageways, between the ribs, and at the extremities of bones that form joints. Fibrocartilage is robust cartilage made of thick fibers. While being the least flexible, this cartilage is also the strongest. It is tough enough to endure blows and steady different body components. Fibrocartilage is found in the meniscus of the knee, in the vertebrae of the spine, and in the discs that support the entirety of the muscle system, tendons, and ligaments. Elastic cartilage is the most flexible kind of cartilage. It provides support for the portions of the body that must bend and move to carry out their duties. Elastic cartilage may resume its original form even after being subjected to high stress. The ear's cartilage may be bent. It can bend and move without injury before returning to its original form. The Eustachian tubes, which carry sound via the auricle into your brain, both of your external ears and the larynx, are bodily components that contain elastic cartilage (voice box) [2]. Diarthrodial joints have a very specific kind of connective tissue known as articular cartilage. Its overall concern is to provide articulation surfaces that are lubricated and smooth, as well as to make it simpler to transfer workloads with low frictional coefficients. The entire biomechanical microenvironment is hostile to articular cartilage due to the absence of blood vessels, lymphatics, and nerves. The capacity of articular cartilage to spontaneously mend and regenerate itself is limited, which is most significant. In this regard, the condition of the articular cartilage is crucial to sustain a healthy joint. It is well known and reported that any sort of damage to articular cartilage causes severe musculoskeletal morbidity [3].

2 Cartilage-Related Diseases

Numerous disorders can have an impact on cartilage, which is a pertinent part of the musculoskeletal system. For instance, defects and changes in cartilage development and ossification are typical characteristics of a group of illnesses known as chondrodystrophies.

2.1 Cartilage Development Disorders

Several ailments and diseases of the cartilage have been reported based on the development and injuries and are detailed in the following sections.

2.1.1 Osteochondritis Dissecans (OD)

Osteochondritis dissecans (OD) is a pathological condition that arises due to inadequate blood supply to the joints, causing the underlying bone to deteriorate. This happens in the knee, elbow, and ankle joints, but it can also happen in other joints [4]. It affects the subchondral articular cartilage of young adults between the ages of 10 and 20 [5]. Sixty-four percent of all knee osteochondral diseases (OCDs) develop mostly on the medial femoral condyle, and they often arise on this side [6]. The main causes of osteochondritis dissecans were found to be swelling in the joint area, disruption of endochondral ossification, cartilage breakdown, and death. Cracks and fissures appeared in the cartilage that had broken down and died. This caused flaps of cartilage to form and then loose bodies [7].

2.1.2 Relapsing Polychondritis

Relapsing polychondritis is an autoimmune condition that affects cartilaginous organs such as the larynx, ears, nose, joints, and major airways [8]. There is a multitude of redness and swelling in these areas. The disease is caused by the immune system's response to collagen type II, which is mostly found in cartilage and sclera. This leads to damage to the chondrocytes, activation of degradative enzymes, viz, matrix metalloproteinase's [MMPs], and generation of free radicals such as reactive oxygen species (ROS) by inflammatory cytokines along with interleukin1 (IL-1) and necrosis factor (TNF-α) in inflammatory cells. Studies have shown the association of relapsing with the activity of matrix metalloproteinases and the HLA-DR4 antigen (matrilin-1) [9].

2.1.3 Chondrocalcinosis

"Chondrocalcinosis" is a condition that happens when calcium pyrophosphate crystals build up in the cartilage. Chondrocalcinosis is a metabolic and progressive disorder that is also termed "pseudogout." If a person has an inherited metabolic disorder like hemochromatosis, hyperthyroidism, or hypomagnesemia, they are more likely to get chondrocalcinosis [10].

2.1.4 Cartilaginous Tumors

The most prevalent kind of predominant bone tumors are cartilage-forming tumors. These tumors have chondrocyte cells and a matrix of a cartilaginous tumor. Since most of these tumors are actual cancers, few are hamartomas or issues with how the body grows. The different shapes of these tumors could be caused by a single multipotent mesenchymal cell that evolves into adult cartilage in a manner analogous to just how fetal cartilage does [11].

2.1.5 Inflammatory Arthropathies

Inflammation in the cartilage is induced by extracellular matrix (ECM) remodeling due to nitric oxide, damage-associated molecular patterns (DAMPs), alarmins, cytokine, and chemokines activity, together with additional reactive oxygen species, are indeed being produced by the enzymes MMPs, aggrecanase, and protease [12]. Matrix metalloproteinases produced by chondrocytes in response to prostaglandins or cytokines promote inflammation [13]. External mediators like proteases, cytokines, and internal cellular dynamics that cause the release of inflammatory mediators without scavenging the oxidized proteins lead to inflammation. These proteins will exacerbate the yield of free radicals (ROS), contributing to the oxidative remuneration that originally triggered the inflammation. Paradoxically, oxidative stress, specifically in chondrocytes, helps accelerate cell aging [14, 15]. Seronegative spondyloarthropathies includes juvenile idiopathic arthritis, psoriatic arthritis, inflammatory bowel illness-subjected arthritis, synovitis, reactive arthritis, and ankylosing spondylitis and its symptoms encompass inflammatory back pain, inflammation peripheral arthritis, eye inflammation, and the places wherein tendons and ligaments integrate [2].

2.2 *Cartilage Injuries*

Injuries in cartilage are very prevalent in the joint areas of the body such as the knee, hip, and ankle, etc. These injuries often occur due to forceful or repetitive impact to the joints during sports activity or accidents or falling from heights, etc. These injuries may also be linked to poor alignment of joints because of congenital abnormalities or defective repair of previous injuries. These cartilage injuries are discussed in the following sections.

2.2.1 Osteoarthritis

Osteoarthritis (OA) has long been conceived as a disease that is caused due to "stress and strain," which ultimately results in cartilage loss. OA was formerly thought that it was the sole outcome of any circumstance that placed additional stress on one joint or made the cartilage matrix more fragile. Osteoarthritis is a disease that causes cartilage to break down over time. It is an inflammatory disease that does not go away easily but rather increases with advancing age, being overweight, having joints that do not line up right, polyarticular diathesis, joint damage microcrystal deposition, and doing a lot of heavy strenuous effort. This is because more free radicals (ROS), nitric oxide, damage-associated molecular patterns (DAMPs), alarmins, cytokines, as well as chemokines are being made by the enzymes MMPs, aggrecanase, and protease [13]. The surface fibrillation, chondrocyte hypertrophy, and matrix fragmentation also begin to show up. The alteration in the bones caused by OA is a thickening of the cortical plate, a shift in bone mass, the formation of cysts in the bones, and the growth of osteophytes [16].

2.2.2 Costochondritis

Costochondritis is a pain in the thorax as a consequence of sternal articulations or costal cartilage inflammation. Most of the time, the pain in the chest wall is expressed as repetitive, aching, sharp, or stabbing. The pain gets worse with deep breathing, moving the upper limbs, and exercise. Costochondritis is most probably triggered by the latest upper respiratory illness or strenuous work. It may take the patient several weeks to several months for the inflammation to subside, making it challenging for them to work and participate in daily activities. Other names for costochondritis include costosternal syndrome, pain in the cartilage (parasternal chondrodynia), and anterior chest wall syndrome [17].

2.2.3 Herniation

A herniation is a condition in which the internal portion of the intervertebral disc (IVD), i.e., nucleus pulposus (NP), oozes out from the outer part known as annular fibrosis and leaks into the spinal canal. When a lumbar disc herniates, it often drags pieces of hyaline cartilage from an adjacent vertebral end plate. Because hyaline cartilage's collagen network strongly restricts inflammatory edema and proteoglycan (PG) loss and any hyaline-cartilage segments that are loose restricts revascularization and deterioration. Because of this, people with cartilaginous herniations are more likely to have sciatica that does not go away. They also have an overall impact on the degeneration of the disc because the bony endplate becomes significantly more porous after the loss of hyaline cartilage. This allows infections and biochemical mediators to migrate from the vertebrae to that of the disc as well as inflammatory molecules from the intervertebral disc toward the vertebral body [18].

2.2.4 Achondroplasia (ACH)

The most common type of dwarfism in people is called achondroplasia (ACH). Ach is a genetic disease that is passed down from parent to child and has 100% penetrance. The limb shortening in ACH, which is also known as rhizomelia, affects the proximal segments out of proportion, accounts for the majority of nanism. Due to issues with cartilage development at the foundation of the skull, the center of the whole face is flattened and the front of the head is large and rounded. Neurosurgery is often required to treat spinal stenosis and foramen magnum constriction, both of which are rather common. The trunk is about the right size, but it is often messed up by too much lumbar lordosis [19]. The ACH gene is located on chromosome 4's short arm, according to genetic linkage studies. Arginine → glycine (R-G) substitution at position 380 (p.Gly380Arg) in the FGFR-3 was discovered to be present in the majority of ACH patients across Europid, Afro, as well as Asian ethnicities (FGFR3). The fact that FGFR3 is reported to be found in the chondrocytes of the growth plate depicted that a mutation in FGFR3 had a direct effect on how the growth plate worked. When wild-type FGFR3 and mutant FGFR3 were put side by side, it was clear that the mutant receptors sent out more signals, which could be made even stronger by fibroblast growth factor (FGF) ligands. This might be because the mutant receptor is less likely to be broken down by lysosomes, which makes the protein more stable [20].

3 Current Treatment Strategies

3.1 Surgical Approaches

The primary objective of the surgical approaches for cartilage is total replacement or the regeneration of cartilage, which is achieved through various techniques, as discussed in the following sections.

3.1.1 Subchondral Drilling

Subchondral drilling is one of the oldest methods intended for stimulating bone marrow that is used to repair articular cartilage with small-sized lesions. The standard procedure is to drill several holes in the layer of bone underneath the cartilage with a particular wire known as Kirschner wire or with a surgical twist drill bit. The diameter of each opening is precisely determined. Subchondral drilling was initially used for the treatment of osteochondral disease (OCDs) and later implicated in the treatment of osteoarthritis [21].

3.1.2 Microfracture

Microfracture is another form of marrow stimulation technique used on smaller sites of defects. Excision of the weathered cartilage is accomplished with the use of a curet, leaving a straight, right-angled crest. The subchondral bone plate is then punctured with many holes (3–4 mm diameter) using an arthroscopic awl. The hole should be deep enough (2–4 mm) for marrow droplets to appear. Microfracture is still a clinically feasible treatment for acute, damaged cartilage lesions, considering its low cost and less invasiveness [22].

3.1.3 Bone Marrow Aspirate Concentrate

Bone marrow mesenchymal stem/stromal cells (BM-MSC), endothelial progenitor cells (EPCs), hematopoietic stem cells (HSC), platelets, and cytokines are readily harvested from the marrow of long bones to create a cell concentrate known as the bone marrow aspirate concentrate (BMAC). It is administered intra-articularly in combination with other therapies for osteochondral defects [23]. This approach is employed in treating bone deformities, cartilage disorders, limb occlusion, osteonecrosis of femur, myocardial infarction, and degenerative disk disease. It is a promising method in cartilage tissue regeneration and rejuvenation, having variety in cell content and ease of production [23].

3.1.4 Osteochondral Autograft Transplantation

It was initially used in 1993 for knee ligament tear [24]. It involves harvesting a piece of subchondral bone and distal femur's fresh cartilage from the region of the less pressured limb. After being sized to fit the defect, it is placed giving a smooth cartilage surface. A quick fixation is achieved for intermediate lesion size (2–4 cm^2). Osteochondral autograft transplantation (OAT) is found to be superior to microfracture for treating knee osteochondritis dissecans [25, 26].

3.1.5 Osteochondral Allograft Transplantation (OCA)

It is also similar to OAT in surgical procedure, the only difference being the graft harvesting from the donor subchondral bone. Osteochondral allograft transplantation (OCA) was first introduced in 1985 for the treatment of knee lesions, which is usually used for large defect size >2 cm^2 [27, 28]. This method provides accurate surface repair with a fast recovery period, but factors such as unavailability, affordability, immunogenic rejection, cartilage delamination, and disease transmission are not favorable [27].

3.1.6 Particulate Juvenile Allograft Cartilage

Particulate juvenile allograft cartilage (PJAC) is a newer technique, replacing osteochondral plugs of adult cartilage with particulate fragments of juvenile cartilage. This particulate fragment of cartilage contains live cells in their purely natural extracellular matrix. The allograft is often collected from 2-year-old neomorts; however, this age range is flexible. This technique is applied to lesions of the knee, ankle, shoulder, elbow, and hip. It is a simple, one-step operation involving a lesion size greater than 5 mm^2. An infamous product named denovoNT was first implemented in a clinical setting for a patella in May 2007. The costs are relatively high because of the scarcity of young and healthy cartilage donors [29]. Allograft putty made from living cartilage (cartimax) is a recently developed approach that makes use of live cartilage fragments, obtained from donor tissues that were allografted and blended with a lyophilized extracellular matrix to produce a clay-like substance that can be shaped to fit any kind of deformity. The putty can be placed in the defect site without a sealant or fibrin glue. It can be used up to 12 months with cryopreservation [30].

3.1.7 Cryopreserved Viable Osteochondral Allograft

The cryopreserved viable osteochondral allograft (CVOCA) is made up of ECM proteins, chondrocytes, and chondrogenic growth factors (CGF) to form a strong and healthy cartilage and a fine cross-sectional covering of bone over cartilage that may be transplanted in a single surgical step [30]. The osseous surface of the implant is directed toward the subchondral bone during implantation. If a loose implant edge is encountered, it can be patched together with a fibrin glue to the cartilage rim. This method is advantageous as the osseous surface permits the integration of bones, which may lessen the possibility of flaking. Cryopreserving an implant extends its life span. The requirement of preoperative patient size and template creation is avoided because the implant with perforation may be shortened and molded to physically fit the defect. There is a significant threat of disease spread, and costs must be taken into account. It is applied in knee defects, patella, and glenoid chondral injuries [31].

3.1.8 Autologous Chondrocyte Implantation and Matrix-Induced Autologous Chondrocyte Implantation

Autologous chondrocyte implantation (ACI) is an established technique to overhaul the remunerated cartilage. The chondrocyte cell covered with a membrane of periosteum is injected at the site of the defect in the first-generation ACI procedure, the second-generation ACI involves collagen membrane patch instead of the periosteum, and the third-generation ACI utilizes chondrocytes-seeded scaffolds. In most cases, the femoral neck is where the chondrocytes are extracted arthroscopically. The target quantity of cartilage to extract is between 200 and 300 mg, after which it will be multiplied in a cell culture dish. The implantation of chondrocytes into the defect is the second step of the ACI process, which is performed 5–9 weeks following the initial procedure for a lesion size of >4cm^2. High early failure rate, high upfront costs, long delay in graft maturation or increased length of rehabilitation, availability, etc., are a few limitations of the method [32]. Matrix-induced autologous chondrocyte implantation (MACI) is a third-generation ACI that has a similar procedure of preparation and application. It can be applied to diseases of the talus, tibiofemoral or patellofemoral joint, and femoral trochlea cartilage injury.

3.1.9 Particulate Autologous Cartilage Implantation

Particulate autologous cartilage implantation (PACI) is a surgical procedure involving a mixture of particulate cartilage and plasma-rich platelets, forming a flexible scaffold that can be placed on chondral defects. Blood is taken from the patient and centrifuged to separate high-platelet plasma. The semisolid CN-biometrix is produced after an incubation period of 30 min, which consists of activated PPP-PRP mixed in a ratio of 50/50 with minced hyaline cartilage chips obtained from the patient. The biometrix is then spread out evenly along the defect, and it is left there for 5 min to ensure good adherence. Pain alleviation, sufficient lesion filling, absence of bone effusion, and improved knee performance are some of the benefits of this technique. Among the potential consequences are a lack of successful graft integration, graft hypertrophy, and graft delamination [33].

3.1.10 Autologous Matrix-Induced Chondrogenesis

Unlike ACI and MACI, autologous matrix-induced chondrogenesis (AMIC) is a single-step process that entails the microfracturing of the damaged cartilage and covers the blood clot. In this method, the bone and remunerated cartilage are detached with the help of a curette. Microfracturing is carried out with a suitable probe. Defects are assessed and the collagen membrane is placed with the help of fibrin glue. The goal of AMIC involves resurfacing the damaged articular cartilage. To eliminate discomfort and restore full physiological motion, it is used on midsize lesions of 1.5 cm^2 [34]. Avoiding the expense and inconvenience of cartilage harvesting and the need for in vitro cell multiplication make this method both practical and affordable. The only possible drawback is malalignment along with other pathological stress [35]. Osteochondral lesions, most commonly found in the knees and ankles, are the most common sites of application for this technique.

3.2 *Tissue Engineering*

Repairing cartilage with tissue engineering is a promising field of study. Building a biomaterial scaffold to house cells and their associated regulatory elements (growth factors (GFs)) is a vital part of tissue engineering. Sponge-like structures, capsules, fibers, microparticles, injectable hydrogels, injectable thermogels, and 3D bioprints are some of the biomaterial scaffolds that have been proposed to maneuver cartilage regeneration [36]. As a result of their hydrophilic nature, unique structure, and efficient in incorporating growth factors, hydrogels have proven successful in tissue engineering applications. Hydrogels are employed as either long-term implants to repair injured cartilage or as cell/growth factor-carrying biomaterial scaffolds to facilitate tissue regeneration. These injectable hydrogels decrease the extent of surgical intervention, minimizing the time required for both the procedure and subsequent recuperation. The flowability makes it simple to fill irregularly shaped defects; the injectable hydrogels often maintain a high water content, are semipermeable to nutrients and metabolites, and exhibit high biocompatibility [37].

In lieu, stimuli-responsive hydrogels are employed in the administration of drug and tissue engineering strategies due to their efficiency to react in response to external stimuli or mild environmental changes. Such dynamic behaviors in the scaffolds have been induced by introducing a wide range of physical (e.g., light intensity, temperature, electrical or magnetic fields, ultrasound), chemical (e.g., pH, ionic interaction, chemical interaction), and biological (e.g., enzymatic) stimuli. The size of a response to such triggers is highly dependent on the stimulus's type and intensity, as well as the materials' susceptibility to those factors [38]. Wang et al. discovered a novel injectable thermogel composed of (4-arm) PEG-PDLA-Chol and (4-arm) PEG-PLLA-Chol to facilitate cartilage tissue regeneration. This scaffold displayed increased chondrocyte adhesion, larger pore size, and a higher mechanical property [39].

3.3 Cell-Based Therapy

Chondrocytes and MSCs differentiated from myriad sources have been the principal cells employed in addressing cartilage abnormalities in the joints. Stem cells (SCs) are cells with inherent efficiency to transform into any defined form of cell. Adult, fetal, or embryonic stem cells (ESCs) may all transform into multiple different cell types. The sole sources of totipotent cells, which can develop into all three distinct major germ layers including organs like the placenta, are zygotic cells, which are present on human embryonic day 4 prior to epiblast production. Embryonic stem cells (eSCs) as well as induced-pluripotent stem cells (iPSCs) may lead up to every diverse kind of cell into an organism with the extra-embryonic cells being the outliers [40].

3.3.1 Mesenchymal Stromal Cells

Mesenchymal stem/stromal cells (MSCs) are multipotent, self-renewing cells that transform into numerous musculoskeletal lineage (osteogenic, chondrogenic, and adipogenic), including neural precursors [41]. MSCs that are induced to become articular chondrocytes have been thought of as having therapeutic promise; however, there are some signs that suggest their epigenetic makeup may be different from that of comparable autologous cartilage tissue. According to a growing body of research, MSCs work by having a trophic influence on endogenous cells. These cells produce a variety of growth factors (GFs) and cytokines along with potent immunomodulatory, anti-inflammatory, and extracellular vesicles (EVs). By preventing T-lymphocyte maturation, in addition to suppressing macrophage activation and immunomodulatory factor production, they seem to minimize localized inflammatory effects inside an afflicted or early-onset osteoarthritis joint [42].

3.3.2 Peripheral Blood Stem Cells (PBSCs)

The quintessential repository for an autologous transplant is PBSCs. While bone marrow was almost exclusively employed until around 10 years ago, peripheral blood stem cells (PBSCs) are presently the most often used source of stem cells (SCs). The widespread use of mobilized PBSCs as a source of stem cells is primarily attributable to the practicality of peripheral collection methods [43, 44]. The HSC component for autologous transplantation has almost entirely been replaced by peripheral blood stem cell (PBSCs) preparations, which are also often utilized for allogeneic transplantation [45]. A study was conducted by Saw et al. to examine the cartilage regeneration capabilities by injecting the patients subjected to articular chondral lesions with and without PBSCs in combination with hyaluronic acid within the patients having grade-III and grade-IV chondral femoral lesions [46]. Another case study of an athlete with grade-IV chondral femoral lesion (lateral) reported by Fu et al. was repaired by treating with activated PBSCs by mobilizing the cells with recombinant-human granulocyte colony-stimulating factor (rhG-CSF). The activated PBSCs not only showed regenerative capabilities but also assisted in patellofemoral malalignment [47].

3.3.3 Bone Marrow Stem Cells

Bone marrow stem cells (BM-MSCs) are widely explored and a promising class of stem cells (SCs) because of their multipotent nature to differentiate within a wide range of adult cell lineages. These particular stem cells have immune-modulating properties that include reducing neutrophil and T-cell proliferation, which they do in part by decreasing proinflammatory cytokine expression and enhancing anti-inflammatory cytokine synthesis. For cartilage remodeling and repairing the damaged cartilaginous tissues, the BM-MSCs need to possess migratory capacities. Moreover, several biological factors have demonstrated their regulatory effects on the migration of the MSCs and chondrocytic cells (Table 3.1) [48–50].

It has been witnessed that employing BM-MSCs to treat damaged cartilage is a safe and efficient procedure. The quality of the cartilage and knee function of OA patients were both improved by administering injections of endogenous BM-MSCs,

Table 3.1 In vitro migratory regulations of MSCs

Factors	Regulatory effects on MSC migration	Citations
Chemokines		
MCP-1/CCL2	Stimulates the migration of H-BM-MSC, H-AD-MSC, mouse BM-MSC	[52–55]
	Nonsignificant results in H-BM-MSC, H-CSPC	[56–58]
MIP-1α/CCL3	Nonsignificant results in H-BM-MSC	[57]
RANTES/CCL5	Stimulates the migration of H-BM-MSC, H-AD-MSC	[57, 59–61]
Eotaxin-1/CCL11	Nonsignificant results in H-BM-MSC	[57]
HCC-4/CCL16	Inhibits the migration of H-BM-MSC	[62]
TARC/CCL17	Stimulates the migration of H-BM-MSC	[59]
PARC/CCL18	Nonsignificant results in H-BM-MSC	[62]
MIP-3β/CCL19	Stimulates the migration of H-BM-MSC, H-AD-MSC	[54, 59, 61]
MIP-3α/CCL20	Stimulates the migration of H-BM-MSC	[53]
	Nonsignificant results in H-BM-MSC	[52]
SLC/CCL21	Stimulates the migration of H-BM-MSC	[59]
MDC/CCL22	Inhibiting the migration of H-CSPC	[62]
	Stimulates the migration of H-BM-MSC	[57]
	Nonsignificant results in R-BM-MSC	[63]
Eotaxin-2/CCL24	Nonsignificant results in H-BM-MSC, H-CSPC	[57, 62]
TECK/CCL25	Stimulates the migration of H-BM-MSC, H-CSPC	[61, 62, 64]
CTACK/CCL27	Inhibits the migration of H-BM-MSC, H-CSPC	[58, 62]
MEC/CCL28	Stimulates the migration of H-BM-MSC	[59]
GRO-α/CXCL1	Nonsignificant results in H-BM-MSC	[57]
ENA-78/CXCL5	Inhibits the migration of H-CSPC	[62]
NAP-2/CXCL7	Nonsignificant results in H-CSPC	[62]
	Stimulates the migration of H-BM-MSC	[65]
IL-8/CXCL8	Stimulates the migration of H-BM-MSC	[53, 64, 66, 67]
IP-10/CXCL10	Stimulates the migration of H-BM-MSC, H-CSPC	[62, 65]
I-TAC/CXCL11	Stimulates the migration of H-BM-MSC	[62]
SDF-1/CXCL12	Stimulates the migration of H-SD-MSC, H-BM-MSC, H-AD-MSC, R-BM-MSC, mouse BM-MSC	[52–55, 57, 61–63, 68–70]
	Nonsignificant results in H-BM-MSC	[59]
	Inhibits the migration of H-CSPC	[62]
BLC/CXCL13	Stimulates the migration of H-BM-MSC, H-AD-MSC	[59–61]
SR-PSOX/CXCL16	Stimulates the migration of H-BM-MSC, H-AD-MSC	[54, 61]
SCM-1α/XCL1	Stimulates the migration of H-CSPC	[62]
	Nonsignificant results in H-BM-MSC	[62]
Fractalkine/CX3CL1	Nonsignificant results in H-BM-MSC	[57]
Growth factors		
PDGF	Stimulates the migration of H-CPC, H-BM-MSC, H-AD-MSC, R-BM-MSC, equine BM-MSC	[52, 57, 63, 67, 71–73]
TGF-β1	Stimulates the migration of H-BM-MSC, H-AD-MSC	[54, 67]

(continued)

Table 3.1 (continued)

Factors	Regulatory effects on MSC migration	Citations
TGF-β3	Stimulates the migration of H-CPC, H-SD-MSC, H-BM-MSC, H-AD-MSC	[70, 74]
VEGF	Stimulates the migration of H-BM-MSC	[67, 68]
	Nonsignificant results in H-BM-MSC, R-BM-MSC	[57, 63]
EGF	Stimulates the migration of H-BM-MSC, H-AD-MSC	[54, 57]
HGF	Stimulates the migration of H-BM-MSC, H-AD-MSC	[52, 54, 57]
IGF-1	Stimulates the migration of H-CPC, H-BM-MSC, H-AD-MSC	[52, 54, 57, 70, 73]
	Nonsignificant results in R-BM-MSC	[63]
NGF	Stimulates the migration of H-CPC	[75]
FGF-2	Stimulates the migration of H-BM-MSC, H-AD-MSC, R-BM-MSC, Rabbit BM-MSC	[54, 63, 68, 76–78]
	Nonsignificant results in H-BM-MSC	[57, 70]
Other cytokines		
IL-1β	Inhibits the migration of H-CPC, equine BM-MSC	[73, 79]
IL-3	Stimulates the migration of H-BM-MSC, H-AD-MSC	[80]
IL-4	Nonsignificant results in R-BM-MSC	[63]
IL-6	Stimulates the migration of H-BM-MSC	[68]
	Nonsignificant results in H-CPC	[73]
INF-γ	Nonsignificant results in R-BM-MSC	[63]
TNF-α	Stimulates the migration of H-AD-MSC, R-BM-MSC	[54, 63]
	Inhibits the migration of H-CPC, equine BM-MSC	[73, 79]
BMP-2	Stimulates the migration of H-BM-MSC, H-CSPC	[70, 81]
BMP-4	Stimulates the migration of H-BM-MSC	[70]
BMP-7	Stimulates the migration of H-BM-MSC	[70]
SCF	Inhibits the migration of mouse BM-MSC	[55]
Complement 5a	Stimulates the migration of mouse BM-MSC	[55]
Other factors		
Platelet-rich plasma (PRP)	Stimulates the migration of H-CSPC, H-SF-MSC, equine BM-MSC	[72, 82–84]
Platelet-poor plasma	Nonsignificant results in H-SF-MSC	[84]
Platelet lysate	Stimulates the migration of H-SF-MSC	[85]
BM aspirate concentrate	Stimulates the migration of equine BM-MSC	[72]
Fibronectin	Stimulates the migration of H-CSPC, mouse CPC	[81, 86–88]
Link protein N-terminal peptide	Stimulates the migration of R-CPC	[89]

Reproduced from [51] with permission from Copyright 2020 Elsevier
H Human, *R* Rat

in accordance with a study conducted in 2013 by Orozco et al. in which 50 patients suffering from severe OA (median age = 52 years) participated in a first phase and second phase clinical investigation and got an endogenous BM-MSC injection. All patients had evidence of cartilage regeneration after a year, according to magnetic resonance imaging (MRI) scans, and their attribute of life had been revamped. Furthermore, Chahal et al. discovered that endogenous BM-MSCs' lower synovial inflammation as demonstrated by a decline in proinflammatory monocytes or macrophages along with IL-12, which is envisaged by a subset of biomarkers of BM-MSCs [90].

3.3.4 Umbilical Cord-Derived Mesenchymal Stem Cells

Umbilical cord-derived mesenchymal stem cells (UC-MSCs) have revealed their propensity for greater self-renewal, differentiation into three distinct germ layers, accumulation in inflamed or injured tissue, stimulation of tissue healing, and immune response modulation. The separation of MSCs from the diverse umbilical cord sections, namely, Wharton's jelly, sub-amnion, may be accomplished using a variety of procedures and culture techniques [91]. Comparatively, UC-MSCs offer the following appealing benefits as MSCs even as UC cells: (1) noninvasive usage, (2) reduced danger of infection, (3) lower incidence of teratomas, and (4) multipotency and lesser immunogenicity along with high immune-suppressive potential [92]. According to a study registered in ClinicalTrial.Gov (reference NCT03800810), subjects in the trial who had been diagnosed with chronic osteoarthritis of the knee received human umbilical cord mesenchymal stem/stromal cells (hUC-MSCs) and hyaluronan injections (HA) three times in total. The individuals further received HA in the trailing second and subsequent week three. The researchers discovered that delivering adult MSCs locally to damaged joints increases the restoration of the meniscal tissue. The exact method through which UC-MSCs alleviate the signs of osteoarthritis is still unknown. Furthermore, the fact that human UC-MSCs contain several soluble factors, including G-CSF, PDGF-BB, and bFGF, may account for the substantial promotion of chondrocyte proliferation as seen in subsequent trials [93].

3.3.5 Adipose Tissue-Derived Mesenchymal Stem Cells

Mesenchymal stem/stromal cells (MSCs) derived from adipose tissues (ASCs) are a population of self-renewing, multipotent adult cells located inside the vascular stroma of adipose tissues. Adipose tissue mesenchymal stem cells (AT-MSCs) are critical for the development, postnatal development, maintenance of homeostasis, repair, and regeneration of the cartilage tissue [66]. Subcutaneous liposuction may easily access a sizeable number of ASCs, and when invigorated properly, they further develop into a variety of cells, viz, adipocytes, osteocytes, and endothelial cells

[94]. A randomized phase-I and phase-II preclinical study, registered under ClinicalTrials.gov (identifier NCT01809769), was maneuvered. The participants in phase-I were categorized into three groups at random: low-dose, mid-dose, and high-subsequent dose. Patients received cell suspension injections in both joints in the third and sixth weeks followed by liposuction and were monitored at 12th, 24th, and 48th weeks later. In Phase I, an additional open clinical study (Phase-II) was done to examine the profound stability and conceivable advantages of human adipose mesenchymal stem cells (haMSCs). Furthermore, a subsequent third injection was administered within this timeframe. This pilot study presented the evidence that haMSCs intra-articular injections significantly alleviated the pain within the knee joints along with the restoration of the function and cartilage volume. The haMSC injections into the articular cartilage also improved the cartilage thickness substantially [95]. ASCs are an effective therapy recourse for a variety of inflammatory and autoimmune disorders, and their therapeutic effectiveness is mostly dependent on immunomodulatory activities mediated by paracrine responses.

3.3.6 Platelet-Rich Plasma

Platelet-rich plasma (PRP) is an analogical product made by the centrifugation of blood samples. Due to an intense accumulation of growth factors (GF), proteins, and cytokines, platelet-rich plasma (PRP) is widely believed to hasten cartilage regeneration at injury sites [96]. PRP is extracted by centrifuging the blood cells across a series of density gradients [97]. PRP has been utilized successfully in many different medical specialties, including orthopedics, sports medicine, ophthalmic, urological, cardiac, pediatric, and plastic surgery [98].

4 Drug Delivery Systems Used in Cartilage Repair and Regeneration

New approaches for drug delivery into cartilage have emerged due to the ardent improvement in drug-delivery systems in recent years. In terms of size, nanomaterials are closely related to biological bodies. The fusion of nanomaterials and biomedical materials is known as nano-biomedical materials. Various composite materials are created by combining nanoparticles and other materials. Nanomaterials have started to permeate several disciplines with more in-depth research, offering tremendous potential for application. Inorganic nanoparticles, liposomes, micelles, polymeric nanoparticles, exosomes, other NP-based drug delivery methods, hydrogel-based and transdermal patches as drug-delivery systems for cartilage are bandied in this section along with their recent advancements and novel uses (Fig. 3.1) [99].

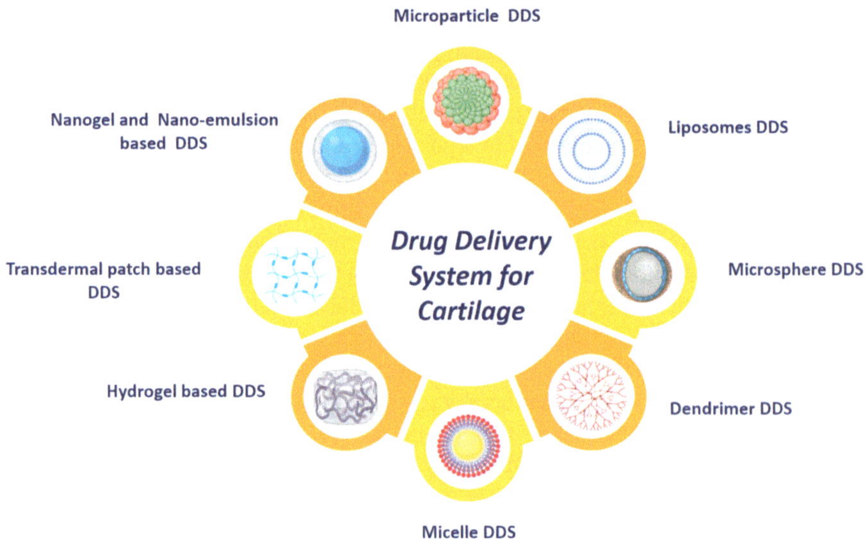

Fig. 3.1 Drug delivery systems in cartilage repair and regeneration

4.1 Nanoparticle-Based DDS

Nanoparticles (NPs) are tiny particles having diameters typically between 1 and 100 nm. NPs have shown tremendous utilization in the areas of tissue repair and regeneration. It is also more challenging to fully predict the behavior of NPs than microparticles. As a result, novel chemical, physical, and biological properties of NPs can eventually be taken advantage of through the control and manipulation of nanostructures. Given that it has a higher surface area–volume ratio, the optimum scale for catalysis, and molecular structures at the nanoscale found in the body, nanotechnology has grown in significance [100]. The fabrication of NPs via the tools of nanotechnology typically uses established top-down or bottom-up approaches. The top-down strategy entails nanofabrication technologies to shrink macro-sized structures and produce nanoscale particles. Contrarily, the bottom-up strategy makes use of physical and chemical processes to combine molecules or even individual atoms into larger nanoscale particles [101]. These nanoparticles demonstrated tremendous potential for controlled delivery of therapeutic agents to the cartilage with the advantages of their design, transport phenomenon, and interactions within cartilaginous tissues. These concepts and novel design strategies for developing multifunctional and smart systems for drug delivery in cartilage are discussed in [102].

4.1.1 Inorganic Nanoparticles

Inorganic materials implicated as nanocarriers have demonstrated tremendous promise in drug-delivery systems. Inorganic nanoparticles are created by combining biodegradable polycations with inorganic particles. Metals, metal oxides, and various carbon-based substances (like those of fullerene, nanotube, and fibers) are the most prevalent forms of inorganic nanoparticles, and perhaps magnetic nanoparticles constructed using super-paramagnetic iron oxide (Fe_2O_3) nanoparticles (SPION) (Fig. 3.2). The inorganic materials acting as the systems' skeletons can load and release medications, maintain blood circulation's structural integrity, and hold good biocompatibility and pharmacological qualities [103]. The inorganic nanocarriers that are examined the most extensively include gold nanoparticles (GNPs), over-graphene oxide (GO), black phosphorous (BP), and mesoporous silica nanoparticles (MSNs). The most preferred silica nanoparticles are mesoporous, which feature homogenous pores, tunable functionalization, good bio-compatibility, large specific-surface area, a high porosity, and the capacity to degrade [104]. Inorganic nanoparticles (NPs) are utilized in theranostic, such as photodynamic therapy (PDT), in addition to applications involving the delivery of drugs. Another inorganic material silica is also widely used in various drug delivery applications including the treatment of arthritis due to their good biocompatibility, degradation, and high payload capacity. Hyaluronan synthase enzyme (HAS2)-loaded silica nanoparticles when tested in a rat model of arthritis demonstrated enhanced production of HA and inhibited inflammation [105].

The inorganic nanoparticles are implicated in developing hydrogels that perform a diverse range of lucrative roles, such as rendering tougher hydrogels and retaining the hydrogel's porosity, antibacterial, and cytocompatibility [106, 107]. Carbon nanomaterials and carbon dots (CDs) are examples of carbon-based nanomaterials used in tissue engineering (TE). CDs prove an important role in increasing cross-linking in carbon-based nanocomposite hydrogels. CDs offer several benefits, including minimal photo-bleaching, red fluorescence, homogeneity, and biocompatibility, which are effective in investigating the deterioration of hydrogels in vivo

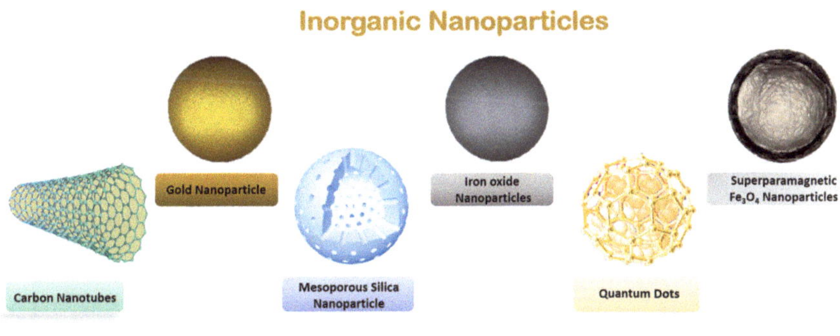

Fig. 3.2 Schematic illustration of inorganic nanoparticles

or tracing drug release [108]. Carbon nanotubes are employed to promote cell growth, chondrogenic differentiation, and repair cartilaginous tissues; these artificial carbon-based three-dimensional (3D) biomaterial-based structures are developed in biomedical research to address bone to cartilage abnormalities [109].

4.1.2 Liposome-Based Drug Delivery Systems

Liposomes are known to be adaptable drug vesicles with the capabilities of site specificity, prolonged or regulated release, safeguarding the pharmaceuticals from deterioration and elimination, increased clinical advantages, and reduced negative adverse signs [110]. The molecules that make up lipids are amphiphilic, meaning that one of its parts prefers (hydrophilic) water while the others do not (hydrophobic). Lipids self-assemble, typically in the form of liposomes, when exposed to water because the hydrophobic parts of the molecule engage negatively with the solvent (Fig. 3.3). A liposome is made up of an encasing lipid bilayer that resembles a membrane and isolates the aqueous core from the surrounding material [111]. Liposomes are biocompatible and biodegradable regardless of whether they are made of synthetic or natural lipids or polymers, making them appropriate for use in biomedical research.

Because of these distinctive qualities, namely, biocompatibility and biodegradability, they are highly desirable as delivery systems. Hydrophilic moieties are confined at the bilayer interface, or hydrophobic drugs infiltrate the bilayer of the liposome from the aqueous core of the liposome. Because of the altered biodistribution, liposomal formulations are more effective at treating patients than traditional formulations in both preclinical animals and humans [112]. Liposomes are often made of biodegradable, biologically inert, and nonimmunogenic lipids; drugs that attach to their membranes should be delivered to the target with minimal side effects and without significant decomposition. Additionally, they have a low level of toxicity, and neither pyrogenic nor antigenic reactions are produced [113]. The size of

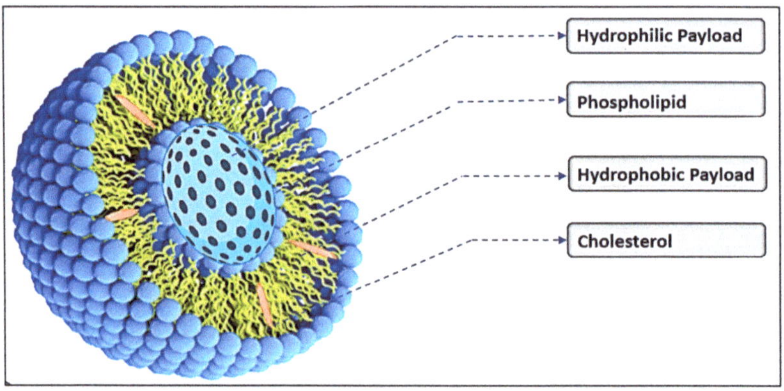

Fig. 3.3 Schematic illustration of the liposome structure

the liposome may vary considerably, from minuscule vesicles (0.025 μm) to large ones (2.5 μm) [114]. Liposomes may also contain either a single layer or two layers of membranes.

The preparation process has a direct impact on the characteristics of generated liposomes. The following elements should be looked into while selecting a method: (1) physicochemical characteristics of the compound to successfully encapsulate and that of liposome constituents, (2) type of media within which the liposomes are uniformly distributed, (3) the effectual level of concentration of the encapsulated compound as well as its foreseeable toxic effect, (4) supplementary application-related techniques (distribution of liposomes), (5) the optimum size, polydispersity-and the life of liposomes for the intended applications, and (6) the capacity to produce safer and biocompatible liposomes [114]. Recent developments have emphasized lipid nanostructure formulation by encapsulating natural substances into liposomes with anti-inflammatory actions derived from medicinal plants as potentially effective systems for OA treatment (Fig. 3.4) [115].

Several studies have reported model systems wherein the liposomes have been implicated as efficient boundary lubricants. Phosphocholination (PC) of liposomes imparts highly hydrated surfaces. These lubricating layers are more efficient in combination with other extracellular matrix components, i.e., hyaluronan, a nonsulfated glycosaminoglycan that has water-holding capacity, thus improving the efficiency of the phosphocholinated liposomes. These PC-liposomes are further loaded with anti-inflammatory and analgesic drugs to alleviate the inflammation and pain that is subjected to chronic lesions in the cartilage structures [116]. Recently, a study reported the development of microfluidic liposomes with a one-step microfluidics technology by immobilizing liposomes onto a polymer hydrogel matrix, which enhances the drug loading, drug release, and extended retention time [117].

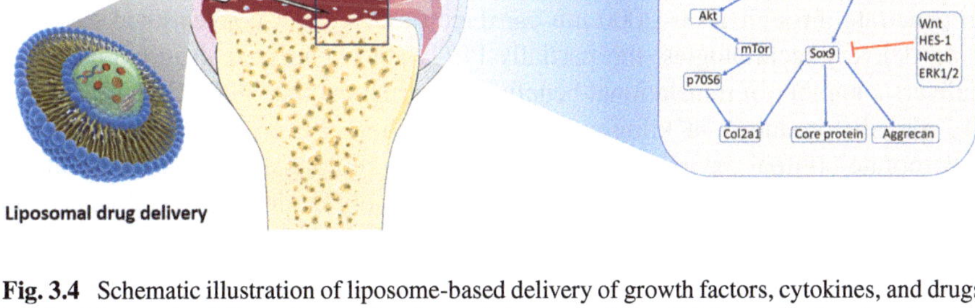

Fig. 3.4 Schematic illustration of liposome-based delivery of growth factors, cytokines, and drugs involved in mechanistic pathways for the maintenance of the GAG matrix for chondrocytic functions

4.1.3 Dendrimer-Based Drug Delivery Systems

Dendrimers are attractive drug delivery systems because these are highly branched 3D macromolecules with internal voids and specific surface activity. In the last two decades, dendrimer-mediated targeted delivery has seen significant advancements [118]. Three distinctive components form dendrimers: an inner core, an internal layer (generation) composed of tandem units that are coupled centrally to the internal core, an outer layer (terminal capability) that is connected till the last domestic generation [119]. The surface has many responsive endings and tight packing which make it simple to load pharmaceuticals, and the interior chambers are ideal for doing so. These dendrimers are virtually monodisperse, extremely branched macromolecules and having flexible surface with severalfunctional groups, are stand out options as nanosized materials for application in drug delivery [120]. Dendrimers are among the most adaptable drug-delivery systems that can have their structure and makeup altered. Through a process called endocytosis-mediated cell internalization, dendrimers can help therapies get through different cell membranes or biological barriers.

One of the main ways that dendrimers improve intracellular absorption is via regulating actin and occludin, two close junction proteins [121]. Dendrimers are taken inside mostly by endocytosis, which is caused by clathrin and caveolae, and partly by macro-pinocytosis. Due to the abundance of both secondary and tertiary amines in dendrimers having pKa values below or equal to physiological pH, the proton sponge reaction takes place. These amines enable the protons generated by ATPase to adhere to them, resulting in osmotic edema and eventually resulting in endosome membrane breaching, which allows the dendrimers to escape [122, 123]. Dendrimers' cellular toxicity is significantly influenced by the amount and kind of surface functional groups they possess. Most cationic dendrimers are very cytotoxic, while most anionic and neutral dendrimers are only mildly cytotoxic or not dangerous at all. It was discovered that dendrimers of a more recent generation and those having positive surface charge groups were much more cytotoxic to cells. To make dendrimers less hazardous to cells, diverse chemical modifications were performed to their exteriors. Dendrimers containing acetyl groups (-$COCH_3$), PEG, and polysaccharides have depicted increased cell survival while maintaining several beneficial characteristics [124]. Cartilage penetration becomes an important aspect when it comes to treating cartilage disorders. The developed delivery systems need to penetrate through 1000–2000 μm cartilage tissues. Over conventional cartilage drug delivery technologies, the partially-PEGylated dendrimer drug combination delivers a number of translational benefits. Dendrimers made of PEG and PAMAM are already produced in kilogram quantities. These PEGylated dendrimers can incorporate optimal payload without affecting the biological activity of the molecule. Clinical studies are conducted safely using polyamidoamine (PAMAM) and albeit other cationic dendrimers (NCT01577537, NCT03500627, and EUDRACT 2016-000877-19) [125]. The PAMAM dendrimers were also used to develop scaffolds encapsulated with stem cells (SCs) with a suitable crosslinker. These PAMAM scaffolds showed cartilage regenerative capabilities [126]. Several PEGylated hydrogels based on carbamate dendrimers were also developed to repair osteochondral defects. The hydrogels displayed the presence of extracellular matrix (ECM) composition such as collagen and glycosaminoglycans at the defected or applied site [127].

4.1.4 Micelle-Based Drug Delivery System

Polymeric micelles are nanosized structures with a core and a shell produced from amphiphilic block copolymers. Micellization is the process by which the amphiphilic molecules join together to form micelles, which are nanoscale structures made up of many molecules. The hydrophobic segments combine to create the micellar core during the micellization stage, whereas the hydrophilic segments build the micellar shell and thus are positioned between both the hydrophobic core and the hydrophilic bulk phase [128]. These micelles however are excellent choices for intracellular delivery, but there are still significant uncertainties regarding how they enter cells. Polymeric micelles are seldom taken into the body in their entirety. They usually disintegrate within the plasma membrane or degrade in lysosomes. Alternatively, micelles may take in the drug as a disaggregated monomer or release it outside of cells, leading the drug to accumulate in various locations much like plasma membrane including different cellular compartments [129, 130]. Micelle-based drug delivery systems are generally implicated for hydrophobic agents. Encapsulation of the secretory-phospholipase (sPLA2) inhibitor within micelles is developed for site-targeted drug delivery for the lesion tissue [131, 132]. siRNAs are encapsulated within the micelles for targeting p56 within the NF-κB pathway [133]. In response to the elevated level of exposure of free radicals in arthritic joints, oxygen species (ROS) responsive micelles were produced, which would aid in reactive oxygen species sensitive-based delivery [134, 135]. Similarly, folic acid, heparin MMP-13, pH-sensing and responsive theranostic micelles, and several other molecules are conjugated into the micellar structures to target the cartilage tissue [136–138]. Recently, a positively charged triblock polymeric micellar system has been developed that has been shown to deliver the hydrophobic fluorescent dye inside porcine knee cartilage explants with deeper penetration and retention [139].

4.1.5 Nanoemulsion-Based Drug Delivery System

Nanoemulsion is an isotropic dispersion system of two nonmiscible liquids. These are thermodynamically unstable. Nanoemulsion droplets tend to be between 20 and 200 nm in size. The primary distinction between emulsion and nanoemulsion in the dispersion media is indeed the size and the form of the particles [140]. Such carriers are amorphous, negatively charged, solid spheres that are lipid-friendly. Site-specificity can be increased by using magnetic nanoparticles. They boost a drug's therapeutic impact while lowering its side effects and harmful responses in order to get it into the body. Major applications include the treatment of reticuloendothelial system (RES) infections, the replacement of liver enzymes, the management of cancer, and the administration of vaccinations. A two-phase system called an emulsion is one that entails the two phases being intimately combined as small droplets with size ranging from 0.1 to 100 nm. This is a thermodynamically unstable system that an emulsifying agent may stabilize (emulgent or emulsifier). The dispersion medium, exterior phase, or perhaps the continuous phase are other names for the dispersed phase. Additionally known as intermediate or interphase, the emulsifier [141]. A "nanoemulsion," alternatively known as "miniemulsion" is an absolutely fine o/w or w/o dispersion kept together via interfacial coating of surface-active agent molecules having droplet sizes around 20 and 600 nm. There are three different forms of nanoemulsions: (a) w/o nanoemulsions, in which water droplets spread out throughout the continuous oil phase (dispersion), and (b) o/w nanoemulsions, in which oil droplets spread throughout the continuous aqueous phase (dispersion) [142, 143]. Emulsion-based targeted delivery of pharmaceuticals can be delivered to various tissue locations [144]. Significant expression of certain receptor proteins along the outer surface of sick cells is a symptom of several illnesses. It is therefore simpler for nanoparticles to penetrate the cell, where the medicine may be released, thanks to the ligand's binding to the receptor. These nanoemulsions have been created to transport hydrophobic medicines and have also been utilized as thrusters for vaccinations. The clinical results led to the development of nanoemulsions for high-end uses including immunotherapy and targeted treatment. According to a study, the researchers developed and reported injectable tolerogenic nanoemulsions co-encapsulated with citrullinated filaggrin β-fibrinogen vimentin and collagen type-II (CitP), a peptide, and rapamycin that substantially alleviated the advancement of chronic and acute arthritic condition [145]. An alternate study reported the development of quercetin-loaded nanoemulsion, which significantly showed inhibitory activity over TNF-α within in vitro cell line studies [146]. These nanoemulsion-based gel formulations of diclofenac diethylamine have shown their efficacy in the treatment of rheumatoid arthritis [147].

4.1.6 Nanogel-Based Drug Delivery System

Nanogels are three-dimensional networks made of polymer chains that are linked together in different ways. Nanogels have a dual nature, possessing both hydrogel and nanoparticle properties simultaneously because they are made up of hydrogel particles that are nanometers in size. They can be prepared from polymeric precursors or by polymerizing monomers. Nanogels are made by cross-linking hydrophilic polymers with a size range of particle between 100 and 200 nm [148]. Due to their smaller size, nanogels may react fast to varying ambient conditions, such as an alteration in pH and temperature. Nanogels may hold a really higher number of activated molecules attributed to their large internal surface-to-area ratio. This allows for the active transport of drugs with a much longer time in circulation and more stability in the biological environment [149]. Even though being hydrophilic in nature, nanogels may easily incorporate a variety of payloads, including drug molecules and biomacromolecules (such as proteins as well as nucleic acids), by altering their structural makeup without compromising their gel-like properties. Nanogels have been the only materials that can retain many bioactive ingredients in a single carrier due to their unique physical characteristics. Dendrimers, micelles, liposomes, and solid lipid nanoparticles, among other forms of nanoparticles, often are not capable of doing this. The enclosed biological particles are shielded by the nanogel network, which may inflate and react to stimuli from being degraded and removed in vivo. Additionally, it enables the delivery procedure to initiate the dispersion of particles at certain locations [104, 150, 151]. The main benefit of nanogels is implicated in its profound ability to transport a wide range of medications, considering those that bind to water and those that do not along with charged solutes. The availability of hydrophobic or hydrophilic moieties in the polymeric nanocomposites, the temperature, the density of the cross-linked gels, the proportion of surfactants, and indeed the kind of such crosslinking within the polymeric system all have a remarkable impact on the characteristics of the nanogel [152]. Nanogels loaded with prednisolone were developed within a study. These nanogels also showed better systemic retention, sustained release, good targeting efficiency at the arthritic site, and high permeability with a significant reduction in inflammation [153]. Furthermore, a study reported the development of a nanogel-based 3D scaffold for regenerative therapy. These scaffolds were intended to treat the cartilage tissue malformation within the crippling jaw disease. The scaffolds were fabricated by crosslinking the freeze-dried gel with interface chemistry and reinforced with mesenchymal stem cells, which substantially showed the efficiency to promote cartilage formation and maintain structural integrity [154]. Another study revealed the development of a methotrexate-based nanogel. The nanogel was designed to deliver methotrexate intracellularly, which significantly reduces proinflammatory cytokine and prevents the deterioration of the cartilage suppressing the arthritic condition [155].

4.2 *Microparticle-Based Drug Delivery Systems*

A tightly regulated delivery system is required to counterpart the limitations of traditional treatments and boost the efficiency of the active molecule by dispensing drugs over a reasonable duration of time to months. MPs are intended to transport the drug at the appropriate time and in the effective dosage to reduce adverse symptoms and get the most therapeutic benefits. Using microparticles is one such strategy that could be viable and meet the aforementioned criteria [156]. By entrapping them in innocuous polymeric shells, solids, liquids, and gases with dimensions between 1 and 1000 μm, they are protected from the hostile microenvironment by a microparticulate delivery system for drugs (MDDS). A number of delivery techniques, including intraperitoneal, intramuscular, subcutaneous, intraorgan, and bronchial administration, are practiced to provide them directly to the target organs. This delivery system, in particular, enables the continuous and regulated release of the encapsulated particles with progressive diffusion over time. Because of their distinctive and alluring characteristics, including biocompatibility, stability, specific targeting, homogeneous encapsulation, and regulated and prolonged blue prints for drug release, they are responsible for lowering toxicity and dose frequency [157]. Vaccines, nucleic acids, and proteins are just a few of the many macromolecules that are contained within microparticles, which also control their release by dispersing them via various release pathways. MPs efficient delivery systems contain either water-immiscible or only minimally water-miscible active ingredients. A core component that makes up MDDS contains the active chemical, and a covering material shields it from the rasping environment outside and delays its release as necessary. To achieve greater therapeutic efficacy, various polymers are utilized by the illness state, target site, and medicine to be integrated [158, 159]. Microspheres and microcapsules are two main categories under which MDDS falls. Microspheres are microparticles with a greater surface area and volume and are spherical in shape, and confine the drug homogenously throughout the matrix. Microspheres can also be divided into solid and hollow varieties. The active component is distributed completely in the polymer matrix during polymer encapsulation; therefore, there is no layer to prevent diffusion and hence the desired release kinetics in microspheres is first-order diffusion. The medicine is positioned centrally inside the polymeric shell of a microcapsule, where it is gradually released by diffusion and disintegration. Inside the microcapsule, a layer covers the medication and is followed by a coating of polymer shell. In accordance with this feature, they are further divided into three groups, viz, monocoated, polycoated, or matrix type, and each one displays the load distribution of the drug in either a single porous chamber, multiple-porous chambers, or perhaps the matrix inside the polymeric shell. The drug is released from microcapsules according to zero order kinetics. The drug to be encapsulated and the chosen polymer will determine the type of microparticle that needs to be created [160, 161]. Utilizing microparticle-based biomaterial technologies, for example, microspheres, microcarriers, and granular hydrogels mentioned (Fig. 3.5), can help promote and hasten the creation of successful products.

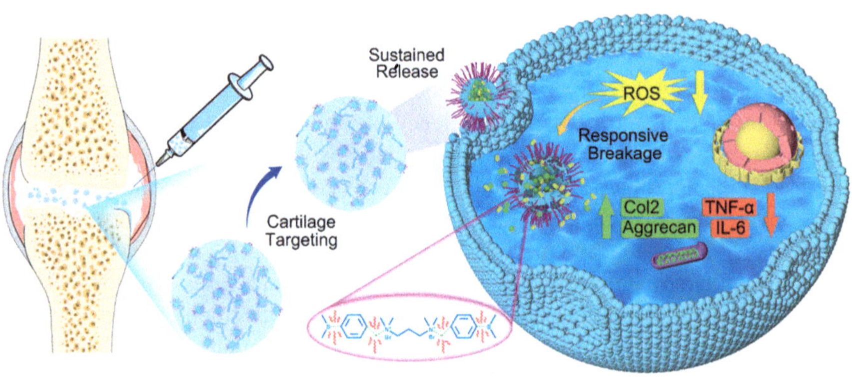

Fig. 3.5 Schematic illustration of microsphere hydrogels showing ROS-responsive and chondrogenic differentiation efficiency for cartilage repair. (Adapted with permission from [162]. Copyright 2022 American Chemical Society)

Drugs that contribute to the regeneration of cartilage tissues, including GFs, corticosteroid drugs, anti-inflammatory molecules, can be delivered via microspheres. Microspheres are a suitable therapy for pain and cartilage degradation due to the localized and sustained release of therapeutics or proteins they offer. In particular, when utilized inside a bioreactor system, microcarriers may be exploited for TE applications by enabling multiscale cell multiplication. The transformation of mesenchymal stem cells on microcarriers has specific relevance to cartilage engineering [163]. A study demonstrated the application of microparticles isolated from BM-MSCs, which reinforced the expression of chondrogenic markers on the contrary, reducing the catabolic markers, viz, matrix-metalloproteinases, ADAMTS 5, and other inflammation markers (iNOS). The microparticles thus showed to be exerting chondrogenic and anti-inflammatory action by inhibiting the macrophages, which leads to the promotion of chondrocytic cell proliferation [164]. The tissue adhesive technique has shown to be another promising clinical practice for cartilage repair, according to a study conducted where the authors developed instant use tissue adhesive paint from silk fibroin microparticle. The paint promotes cell adhesion migration and proliferation. The developed tissue adhesive paint can directly adhere to the surface of the cartilage imparting a polished-coated surface and protecting the cartilage from friction, erosion, and degradation along with wear and tear properties [165]. Microspheres of polyester poly(lactide-co-glycolide) (PLGA) and nonuniform microparticles of size 47 and 210 μm were developed in a study that aimed to regenerate cartilage tissue. Chondrocytes were co-cultured to investigate the development of cartilage tissue that resembled hyaline-like cartilaginous structure [166].

4.3 Hydrogel-Based Drug Delivery Systems

Hydrogels are widely preferred for delivery in different ways because of their different sizes, shapes, and functions. Components within the hydrogels vary in size from centimeters to nanometers. Most of the ways that hydrogels can get into the human body depend on their large-scale design. Hydrogels can be shaped and molded into almost any size or shape. On a scale of several nanometers, a covalently bonded polymeric network envelops the water inside the hydrogel network. The drug-binding sites in the polymer chains may be created in advance utilizing a variety of physical and chemical techniques [167]. The mesh size, molecular scale, or even atomic scale are all crucial for regulated medication release. In general, when developing hydrogel delivery systems, the bioactivity of the drug must be kept during packing, transit, and storage. The hydrogel must therefore be chemically as well as physically stable. There are many methods to administer hydrogels within the body, including surgery, a localized needle injection, or even a systemic intravenous (IV) infusion. How the drug is released from the hydrogel is often key to getting the intended therapeutic effects. Depending on the drug's application, the drug needs to be available for a short or long time, and its release needs to be continuous or pulsatile. When the drug is released after the intended period, the hydrogel should either dissolve so it does not need to be surgically removed [168–170]. When

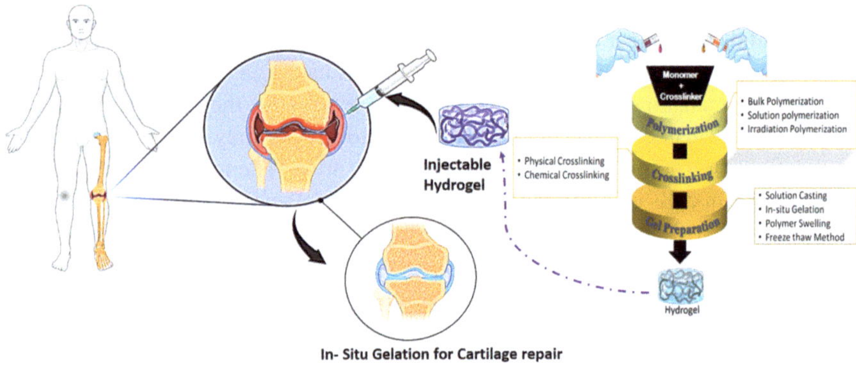

Fig. 3.6 Schematic illustration of the application and development of in situ gel forming injectable hydrogel. (Created with BioRender.com)

hydrogels are put inside the body, they bypass the epithelial layers and deliver drugs where they are intended. Surgical implantation, on the other hand, is invasive and can cause pain and other problems for the patient. To address this issue, there has been a significant emphasis on making macroscale hydrogels that can be injected. These injectable hydrogel systems are either in situ-gelling hydrogels, shear-thinning hydrogels, or shape-memory hydrogels (Fig. 3.6) [171, 172].

Due to their adaptable characteristics and controllable breakdown, hydrogels are gaining interest in potential local drug delivery methods, hence protecting drugs that break down or degrade easily [173]. Several hydrogels are developed, studied, and reported as well with repair and regenerative capabilities (Table 3.2).

4.4 Transdermal Drug Delivery Systems

In the coming years, transdermal administration may be an efficient alternative to hypodermic needles in the same way it now replaces oral medications. Transdermal delivery offers numerous advantages over oral delivery. It is utilized in particular whenever the liver has quite a significant first-pass impact that might cause medications to break down too fast. Transdermal delivery is also better than hypodermic injections, which hurt, make dangerous medical waste, and put people at risk of getting sick by sharing needles, especially in developing countries. Transdermal systems are also noninvasive and easy to use by yourself. They can make you feel better for a long time (up to 1 week). They also make it easier for patients to follow directions, and the devices are usually cheap [175]. The microneedle-based transdermal delivery systems (MNs) can deliver the drug out from the stratum corneum (SC) toward the viable epidermis, from where they may enter readily the dermis and blood vessels (Fig. 3.7).

Table 3.2 Summary of selected hydrogels based on PEG, CS, HA, and gelatin for cartilage regeneration

Hydrogel	Crosslinkers	Time point	Cells	Findings	Viability, ECM production (μg μg^{-1}), compressive modulus
CS-based (with graphene)	EDC/NHS	3 weeks	MSCs	EDAG-CS porous hydrogels exhibit higher mechanical properties, but degrade slower than CS-CS. EDAG-CS and CS-CS produce type II collagen and MSCs do not differentiate to endochondral ossification	Cell viability and ECM production not reported Compressive modulus: 1.12 kPa
Carboxymethylated pullulan CMP-TA/ CS-TA	Enzymatic crosslinking (horseradish peroxides and H_2O_2)	14 days	Porcine auricular chondrocytes	Hydrogel with a CMP-TA/CS-TA weight ratio of 3/1 provided best host tissue-mimetic microenvironment	Viability: 80–95% (CMP-TA), >95% (CMP-TA/ CS-TA) ECM production not reported Compressive modulus (CMP-TA): 0.96 kPa
GEL-based	UV photo-polymerization	3 days, 4 weeks	Chondrocytes	Supplementation of HA to increase viscosity and PCL polymer for structural support enables the build-up of a layered hydrogel structure. Cell viability detected until 28 days using GAG/DNA	Viability: 73 ± 2% (GEL-MA), 82 ± 8% (GEL-MA-HA) (day 3) GAG/DNA: 7.1 ± 3.3 (GEL-MA), 10.5 ± 5.2 (GEL-MA-HA) (week 4) Compressive modulus: 32.28–53.54 kPa (GEL-MA)

(continued)

Table 3.2 (continued)

Hydrogel	Crosslinkers	Time point	Cells	Findings	Viability, ECM production (μg μg^{-1}), compressive modulus
CS-PEG	UV photo-polymerization	24 h, 3 weeks	Juvenile, adult, and osteoarthritic chondrocytes	7% w/v of PEGDA and 3% w/v. Deposition of ECM and long-term cell viability are reported	Viability: 94%
					ECM production not reported
					Compressive modulus: 0.73–1.04 kPa (day 1, week 3)
CS-PEG	Photo-polymerization	3 weeks	Neonatal bovine chondrocytes	Dual gradient hydrogel with interconnected chambers (20% w/v PEG, 3% w/v CS; zone 1; to 2% w/v PEG, 8% w/v CS; zone 5) with better mechanical properties, cell proliferation, and matrix deposition than CS hydrogel gradient alone	Viability: >90% in all zones (7 days)
					GAG/DNA: 40.58–53.72 kPa (zone 1–zone 5)
					Compressive modulus: 9.2 kPa–124.7 kPa (zone 1–zone 5)
CS-PEG	UV photo-polymerization	2, 4, 6 weeks	D1 mouse bone marrow progenitor cell	Higher collagen II content was detected in the superficial layer (PEG: CS: MMP-pep), collagen X content and GAG production were increasing from the superficial layer, transitional (PEG: CS) to the deep layer (PEG: HA) mimicking the cartilage layer	Viability: 86% (PEG-CS-MMP-pep), >95% (CS-PEG), 84% (PEG-HA)
PEG-CS-MMP-pep					GAG/DNA: PEG-CS-MMP-pep: 2.6; CS-PEG: 2.8; PEG-HA: 6.6 (2 weeks)
PEG-HA					Compressive modulus: PEG-CS-MMP-pep: 199.71 (120–209 kPa), CS-PEG: 203.95 (121–270 kPa); PEG-HA: 942.83 (271–1227 kPa)

CS-PEG	EDC/NHS	24 h (in vitro)	Bovine articular chondrocytes	Cells within hydrogel retain viability and metabolically active, minimal inflammatory response compared to PEG, stronger adhesion compared with fibrin in vivo	Viability: 93–96%
		4 weeks (in vivo)			ECM production not reported
					Compressive modulus: 11.31–22.28 kPa
CS-PEG	UV photo-polymerization	24 h, 3, 6 weeks	MSCs	PEG-CS gels have 10 times stronger adhesion to cartilage tissue compared to fibrin, and decreased inflammatory response compared with PEG	Viability: 79% (PEG), 84% (CS-PEG) (24 h)
					GAGs/DNA: PEG: 10.75 (week 3), 20.12 (week 6). CS-PEG: 18.11 (week 3), 36.18 (week 6)
					Compressive modulus not reported
PEG-LA-DM	UV photo-polymerization	6 weeks (in vitro)	Juvenile swine articular chondrocytes	Encapsulated chondrocytes in 60/40 and 70/30 PEG-LA-DM/PEG-DM polymer ratios	Cell viability and ECM production not reported
PEG-DM		6, 12, 18 weeks (in vivo)		Improved cartilage matrix formation of aggrecan and collagen type II/VI	Compressive modulus: 60–670 kPa
PEG-HA	Enzymatic crosslinking (HRP and H_2O_2) and Diels–Alder click chemistry	7 days	Chondrogenic teratocarcinoma cells	Enzymatic crosslinking hydrogels result in 5 min of gelation time, have great shape memory and antifatigue properties	Viability: >95%
					No ECM production reported
					Compressive modulus: 60.7–109.4 kPa

(continued)

Table 3.2 (continued)

Hydrogel	Crosslinkers	Time point	Cells	Findings	Viability, ECM production (μg $μg^{-1}$), compressive modulus
PEG-GelMa	Photo-polymerization	7 days, 3 weeks	Human MSCs	Inkjet-bioprinted MSCs of PEG-MA show better mechanical properties than PEG, and increased 63% under chondrogenic differentiation after 21 days	Viability: 85.87% (PEG-GelMa), 82.43% (PEG)
					GAGs/DNA: 73.81 (PEG-GelMa), 67.64 (PEG) (7 days)
PEG					Compressive modulus: 36.70 kPa (PEG-GelMa), 45.31 kPa (PEG)
GEL-CS	EDC/NHS and click chemistry	1, 3 and 7 days	Fibroblasts and macrophages	Overall good cytocompatibility, mimicking cartilage ECM composition	Viability: 10% CS–gelatin (79% MTT);
					Compressive modulus: not measured
					ECM production: not measured
GEL-CS	Chemical crosslinking (glutaraldehyde, dithiothreitol/2-propanol); and photo-crosslinking (UV-light)	21 days	Human mesenchymal stem cells (MSCs)	Spatially patterned μRB scaffold enabled MSCs to produce cartilage with biomimetic zonal biochemical and mechanical properties	Viability: not measured quantitatively
					Compressive modulus: trilayer μRB up to 456 kPa
					ECM production: sGAG per d.w trilayer μRB up to 0.08 μg $μg^{-1}$ (1.5–7.5% dry weight)

GEL-HA	Photo-crosslinking (UV-light)	Up to 8 weeks	Human bone marrow–derived mesenchymal stem cells (hBMSCs) and in vivo (New Zealand white rabbits)	MSCs chondrogenesis scaffolds at 9:1 (% w/v) ratio mGL/mHA lowest hBMSC hypertrophy and highest glycosaminoglycan production, supported cartilage formation in vivo after 12 weeks	Viability: not measured quantitatively Compressive modulus: up to 25 kPa (8.5/1.5 ratio mGL/mHA) ECM production: up to 17 μg GAG/construct
GEL-HA	Photo-crosslinking (UV-light)	1, 4, 7 days	Chondrocytes and in vivo (nude mice)	At a relatively low cell-seeding concentration, homogeneous tubular cartilage regenerated with a tracheal shape, mechanical strength, and cartilage ECM deposition	Cell viability: about 98%. Storage modulus (G') = 1 kPa ECM production: no reported in vitro but in vivo (about 70 mg g^{-1} LDTM scaffold and 200 mg g^{-1} for Exp)
GEL-HA	Enzymatic crosslinking (HRP and H_2O_2)	Up to 14 days	Mesenchymal stem cells (MSCs)	70% HA and 30% gel are most promising for chondrogenesis, best mechanical properties	Viability is not measured quantitatively; cell proliferation decreases in hydrogels rich in HA. Compressive modulus (G') up to 789 ± 220 Pa ECM production = GAGs about 80% in the 30/70 mixture and HA gel

(continued)

Table 3.2 (continued)

Hydrogel	Crosslinkers	Time point	Cells	Findings	Viability, ECM production (μg μg^{-1}), compressive modulus
Collagen-CS-HA (CCH)	Chemical crosslinking (Genipin, 0.75 mM)	4 weeks	Allogeneic chondrocytes (in vivo Rabbits model)	CCH hydrogel provides structural and mechanical similarity to cartilage; is used as a carrier of chondrocytes; helps cell growth and matrix synthesis	Viability is not measured quantitatively; cell proliferation increases in CCH hydrogels
					Compressive modulus increases (CCH about 51.53 ± 0.54 kPa.)
					GAG/DNA ratio about 4 (7 days, approx. 6 μg μl^{-1})
HA-GEL-CS	Diels–Alder click chemistry	N/A	N/A	HA-furan GEL-furan, MAL-PEG-MAL	Cell viability and ECM production not reported or tested
				HA-GEL-CS with EDC NHS	Compressive modulus: 1.62–2.21 kPa
HA-GEL-CS	Click chemistry (Cu(I) as the catalyst)	Up to 7 days	Chondrocytes	Good reactivity under very mild conditions by click-chemistry. In vitro cell culture confirmed that chondrocytes could adhere and proliferate on the hydrogel	Viability is not measured quantitatively; cell number increases over time
					Compressive modulus ~7 kPa
					ECM production: not measured

GEL-MA	Photo-crosslinking (UV light)	1 day, 5 and 8 weeks	Human chondrocytes	The addition of CS and HA to GEL-MA can enhance the differentiation and improve mechanical properties and ECM distribution. Significant COL2A1 expression	Viability: 80–90%
GEL-HA					GAGs/DNA: 110.9 (GEL-MA), 140 (GEL-HA), 92.7 (GEL-CS), 125.3 (GEK-HA-CS) (week 8)
GEL-CS					Compressive modulus: 20.17 kPa (GEL-MA), 31.08 kPa (GEL-HA), 21.66 kPa (GEL-CS), 32.57 kPa (GEL-HA-CS)
GEL-HA-CS					

Reproduced from [174] with permission
CS Chondroitin sulfate, *CMP* Carboxymethyl pullulan, *DM* Dimethacrylate, *EDAG* Ethylenediamine graphene, *GEL* Gelatin, *GELMA* Gelatin methacrylates, *HA* Hyaluronic acid, *LA* 4, 5 Lactic acid, *MA* Methacrylates, *MMPpep* Matrix metalloproteinase-sensitive peptides, *PEG* Polyethylene glycol, *TA* Tyramine

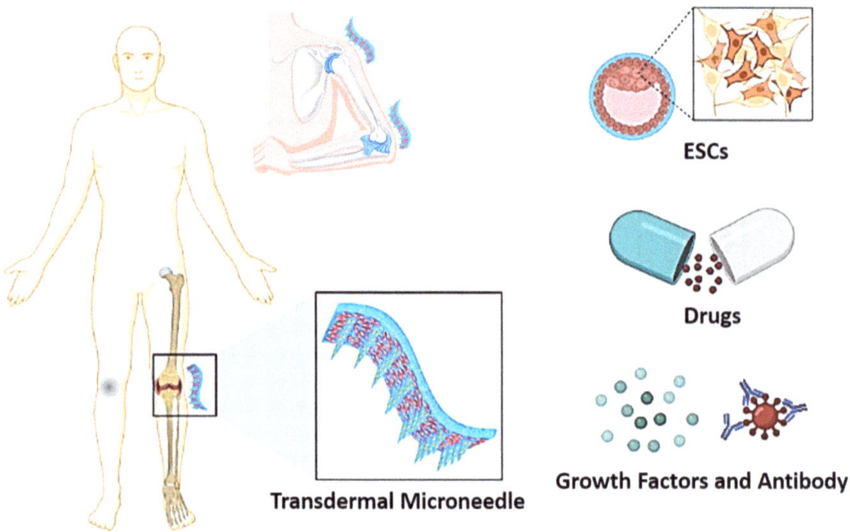

Fig. 3.7 Schematic illustration of the application of transdermal microneedle in cartilage repair. (Created with BioRender.com)

In an attempt for such type of MN delivery systems to operate, the MNs need to reside in the skin for longer periods, attributed to the fact that they are designed to be utilized transdermally or with the drug enclosed in the MNs' structure, where the hydrogel material's composition allows for control over the rate of release [176–178]. Triptolide (TP), a potential anti-inflammatory medication for treating osteoarthritis, has limited its practical applicability due to substantial toxicity, lower solubility, and high clearance into the body. To diversify its utilization possibilities for osteoarthritis therapy, the authors in a study have constructed a dissolving microneedle (DMN) device that is filled with liposomes to dispense poorly water-soluble triptolide and ameliorate osteoarthritis symptoms. The DMNs significantly alleviate the damage to cartilage and symptoms of arthritis [179]. For treating the pain raised as a consequence of cartilaginous lesions or arthritic symptoms, another transdermal microneedle (TDM) approach in conjugation with NSAIDs was developed. The TDM significantly decreased the pain score and alleviated synovitis [180].

4.5 Delivery Molecules in Cartilage Repair and Regeneration

The progression of the therapeutic modalities has forced the drug delivery systems to advance toward small molecules including nucleic acids, peptides, proteins, etc.

4.5.1 Growth Factors and Peptides

A class of physiologically active polypeptides known as growth factors are created by the body and have the ability to promote cellular differentiation, growth, and division. Numerous growth factors intervene in articular cartilage to control its growth and homeostasis during the course of a person's life. Therefore, growth factors provide potential therapies for augmenting cartilage regeneration in cases of isolated articular cartilage abnormalities or more severe cartilage loss, such as osteoarthritis (OA). When pondering about a maimed OA joint, the repercussions of treatment, such as growth factors, should be addressed on that same cartilage, muscles, tendons and ligaments, meniscus, any unprotected subchondral bone, in addition to mesenchymal stem/stromal cells (MSCs) which then enter the articular environment. A myriad of anabolic growth factors promote the production of collagen (Col-2), aggrecan (AGs), and proteoglycans (PGs) in chondrocytes, synoviocytes, and MSCs and promote the transformation of MSCs into chondrogenic cells, which lessen the implications of cytokines like IL-1 and MMPs. In regard to pro-anabolic and anticatabolic actions, an exemplary growth factor (GF) for broad sense tissue repair or the regeneration of the cartilage disease is impactful against the age of the patient. Transforming growth factor (TGF-β) appears to be in three distinct forms (i.e., TGF-β1, β2, and β3) in the ECM of the tissue. These are regulated by the SMAD signaling pathway [181]. Bone morphogenic proteins (BMPs) form the subclass of TGF acting through similar signaling pathways. Among 15 BMPs discovered till now, BMP2, BMP3, BMP4, and BMP7 are vastly characterized by cartilage defect [182]. IGF is a part of the protein family that is physically homologous to proinsulin and has the ability to increase matrix formation and stimulate chondrocyte growth [183]. The FGF family is composed of 23 members, of which the most explored are, viz, FGF2, FGF9, and FGF18. VEGF is a member of a homo-dimeric protein family consisting of six members, viz, VEGF-A, B, C, D, E, and placental growth factor. VEGF regulates vascular permeability, vasodilation, and angiogenesis of endothelial cells [184]. VEGF has a critical function in regulating the osteoblast maturation of ADSC in response to hypoxic conditions [185]. Kartogenin and TGF-β3 exhibited an additive effect on the MSC found in synovia and umbilical cord [186, 187]. The detailed list of growth factors and peptides used in the drug delivery system for cartilage is tabulated in Table 3.3.

5 Challenges for Drug Delivery System

Numerous cutting-edge cartilage drug delivery strategies have been introduced during the past 20 years. Anti-inflammatory therapies have evolved into an effective and ground-breaking way to stop inflammatory arthropathies including psoriatic as well as rheumatoid arthritis from progressing. But with the majority of diseases that impact cartilage, there are still no efficient disease-modifying drugs. This is because the pathophysiology of the disease is not fully understood. It also takes a longer

Table 3.3 Growth factor and peptides with drug delivery system in cartilage disease

Sr. no.	Growth factors	Cartilage disease	Drug delivery system	References
1.	TGF-β and ghrelin	Osteoarthritis	Microsphere/hydrogel	[188]
2.	TGF-β1 and BMP-2	Articular cartilage defect	Chitosan nanoparticle	[189]
3.	TGF-β1	Osteochondral defect	PCEC hydrogel	[190]
4.	TGF-β1 and SDF-1α	Cartilage defect	Silk-fibroin porous-gelatin scaffold	[191]
5.	Ginsenoside, Rb1/TGF-β1	Cartilage defect	Silk-fibroin gelatin-porous scaffold	[192]
6.	TGF-β1	Articular cartilage disease	Self-assembling peptide hydrogel	[193]
7.	BMP-2	Chondral defect	Injectable hydrogel	[194]
8.	BMP-2	Osteochondral defect	Hap/Col scaffold	[195]
9.	BMP-7	Articular cartilage injuries	Collagen sponge	[196]
10.	BMP-7	Articular cartilage damage	Cryogel	[197]
11.	BMP-4	Articular cartilage defect	Gene therapy	[198]
12.	BMP-7	Osteochondral defect	Nanoparticles	[199]
13.	BMP-7	Small cartilage defects	Microspheres	[200]
14.	IGF-1	Osteoarthritis	Modified RNAs	[201]
15.	IGF-1	Osteochondral defects	Composite hydrogel	[202]
16.	IGF-1	Osteochondral defects	Gene therapy	[203]
17.	IGF-1 and BMP-7	Knee injury	Gene therapy	[204]
18.	Human IGF-1 pro peptide	Articular cartilage defect	Gene therapy	[205]
19.	IGF-1 and TGF-β	Osteochondral defects	Laminin gel	[206]
20.	IGF-1 and HA	Osteochondral injury	Absorbable gelatin sponge	[207]
21.	FGF-18 (sprifermin)	Rotator cuff injury	Gene therapy	[208]
22.	FGF-2	Calvarial bone defects	Gene therapy	[209]
23.	hFGF-2	Osteochondral defect	Gene therapy	[210]
24.	FGF-9	Osteoarthritis	Intra-articular injection	[211]
25.	IGF-1 and FGF-2	Osteochondral defect	Gene therapy	[212, 213]
26.	FGF-2	Osteochondral defect	Hap/col scaffold	[214]
27.	FGF-2	Rotator cuff injury	ADM graft	[215]
28.	VEGF attenuated PRP	Osteoarthritis	Microsphere	[216]
29.	Bevacizumab (anti-VEGF)	Osteochondral defect	Intra-articular injection	[217]
30.	Bevacizumab	Growth plate injury	Intraperitoneal injection	[218]
31.	Bevacizumab	Osteochondral defect	Intravenous injection	
32.	VEGF	Radius defects	Gene-activated matrix	[219]
33.	VEGF	Meniscus lacerations	mRNA therapy	[220]

(continued)

Table 3.3 (continued)

Sr. no.	Growth factors	Cartilage disease	Drug delivery system	References
34.	VEGF-A	Bone defect	Gene therapy	[221]
35.	Exogenous bFGF	Articular cartilage defect	Nanoparticles	[222]
36.	Kartogenin and poly-l-lysine	Articular cartilage defect	Nanoparticles	[223]
37.	Kartogenin and PLGA	Articular cartilage defect	Microparticles	[224]
38.	Kartogenin	Osteoarthritis	Intra-articular injection	[225]

time to diagnose and treat the disease because cartilage does not have nerves, and it is hard to deliver potential therapies because cartilage does not have blood vessels. Mechanistic investigations are identifying pharmacological targets for osteoarthritis specifically and opening the door for the development of potential therapies. But there are still challenges associated with the effective drug delivery system for cartilage. Cationic carriers for drug delivery to cartilage are starting to look like a good way to help find potential DMOADs and make them work better. Such delivery methods have the added benefit of being able to treat a small area for a long time with low doses that are still effective. The same principles might be utilized to administer medications for various disorders that damage cartilage since all these charged carriers enhance distribution by interacting mostly with matrix proteoglycans. The acquisition of gene therapies and/or medications that access proinflammatory mediators to chondrocytes in individuals with hereditary diseases that impact cartilage are two examples of potential uses. Another is the allocation of harmful chemotherapeutic drugs to cartilaginous tumors in the bone and joints without negatively impacting normal healthy tissues. This is a big step toward developing a DDS that can slow or inhibit the advancement of cartilage-related diseases. One potential treatment may work better with a certain method of delivery than with another. Candidate drug selection is a vital part of the scientific progress of a cartilage drug delivery system [226]. When designing novel drug delivery systems, it is crucial to consider both the drug's release mechanism and its effectiveness in reaching its intended target. Taking into consideration the complexity of physiological variables and tissue structure, making multiple targeted designs fosters a big difference in how efficiently materials can be guided. To justify that these strategies can be feasible, more tests of how reactions change every stimulus in vitro and even in vivo are needed. Hence, a significant amount of progress has been made in the research of new materials. Even so, research on improved and more sophisticated multiple-responsive carriers is still mostly practiced in the sector of cartilage research, and the advancement of multiple-response drug-delivery methods is still in its preliminary phase. But in a few aspects that are related to the disorder, they are still very important for the development of drug carriers. Such flaws might not be reported until more preclinical evidence is collected, it is imperative to consider both the drug's release mechanism and its effectiveness as a target when designing novel drug delivery materials. In order to make it more straightforward for modern approaches to be assessed in clinical settings, to ensure the successful completion of clinical studies, to ensure that regulatory practices for materials acknowledge both healthcare needs and market needs, facilitate the adoption of clinical trials, and find an appropriate balance between the advantages and risks. It may be helpful to always have close collaboration and understanding with people who specialize in the area of particle toxicology.

6 Conclusion

In the field of medicine, drug carriers are innovative methods to deliver drugs to wherever it is targeted. Even though a plethora of polymers have been utilized in different studies to make the carriers, the most promising drug delivery system, which is cheap, easy to make, and capable of delivering various drugs, is to be evaluated. Many gel-based systems have been researched, designed, and protected by intellectual property. It seems that only some of these products have successfully entered the market. This elucidates that more research with better performance of existing systems and creating newer systems is the need of the hour. This chapter discussed various drug delivery systems from different points of view, like how they work (loading, releasing, and targeting), how they look (shape and route of administration), and what kinds of drugs they deliver. An improved comprehensive study of this phenomenon and uptake strategies, in general, brings one move toward being able to regulate the mechanics of nanocarrier internalization by design and will result in superior DDS with improved drug effectiveness.

References

1. L. Wachsmuth, S. Söder, Z. Fan, F. Finger, T. Aigner, Immunolocalization of matrix proteins in different human cartilage subtypes, Histol. Histopathol. 21 (2006) 477–485. https://doi.org/10.14670/HH-21.477.
2. Y. Krishnan, A.J. Grodzinsky, Cartilage diseases, Matrix Biol. 71–72 (2018) 51–69. https://doi.org/10.1016/j.matbio.2018.05.005.
3. A.J. Sophia Fox, A. Bedi, S.A. Rodeo, The basic science of articular cartilage: Structure, composition, and function, Sports Health. 1 (2009) 461–468. https://doi.org/10.1177/1941738109350438.
4. K.L. Bauer, J.D. Polousky, Management of Osteochondritis Dissecans Lesions of the Knee, Elbow and Ankle, Clin. Sports Med. 36 (2017) 469–487. https://doi.org/10.1016/j.csm.2017.02.005.
5. L. Andriolo, D.C. Crawford, D. Reale, S. Zaffagnini, C. Candrian, A. Cavicchioli, G. Filardo, Osteochondritis Dissecans of the Knee: Etiology and Pathogenetic Mechanisms. A Systematic Review, Cartilage. 11 (2020) 273–290. https://doi.org/10.1177/1947603518786557.
6. J.J. Nepple, M.D. Milewski, K.G. Shea, Research in Osteochondritis Dissecans of the Knee: 2016 Update, J. Knee Surg. 29 (2016) 533–538. https://doi.org/10.1055/s-0036-1586723.

7. B. Ytrehus, C.S. Carlson, S. Ekman, Etiology and pathogenesis of osteochondrosis, Vet. Pathol. 44 (2007) 429–448. https://doi.org/10.1354/vp.44-4-429.
8. T. Lahmer, M. Treiber, A. von Werder, F. Foerger, A. Knopf, U. Heemann, K. Thuermel, Relapsing polychondritis: An autoimmune disease with many faces, Autoimmun. Rev. 9 (2010) 540–546. https://doi.org/10.1016/j.autrev.2010.02.016.
9. L. Longo, A. Greco, A. Rea, V.R. Lo Vasco, A. De Virgilio, M. De Vincentiis, Relapsing polychondritis: A clinical update, Autoimmun. Rev. 15 (2016) 539–543. https://doi.org/10.1016/j.autrev.2016.02.013.
10. P. Richette, T. Bardin, M. Doherty, An update on the epidemiology of calcium pyrophosphate dihydrate crystal deposition disease, Rheumatology. 48 (2009) 711–715. https://doi.org/10.1093/rheumatology/kep081.
11. S.A. Qasem, B.R. DeYoung, Cartilage-forming tumors, Semin. Diagn. Pathol. 31 (2014) 10–20. https://doi.org/10.1053/j.semdp.2014.01.006.
12. S. Zhou, T.S. Thornhill, F. Meng, L. Xie, J. Wright, J. Glowacki, Influence of osteoarthritis grade on molecular signature of human cartilage, J. Orthop. Res. 34 (2016) 454–462. https://doi.org/10.1002/jor.23043.
13. F. Berenbaum, Osteoarthritis as an inflammatory disease (osteoarthritis is not osteoarthrosis!), Osteoarthr. Cartil. 21 (2013) 16–21. https://doi.org/10.1016/j.joca.2012.11.012.
14. Y. Li, D. Tong, P. Liang, E. Lönnblom, J. Viljanen, B. Xu, K.S. Nandakumar, R. Holmdahl, Cartilage-binding antibodies initiate joint inflammation and promote chronic erosive arthritis, Arthritis Res. Ther. 22 (2020) 1–14. https://doi.org/10.1186/s13075-020-02169-0.
15. H. Huang, Z. Lou, S. Zheng, J. Wu, Q. Yao, R. Chen, L. Kou, D. Chen, Intra-articular drug delivery systems for osteoarthritis therapy: shifting from sustained release to enhancing penetration into cartilage, Drug Deliv. 29 (2022) 767–791. https://doi.org/10.1080/10717544.2022.2048130.
16. J. Fábio dos Santos Duarte Lana, B. Lima Rodrigues, Osteoarthritis as a Chronic Inflammatory Disease: A Review of the Inflammatory Markers, Osteoarthr. Biomarkers Treat. (2019). https://doi.org/10.5772/intechopen.82565.
17. J. Barranco-Trabi, V. Mank, J. Roberts, D.P. Newman, Atypical Costochondritis: Complete Resolution of Symptoms After Rib Manipulation and Soft Tissue Mobilization, Cureus. 13 (2021) 1–10. https://doi.org/10.7759/cureus.14369.
18. P. Lama, U. Zehra, C. Balkovec, H.A. Claireaux, L. Flower, I.J. Harding, P. Dolan, M.A. Adams, Significance of cartilage endplate within herniated disc tissue, Eur. Spine J. 23 (2014) 1869–1877. https://doi.org/10.1007/s00586-014-3399-3.
19. R.M. Pauli, Achondroplasia: A comprehensive clinical review, Orphanet Journal of Rare Diseases, 2019. https://doi.org/10.1186/s13023-018-0972-6.
20. D.M. Ornitz, L. Legeai-mallet, F. Growth, F. Receptor, D. Ornitz, L. Legeai-mallet, Review Developmental Dynamics DOI 10.1002/dvdy. 24479 Achondroplasia: Development, Pathogenesis, and Therapy, (2016) 1–41. https://doi.org/10.1002/dvdy.
21. L. Gao, L.K.H. Goebel, P. Orth, M. Cucchiarini, H. Madry, Subchondral drilling for articular cartilage repair: A systematic review of translational research, DMM Dis. Model. Mech. 11 (2018). https://doi.org/10.1242/DMM.034280/265026/AM/SUBCHONDRAL-DRILLING-FOR-ARTICULAR-CARTILAGE.
22. M.F. Sommerfeldt, R.A. Magnussen, T.E. Hewett, C.C. Kaeding, D.C. Flanigan, Microfracture of articular cartilage, JBJS Rev. 4 (2016). https://doi.org/10.2106/JBJS.RVW.15.00005.
23. W. Xing, D. Mu, Q. Wang, S. Fu, M. Xin, J. Luan, Improvement of Fat Graft Survival with Autologous Bone Marrow Aspirate and Bone Marrow Concentrate: A One-Step Method, Plast. Reconstr. Surg. 137 (2016) 676e-686e. https://doi.org/10.1097/PRS.0000000000001993.
24. Y. Matsusue, T. Yamamuro, H. Hama, Arthroscopic multiple osteochondral transplantation to the chondral defect in the knee associated with anterior cruciate ligament disruption, Arthrosc. J. Arthrosc. Relat. Surg. 9 (1993) 318–321. https://doi.org/10.1016/S0749-8063(05)80428-1.
25. D.L. Richter, J.A. Tanksley, M.D. Miller, Osteochondral Autograft Transplantation: A Review of the Surgical Technique and Outcomes, Sports Med. Arthrosc. 24 (2016) 74–78. https://doi.org/10.1097/JSA.0000000000000099.

26. R. Gudas, R. Simonaityte, E. Čekanauskas, R. Tamošiunas, A prospective, randomized clinical study of osteochondral autologous transplantation versus microfracture for the treatment of osteochondritis dissecans in the knee joint in children, J. Pediatr. Orthop. 29 (2009) 741–748. https://doi.org/10.1097/BPO.0B013E3181B8F6C7.
27. A. Hennig, J. Abate, Osteochondral allografts in the treatment of articular cartilage injuries of the knee, Sports Med. Arthrosc. 15 (2007) 126–132. https://doi.org/10.1097/JSA.0B013E31812E5373.
28. F. Yamashita, K. Sakakida, F. Suzu, S. Takai, The transplantation of an autogeneic osteochondral fragment for osteochondritis dissecans of the knee., Clin. Orthop. Relat. Res. NO. 201 (1985) 43–50. https://doi.org/10.1097/00003086-198512000-00007.
29. C.M. Wixted, T.J. Dekker, S.B. Adams, Particulated juvenile articular cartilage allograft transplantation for osteochondral lesions of the knee and ankle, https://doi.org/10.1080/17434440.2020.1733973. 17 (2020) 235–244. https://doi.org/10.1080/17434440.2020.1733973.
30. B.B. Hinckel, D. Thomas, E.E. Vellios, K.J. Hancock, J.G. Calcei, S.L. Sherman, C.D. Eliasberg, T.L. Fernandes, J. Farr, C. Lattermann, A.H. Gomoll, Algorithm for Treatment of Focal Cartilage Defects of the Knee: Classic and New Procedures, Cartilage. 13 (2021) 473S–495S. https://doi.org/10.1177/1947603521993219/ASSET/IMAGES/LARGE/10.1177_1947603521993219-FIG2.JPEG.
31. J.M. Woodmass, H.P. Melugin, I.T. Wu, D.B.F. Saris, M.J. Stuart, A.J. Krych, Viable Osteochondral Allograft for the Treatment of a Full-Thickness Cartilage Defect of the Patella, Arthrosc. Tech. 6 (2017) e1661–e1665. https://doi.org/10.1016/J.EATS.2017.06.034.
32. M. Krill, N. Early, J.S. Everhart, D.C. Flanigan, Autologous Chondrocyte Implantation (ACI) for Knee Cartilage Defects, JBJS Rev. 6 (2018) E5. https://doi.org/10.2106/JBJS.RVW.17.00078.
33. R. Cugat, G. Samitier, G. Vinagre, M. Sava, E. Alentorn-Geli, M. García-Balletbó, X. Cuscó, R. Seijas, D. Barastegui, J. Navarro, P. Laiz, Particulated Autologous Chondral–Platelet-Rich Plasma Matrix Implantation (PACI) for Treatment of Full-Thickness Cartilage Osteochondral Defects, Arthrosc. Tech. 10 (2021) e539–e544. https://doi.org/10.1016/J.EATS.2020.10.038.
34. J.P. Benthien, P. Behrens, Autologous matrix-induced chondrogenesis (AMIC): Combining microfracturing and a collagen I/III matrix for articular cartilage resurfacing, Cartilage. 1 (2010) 65–68. https://doi.org/10.1177/1947603509360044/ASSET/IMAGES/LARGE/10.1177_1947603509360044-FIG2.JPEG.
35. M. Wiewiorski, A. Barg, V. Valderrabano, Autologous matrix-induced chondrogenesis in osteochondral lesions of the talus, Foot Ankle Clin. 18 (2013) 151–158. https://doi.org/10.1016/j.fcl.2012.12.009.
36. P. Gupta, K. Vermani, S. Garg, Hydrogels: from controlled release to pH-responsive drug delivery, Drug Discov. Today. 7 (2002) 569–579. https://doi.org/10.1016/S1359-6446(02)02255-9.
37. J. Radhakrishnan, A. Subramanian, U.M. Krishnan, S. Sethuraman, Injectable and 3D Bioprinted Polysaccharide Hydrogels: From Cartilage to Osteochondral Tissue Engineering, Biomacromolecules. 18 (2017) 1–26. https://doi.org/10.1021/ACS.BIOMAC.6B01619/ASSET/IMAGES/LARGE/BM-2016-01619E_0007.JPEG.
38. M.A. Mohamed, A. Fallahi, A.M.A. El-Sokkary, S. Salehi, M.A. Akl, A. Jafari, A. Tamayol, H. Fenniri, A. Khademhosseini, S.T. Andreadis, C. Cheng, Stimuli-responsive hydrogels for manipulation of cell microenvironment: From chemistry to biofabrication technology, Prog. Polym. Sci. 98 (2019) 101147. https://doi.org/10.1016/J.PROGPOLYMSCI.2019.101147.
39. C. Wang, N. Feng, F. Chang, J. Wang, B. Yuan, Y. Cheng, H. Liu, J. Yu, J. Zou, J. Ding, X. Chen, Injectable Cholesterol-Enhanced Stereocomplex Polylactide Thermogel Loading Chondrocytes for Optimized Cartilage Regeneration, Adv. Healthc. Mater. 8 (2019) 1900312. https://doi.org/10.1002/ADHM.201900312.
40. C.H. Hulme, J. Perry, H.S. McCarthy, K.T. Wright, M. Snow, C. Mennan, S. Roberts, Cell therapy for cartilage repair, Emerg. Top. Life Sci. 5 (2021) 575–589. https://doi.org/10.1042/ETLS20210015.

41. M.F. Pittenger, D.E. Discher, B.M. Péault, D.G. Phinney, J.M. Hare, A.I. Caplan, Mesenchymal stem cell perspective: cell biology to clinical progress, Npj Regen. Med. 4 (2019). https://doi.org/10.1038/s41536-019-0083-6.
42. J.B. Richardson, K.T. Wright, J. Wales, J.H. Kuiper, H.S. McCarthy, P. Gallacher, P.E. Harrison, S. Roberts, Efficacy and safety of autologous cell therapies for knee cartilage defects (autologous stem cells, chondrocytes or the two): Randomized controlled trial design, Regen. Med. 12 (2017) 493–501. https://doi.org/10.2217/rme-2017-0032.
43. Z. Wang, G. Schuch, J.K. Williams, S. Soker, Peripheral Blood Stem Cells, Essentials Stem Cell Biol. Third Ed. (2014) 227–244. https://doi.org/10.1016/B978-0-12-409503-8.00017-2.
44. S. Kulkarni, J. Treleaven, Patient selection: Preliminary interview, screening of patient and donor, Elsevier Ltd, 2009. https://doi.org/10.1016/B978-0-443-10147-2.50024-2.
45. T. Mori, T. Osumi, Hematopoietic Stem Cell Transplantation, Non-Hodgkin's Lymphoma Child. Adolesc. (2019) 305–313. https://doi.org/10.1007/978-3-030-11769-6_25.
46. K.Y. Saw, A. Anz, C. Siew-Yoke Jee, S. Merican, R. Ching-Soong Ng, S.A. Roohi, K. Ragavanaidu, Articular cartilage regeneration with autologous peripheral blood stem cells versus hyaluronic acid: A randomized controlled trial, Arthrosc. – J. Arthrosc. Relat. Surg. 29 (2013) 684–694. https://doi.org/10.1016/j.arthro.2012.12.008.
47. W.L. Fu, Y.F. Ao, X.Y. Ke, Z.Z. Zheng, X. Gong, D. Jiang, J.K. Yu, Repair of large full-thickness cartilage defect by activating endogenous peripheral blood stem cells and autologous periosteum flap transplantation combined with patellofemoral realignment, Knee. 21 (2014) 609–612. https://doi.org/10.1016/j.knee.2013.10.010.
48. Z. Wang, L. Han, T. Sun, J. Ma, S. Sun, L. Ma, B. Wu, Extracellular matrix derived from allogenic decellularized bone marrow mesenchymal stem cell sheets for the reconstruction of osteochondral defects in rabbits, Acta Biomater. 118 (2020) 54–68. https://doi.org/10.1016/J.ACTBIO.2020.10.022.
49. A.J. Favreau-Lessard, D.B. Sawyer, Cardiac Regeneration and Stem Cells as Therapy for Heart Disease, Elsevier Inc., 2018. https://doi.org/10.1016/b978-0-12-809657-4.65740-x.
50. S. Bhat, P. Viswanathan, S. Chandanala, S.J. Prasanna, R.N. Seetharam, Expansion and characterization of bone marrow derived human mesenchymal stromal cells in serum-free conditions, Sci. Rep. 11 (2021) 1–18. https://doi.org/10.1038/s41598-021-83088-1.
51. S. Zhang, B. Hu, W. Liu, P. Wang, X. Lv, S. Chen, H. Liu, Z. Shao, Articular cartilage regeneration: The role of endogenous mesenchymal stem/progenitor cell recruitment and migration, Semin. Arthritis Rheum. 50 (2020) 198–208. https://doi.org/10.1016/j.semarthrit.2019.11.001.
52. J.M. Lee, B.S. Kim, H. Lee, G. Il Im, In vivo tracking of mesechymal stem cells using fluorescent nanoparticles in an osteochondral repair model, Mol. Ther. 20 (2012) 1434–1442. https://doi.org/10.1038/mt.2012.60.
53. M.S. Park, Y.H. Kim, Y. Jung, S.H. Kim, J.C. Park, D.S. Yoon, S.H. Kim, J.W. Lee, In Situ Recruitment of Human Bone Marrow-Derived Mesenchymal Stem Cells Using Chemokines for Articular Cartilage Regeneration, Cell Transplant. 24 (2015) 1067–1083. https://doi.org/10.3727/096368914X681018.
54. S.J. Baek, S.K. Kang, J.C. Ra, In vitro migration capacity of human adipose tissue-derived mesenchymal stem cells reflects their expression of receptors for chemokines and growth factors, Exp. Mol. Med. 43 (2011) 596–603. https://doi.org/10.3858/emm.2011.43.10.069.
55. J.D. Miller, S.M. Lankford, K.B. Adler, A.R. Brody, Mesenchymal stem cells require MARCKS protein for directed chemotaxis in vitro, Am. J. Respir. Cell Mol. Biol. 43 (2010) 253–258. https://doi.org/10.1165/rcmb.2010-0015RC.
56. J. Ringe, S. Strassburg, K. Neumann, M. Endres, M. Notter, G.R. Burmester, C. Kaps, M. Sittinger, Towards in situ tissue repair: Human mesenchymal stem cells express chemokine receptors CXCR1, CXCR2 and CCR2, and migrate upon stimulation with CXCL8 but not CCL2, J. Cell. Biochem. 101 (2007) 135–146. https://doi.org/10.1002/jcb.21172.

57. A.L. Ponte, E. Marais, N. Gallay, A. Langonné, B. Delorme, O. Hérault, P. Charbord, J. Domenech, The In Vitro Migration Capacity of Human Bone Marrow Mesenchymal Stem Cells: Comparison of Chemokine and Growth Factor Chemotactic Activities, Stem Cells. 25 (2007) 1737–1745. https://doi.org/10.1634/stemcells.2007-0054.
58. M. Endres, K. Andreas, G. Kalwitz, U. Freymann, K. Neumann, J. Ringe, M. Sittinger, T. Häupl, C. Kaps, Chemokine profile of synovial fluid from normal, osteoarthritis and rheumatoid arthritis patients: CCL25, CXCL10 and XCL1 recruit human subchondral mesenchymal progenitor cells, Osteoarthr. Cartil. 18 (2010) 1458–1466. https://doi.org/10.1016/j.joca.2010.08.003.
59. R.F. Martin, V. Murray, G. D'Cunha, M. Pardee, E. Kampouris, A. Haigh, D.P. Kelly, G.S. Hodgson, Rapid communication, Int. J. Radiat. Biol. 57 (1990) 939–946. https://doi.org/10.1080/09553009014551061.
60. H. Brisby, N. Papadimitriou, E. Runesson, N. Sasaki, A. Lindahl, H.B. Henriksson, Moderate physical exercise results in increased cell activity in articular cartilage of the knee joint in rats, Cells Tissues Organs. 198 (2013) 237–248. https://doi.org/10.1159/000355919.
61. M. Honczarenko, Y. Le, M. Swierkowski, I. Ghiran, A.M. Glodek, L.E. Silberstein, Human Bone Marrow Stromal Cells Express a Distinct Set of Biologically Functional Chemokine Receptors, Stem Cells. 24 (2006) 1030–1041. https://doi.org/10.1634/stemcells.2005-0319.
62. G. Kalwitz, K. Andreas, M. Endres, K. Neumann, M. Notter, J. Ringe, M. Sittinger, C. Kaps, Chemokine profile of human serum from whole blood: Migratory effects of CXCL-10 and CXCL-11 on human mesenchymal stem cells, Connect. Tissue Res. 51 (2010) 113–122. https://doi.org/10.3109/03008200903111906.
63. Y. Naaldijk, A.A. Johnson, S. Ishak, H.J. Meisel, C. Hohaus, A. Stolzing, Migrational changes of mesenchymal stem cells in response to cytokines, growth factors, hypoxia, and aging, Exp. Cell Res. 338 (2015) 97–104. https://doi.org/10.1016/j.yexcr.2015.08.019.
64. M. Ullah, J. Eucker, M. Sittinger, J. Ringe, Mesenchymal stem cells and their chondrogenic differentiated and dedifferentiated progeny express chemokine receptor CCR9 and chemotactically migrate toward CCL25 or serum, Stem Cell Res. Ther. 4 (2013) 1–16. https://doi.org/10.1186/scrt310.
65. G. Kalwitz, M. Endres, K. Neumann, K. Skriner, J. Ringe, O. Sezer, M. Sittinger, T. Häupl, C. Kaps, Gene expression profile of adult human bone marrow-derived mesenchymal stem cells stimulated by the chemokine CXCL7, Int. J. Biochem. Cell Biol. 41 (2009) 649–658. https://doi.org/10.1016/j.biocel.2008.07.011.
66. Y. Wu, M.J. Hoogduijn, C.C. Baan, S.S. Korevaar, R. De Kuiper, L. Yan, L. Wang, N.M.V. Besouw, Adipose Tissue-Derived Mesenchymal Stem Cells Have a Heterogenic Cytokine Secretion Profile, Stem Cells Int. 2017 (2017). https://doi.org/10.1155/2017/4960831.
67. Y. Mishima, M. Lotz, Chemotaxis of Human Articular Chondrocytes and Mesenchymal Stem Cells, J. Orthop. Res. 26 (2008) 1407–1412. https://doi.org/10.1002/jor.20668.
68. A. Schmidt, D. Ladage, T. Schinköthe, U. Klausmann, C. Ulrichs, F. Klinz, K. Brixius, S. Arnhold, B. Desai, U. Mehlhorn, R.H.G. Schwinger, P. Staib, K. Addicks, W. Bloch, Basic Fibroblast Growth Factor Controls Migration in Human Mesenchymal Stem Cells, Stem Cells. 24 (2006) 1750–1758. https://doi.org/10.1634/stemcells.2005-0191.
69. Y. Wang, J. Chen, W. Fan, J. Zhang, B. Hua, B. Sun, L. Zhu, X. Niu, Z. Yan, C. Guo, Stromal cell-derived factor-1α and transforming growth factor-β1 synergistically facilitate migration and chondrogenesis of synovium-derived stem cells through MAPK pathways, Am. J. Transl. Res. 9 (2017) 2656–2667.
70. A. Mendelson, E. Frank, C. Allred, E. Jones, M. Chen, W. Zhao, J.J. Mao, Chondrogenesis by chemotactic homing of synovium, bone marrow, and adipose stem cells in vitro, FASEB J. 25 (2011) 3496–3504. https://doi.org/10.1096/fj.10-176305.
71. R. Rollín, F. Marco, E. Camafeita, E. Calvo, L. López-Durán, J.Á. Jover, J.A. López, B. Fernández-Gutiérrez, Differential proteome of bone marrow mesenchymal stem cells from osteoarthritis patients, Osteoarthr. Cartil. 16 (2008) 929–935. https://doi.org/10.1016/j.joca.2007.12.006.

72. H.L. Holmes, B. Wilson, J.P. Goerger, J.L. Silverberg, I. Cohen, W.R. Zipfel, L.A. Fortier, Facilitated recruitment of mesenchymal stromal cells by bone marrow concentrate and platelet rich plasma, PLoS One. 13 (2018) 1–12. https://doi.org/10.1371/journal.pone.0194567.
73. H. Joos, A. Wildner, C. Hogrefe, H. Reichel, R.E. Brenner, Interleukin-1 beta and tumor necrosis factor alpha inhibit migration activity of chondrogenic progenitor cells from non-fibrillated osteoarthritic cartilage, Arthritis Res. Ther. 15 (2013) 1–13. https://doi.org/10.1186/ar4299.
74. S. Koelling, J. Kruegel, M. Irmer, J.R. Path, B. Sadowski, X. Miro, N. Miosge, Migratory Chondrogenic Progenitor Cells from Repair Tissue during the Later Stages of Human Osteoarthritis, Cell Stem Cell. 4 (2009) 324–335. https://doi.org/10.1016/j.stem.2009.01.015.
75. Y. Jiang, C. Hu, S. Yu, J. Yan, H. Peng, H.W. Ouyang, R.S. Tuan, Cartilage stem/progenitor cells are activated in osteoarthritis via interleukin-1β/nerve growth factor signaling, Arthritis Res. Ther. 17 (2015) 1–13. https://doi.org/10.1186/s13075-015-0840-x.
76. J.S. Lee, S.K. Kim, B.J. Jung, S.B. Choi, E.Y. Choi, C.S. Kim, Enhancing proliferation and optimizing the culture condition for human bone marrow stromal cells using hypoxia and fibroblast growth factor-2, Stem Cell Res. 28 (2018) 87–95. https://doi.org/10.1016/j.scr.2018.01.010.
77. H. Chuma, H. Mizuta, S. Kudo, K. Takagi, Y. Hiraki, One day exposure to FGF-2 was sufficient for the regenerative repair of full-thickness defects of articular cartilage in rabbits, Osteoarthr. Cartil. 12 (2004) 834–842. https://doi.org/10.1016/j.joca.2004.07.003.
78. B. Awan, D. Turkov, C. Schumacher, A. Jacobo, A. McEnerney, A. Ramsey, G. Xu, D. Park, S. Kalomoiris, W. Yao, L.E. Jao, M.L. Allende, C.B. Lebrilla, F.A. Fierro, FGF2 Induces Migration of Human Bone Marrow Stromal Cells by Increasing Core Fucosylations on N-Glycans of Integrins, Stem Cell Reports. 11 (2018) 325–333. https://doi.org/10.1016/j.stemcr.2018.06.007.
79. H.L. Reesink, R.M. Sutton, C.R. Shurer, R.P. Peterson, J.S. Tan, J. Su, M.J. Paszek, A.J. Nixon, Galectin-1 and galectin-3 expression in equine mesenchymal stromal cells (MSCs), synovial fibroblasts and chondrocytes, and the effect of inflammation on MSC motility, Stem Cell Res. Ther. 8 (2017) 1–12. https://doi.org/10.1186/s13287-017-0691-2.
80. A. Barhanpurkar-Naik, S.T. Mhaske, S.T. Pote, K. Singh, M.R. Wani, Interleukin-3 enhances the migration of human mesenchymal stem cells by regulating expression of CXCR4, Stem Cell Res. Ther. 8 (2017) 1–15. https://doi.org/10.1186/s13287-017-0618-y.
81. C. Jiang, P. Ma, B. Ma, Z. Wu, G. Qiu, X. Su, Z. Xia, Z. Ye, Y. Wang, Plasma-derived fibronectin stimulates chondrogenic differentiation of human subchondral cortico-spongious progenitor cells in late-stage osteoarthritis, Int. J. Mol. Sci. 16 (2015) 19477–19489. https://doi.org/10.3390/ijms160819477.
82. J.P. Krüger, S. Hondke, M. Endres, A. Pruss, A. Siclari, C. Kaps, Human platelet-rich plasma stimulates migration and chondrogenic differentiation of human subchondral progenitor cells, J. Orthop. Res. 30 (2012) 845–852. https://doi.org/10.1002/jor.22005.
83. P.C. Kreuz, J.P. Krüger, S. Metzlaff, U. Freymann, M. Endres, A. Pruss, W. Petersen, C. Kaps, Platelet-Rich Plasma Preparation Types Show Impact on Chondrogenic Differentiation, Migration, and Proliferation of Human Subchondral Mesenchymal Progenitor Cells, Arthrosc. – J. Arthrosc. Relat. Surg. 31 (2015) 1951–1961. https://doi.org/10.1016/j.arthro.2015.03.033.
84. J. Amrichová, T. Špaková, J. Rosocha, D. Harvanová, D. Bačenková, M. Lacko, S. Horňák, Effect of PRP and PPP on proliferation and migration of human chondrocytes and synoviocytes in vitro, Cent. Eur. J. Biol. 9 (2014) 139–148. https://doi.org/10.2478/s11535-013-0255-0.
85. A. Damerau, A. Lang, M. Pfeiffenberger, F. Buttgereit, T. Gaber, FRI0002 Development of an in vitro multi-component 3d joint model to simulate the pathogenesis of arthritis, (2017) 480.2–480. https://doi.org/10.1136/annrheumdis-2017-eular.5889.
86. R. Kulawig, J.P. Krüger, O. Klein, Z. Konthur, H. Schütte, J. Klose, C. Kaps, M. Endres, Identification of fibronectin as a major factor in human serum to recruit subchondral mesenchymal progenitor cells, Int. J. Biochem. Cell Biol. 45 (2013) 1410–1418. https://doi.org/10.1016/j.biocel.2013.04.016.

87. R.H. Kalkreuth, J.P. Krüger, S. Lau, P. Niemeyer, M. Endres, P.C. Kreuz, C. Kaps, Fibronectin stimulates migration and proliferation, but not chondrogenic differentiation of human subchondral progenitor cells, Regen. Med. 9 (2014) 759–773. https://doi.org/10.2217/rme.14.40.
88. T. Tao, Y. Li, C. Gui, Y. Ma, Y. Ge, H. Dai, K. Zhang, J. Du, Y. Guo, Y. Jiang, J. Gui, Fibronectin enhances cartilage repair by activating progenitor cells through integrin $\alpha5\beta1$ Receptor, 2018. https://doi.org/10.1089/ten.tea.2017.0322.
89. R. He, B. Wang, M. Cui, Z. Xiong, H. Lin, L. Zhao, Z. Li, Z. Wang, S. Peggrem, Z. Xia, Z. Shao, Link Protein N-Terminal Peptide as a Potential Stimulating Factor for Stem Cell-Based Cartilage Regeneration, Stem Cells Int. 2018 (2018). https://doi.org/10.1155/2018/3217895.
90. K. Zha, X. Li, Z. Yang, G. Tian, Z. Sun, X. Sui, Y. Dai, S. Liu, Q. Guo, Heterogeneity of mesenchymal stem cells in cartilage regeneration: from characterization to application, Npj Regen. Med. 6 (2021). https://doi.org/10.1038/s41536-021-00122-6.
91. T. Nagamura-Inoue, Umbilical cord-derived mesenchymal stem cells: Their advantages and potential clinical utility, World J. Stem Cells. 6 (2014) 195. https://doi.org/10.4252/wjsc.v6.i2.195.
92. Y. Shang, H. Guan, F. Zhou, Biological Characteristics of Umbilical Cord Mesenchymal Stem Cells and Its Therapeutic Potential for Hematological Disorders, Front. Cell Dev. Biol. 9 (2021) 1–11. https://doi.org/10.3389/fcell.2021.570179.
93. I.H. Dilogo, A.F. Canintika, A.L. Hanitya, J.A. Pawitan, I.K. Liem, J. Pandelaki, Umbilical cord-derived mesenchymal stem cells for treating osteoarthritis of the knee: a single-arm, open-label study, Eur. J. Orthop. Surg. Traumatol. 30 (2020) 799–807. https://doi.org/10.1007/s00590-020-02630-5.
94. S. Ceccarelli, P. Pontecorvi, E. Anastasiadou, C. Napoli, C. Marchese, Immunomodulatory Effect of Adipose-Derived Stem Cells: The Cutting Edge of Clinical Application, Front. Cell Dev. Biol. 8 (2020) 1–12. https://doi.org/10.3389/fcell.2020.00236.
95. Y. Song, H. Du, C. Dai, L. Zhang, S. Li, D.J. Hunter, L. Lu, C. Bao, Human adipose-derived mesenchymal stem cells for osteoarthritis: A pilot study with long-term follow-up and repeated injections, Regen. Med. 13 (2018) 295–307. https://doi.org/10.2217/rme-2017-0152.
96. F. Slimi, W. Zribi, M. Trigui, R. Amri, N. Gouiaa, C. Abid, M.A. Rebai, T. Boudawara, S. Jebahi, H. Keskes, The effectiveness of platelet-rich plasma gel on full-thickness cartilage defect repair in a rabbit model, https://doi.org/10.1302/2046-3758.103.BJR-2020-0087.R2. 10 (2021) 192–202. https://doi.org/10.1302/2046-3758.103.BJR-2020-0087.R2.
97. S.G. Boswell, B.J. Cole, E.A. Sundman, V. Karas, L.A. Fortier, Platelet-Rich Plasma: A Milieu of Bioactive Factors, Arthrosc. J. Arthrosc. Relat. Surg. 28 (2012) 429–439. https://doi.org/10.1016/J.ARTHRO.2011.10.018.
98. R. Alves, R. Grimalt, A Review of Platelet-Rich Plasma: History, Biology, Mechanism of Action, and Classification, Ski. Appendage Disord. 4 (2018) 18–24. https://doi.org/10.1159/000477353.
99. G.Z. Jin, Current nanoparticle-based technologies for osteoarthritis therapy, Nanomaterials. 10 (2020) 1–20. https://doi.org/10.3390/nano10122368.
100. J. Jeevanandam, A. Barhoum, Y.S. Chan, A. Dufresne, M.K. Danquah, Review on nanoparticles and nanostructured materials: History, sources, toxicity and regulations, Beilstein J. Nanotechnol. 9 (2018) 1050–1074. https://doi.org/10.3762/bjnano.9.98.
101. A. Biswas, I.S. Bayer, A.S. Biris, T. Wang, E. Dervishi, F. Faupel, Advances in top-down and bottom-up surface nanofabrication: Techniques, applications & future prospects, Adv. Colloid Interface Sci. 170 (2012) 2–27. https://doi.org/10.1016/j.cis.2011.11.001.
102. X. Li, B. Dai, J. Guo, L. Zheng, Q. Guo, J. Peng, J. Xu, L. Qin, Nanoparticle–Cartilage Interaction: Pathology-Based Intra-articular Drug Delivery for Osteoarthritis Therapy, 2021. https://doi.org/10.1007/s40820-021-00670-y.
103. Z. Shi, Y. Zhou, T. Fan, Y. Lin, H. Zhang, L. Mei, Inorganic nano-carriers based smart drug delivery systems for tumor therapy, Smart Mater. Med. 1 (2020) 32–47. https://doi.org/10.1016/j.smaim.2020.05.002.

104. L. Valot, M. Maumus, L. Brunel, J. Martinez, M. Amblard, D. Noël, A. Mehdi, G. Subra, A collagen-mimetic organic-inorganic hydrogel for cartilage engineering, Gels. 7 (2021) 1–15. https://doi.org/10.3390/gels7020073.
105. M. Zheng, H. Jia, H. Wang, L. Liu, Z. He, Z. Zhang, W. Yang, L. Gao, X. Gao, F. Gao, Application of nanomaterials in the treatment of rheumatoid arthritis, RSC Adv. 11 (2021) 7129–7137. https://doi.org/10.1039/d1ra00328c.
106. J. Yang, C.R. Han, J.F. Duan, F. Xu, R.C. Sun, Insitu grafting silica nanoparticles reinforced nanocomposite hydrogels, Nanoscale. 5 (2013) 10858–10863. https://doi.org/10.1039/c3nr04252a.
107. M.A. Bonifacio, A. Cochis, S. Cometa, A. Scalzone, P. Gentile, G. Procino, S. Milano, A.C. Scalia, L. Rimondini, E. De Giglio, Advances in cartilage repair: The influence of inorganic clays to improve mechanical and healing properties of antibacterial Gellan gum-Manuka honey hydrogels, Mater. Sci. Eng. C. 108 (2020) 110444. https://doi.org/10.1016/j.msec.2019.110444.
108. Y. Ao, E. Zhang, Y. Liu, L. Yang, J. Li, F. Wang, Advanced Hydrogels With Nanoparticle Inclusion for Cartilage Tissue Engineering, Front. Bioeng. Biotechnol. 10 (2022) 1–13. https://doi.org/10.3389/fbioe.2022.951513.
109. N. Amiryaghoubi, M. Fathi, A. Barzegari, J. Barar, H. Omidian, Y. Omidi, Recent advances in polymeric scaffolds containing carbon nanotube and graphene oxide for cartilage and bone regeneration, Mater. Today Commun. 26 (2021) 102097. https://doi.org/10.1016/j.mtcomm.2021.102097.
110. P. Liu, G. Chen, J. Zhang, A Review of Liposomes as a Drug Delivery System: Current Status of Approved Products, Regulatory Environments, and Future Perspectives, Molecules. 27 (2022). https://doi.org/10.3390/molecules27041372.
111. G. Wen, H.F. Deng, W.Z. Bing, S.Z. Luo, Preliminarily experimental study on biological properties of 153Sm-citrate-nano-hydroxyapatite, J. Sichuan Univ. (Medical Sci. Ed.) 38 (2007) 1017–1020.
112. G. Gregoriadis, A.T. Florence, Liposomes in Drug Delivery, Drugs. 45 (1993) 15–28. https://doi.org/10.2165/00003495-199345010-00003.
113. W. Deng, W. Chen, S. Clement, A. Guller, Z. Zhao, A. Engel, E.M. Goldys, Controlled gene and drug release from a liposomal delivery platform triggered by X-ray radiation, Nat. Commun. 9 (2018) 1–11. https://doi.org/10.1038/s41467-018-05118-3.
114. A. Akbarzadeh, R. Rezaei-Sadabady, S. Davaran, S.W. Joo, N. Zarghami, Y. Hanifehpour, M. Samiei, M. Kouhi, K. Nejati-Koshki, Liposome: Classification, preparation, and applications, Nanoscale Res. Lett. 8 (2013) 1. https://doi.org/10.1186/1556-276X-8-102.
115. O. Craciunescu, M. Icriverzi, P.E. Florian, A. Roseanu, M. Trif, Mechanisms and pharmaceutical action of lipid nanoformulation of natural bioactive compounds as efficient delivery systems in the therapy of osteoarthritis, Pharmaceutics. 13 (2021) 1–24. https://doi.org/10.3390/pharmaceutics13081108.
116. W. Lin, R. Goldberg, J. Klein, Poly-phosphocholination of liposomes leads to highly-extended retention time in mice joints, J. Mater. Chem. B. 10 (2022) 2820–2827. https://doi.org/10.1039/d1tb02346b.
117. J. Yang, Y. Zhu, F. Wang, L. Deng, X. Xu, W. Cui, Microfluidic liposomes-anchored microgels as extended delivery platform for treatment of osteoarthritis, Chem. Eng. J. 400 (2020) 126004. https://doi.org/10.1016/j.cej.2020.126004.
118. R. Gauro, M. Nandave, V.K. Jain, K. Jain, Advances in dendrimer-mediated targeted drug delivery to the brain, J. Nanoparticle Res. 23 (2021). https://doi.org/10.1007/s11051-021-05175-8.
119. K. Jain, P. Kesharwani, U. Gupta, N.K. Jain, Dendrimer toxicity: Let's meet the challenge, Int. J. Pharm. 394 (2010) 122–142. https://doi.org/10.1016/j.ijpharm.2010.04.027.
120. M. Mariyam, K. Ghosal, S. Thomas, N. Kalarikkal, M.S. Latha, Dendrimers: General Aspects, Applications and Structural Exploitations as Prodrug/Drug-delivery Vehicles in Current Medicine, Mini-Reviews Med. Chem. 18 (2017) 439–457. https://doi.org/10.2174/1389557517666170512095151.

121. S. Sadekar, H. Ghandehari, Transepithelial transport and toxicity of PAMAM dendrimers: Implications for oral drug delivery, Adv. Drug Deliv. Rev. 64 (2012) 571–588. https://doi.org/10.1016/j.addr.2011.09.010.
122. L.P. Mendes, J. Pan, V.P. Torchilin, Dendrimers as nanocarriers for nucleic acid and drug delivery in cancer therapy, Molecules. 22 (2017) 1–21. https://doi.org/10.3390/molecules22091401.
123. C.A. Razzino, V. Serafín, M. Gamella, M. Pedrero, A. Montero-Calle, R. Barderas, M. Calero, A.O. Lobo, P. Yáñez-Sedeño, S. Campuzano, J.M. Pingarrón, An electrochemical immunosensor using gold nanoparticles-PAMAM-nanostructured screen-printed carbon electrodes for tau protein determination in plasma and brain tissues from Alzheimer patients, Biosens. Bioelectron. 163 (2020) 112238. https://doi.org/10.1016/j.bios.2020.112238.
124. A. Janaszewska, J. Lazniewska, P. Trzepiński, M. Marcinkowska, B. Klajnert-Maculewicz, Cytotoxicity of dendrimers, Biomolecules. 9 (2019) 1–23. https://doi.org/10.3390/biom9080330.
125. B.C. Geiger, S. Wang, R.F. Padera, A.J. Grodzinsky, P.T. Hammond, Cartilage-penetrating nanocarriers improve delivery and efficacy of growth factor treatment of osteoarthritis, Sci. Transl. Med. 10 (2018) 1–13. https://doi.org/10.1126/scitranslmed.aat8800.
126. F. Liu, X. Wang, Y. Li, M. Ren, P. He, L. Wang, J. Xu, S. Yang, P. Ji, Dendrimer-modified gelatin methacrylate hydrogels carrying adipose-derived stromal/stem cells promote cartilage regeneration, Stem Cell Res. Ther. 13 (2022) 1–15. https://doi.org/10.1186/s13287-022-02705-6.
127. L. Degoricija, P.N. Bansal, S.H.M. Söntjens, N.S. Joshi, M. Takahashi, B. Snyder, M.W. Grinstaff, Hydrogels for osteochondral repair based on photocrosslinkable carbamate dendrimers, Biomacromolecules. 9 (2008) 2863–2872. https://doi.org/10.1021/bm800658x.
128. H.M. Aliabadi, A. Lavasanifar, Polymeric micelles for drug delivery, Expert Opin. Drug Deliv. 3 (2006) 139–162. https://doi.org/10.1517/17425247.3.1.139.
129. L.C. Nelemans, L. Gurevich, Drug delivery with polymeric nanocarriers-cellular uptake mechanisms, Materials (Basel). 13 (2020) 1–21. https://doi.org/10.3390/ma13020366.
130. D. Maysinger, J. Lovrić, A. Eisenberg, R. Savić, Fate of micelles and quantum dots in cells, Eur. J. Pharm. Biopharm. 65 (2007) 270–281. https://doi.org/10.1016/j.ejpb.2006.08.011.
131. Y. Wei, L. Yan, L. Luo, T. Gui, B. Jang, A. Amirshaghaghi, T. You, A. Tsourkas, L. Qin, Z. Cheng, Phospholipase A2 inhibitor-loaded micellar nanoparticles attenuate inflammation and mitigate osteoarthritis progression, Sci. Adv. 7 (2021). https://doi.org/10.1126/SCIADV.ABE6374.
132. L. Ma, X. Zheng, R. Lin, A.R. Sun, J. Song, Z. Ye, D. Liang, M. Zhang, J. Tian, X. Zhou, L. Cui, Y. Liu, Y. Liu, Knee Osteoarthritis Therapy: Recent Advances in Intra-Articular Drug Delivery Systems, (2022).
133. Q. Wang, H. Jiang, Y. Li, W. Chen, H. Li, K. Peng, Z. Zhang, X. Sun, Targeting NF-kB signaling with polymeric hybrid micelles that co-deliver siRNA and dexamethasone for arthritis therapy, Biomaterials. 122 (2017) 10–22. https://doi.org/10.1016/j.biomaterials.2017.01.008.
134. X. Wu, P. Li, J. Cheng, Q. Xu, B. Lu, C. Han, W. Huo, ROS-Sensitive Nanoparticles Co-delivering Dexamethasone and CDMP-1 for the Treatment of Osteoarthritis Through Chondrogenic Differentiation Induction and Inflammation Inhibition, Front. Bioeng. Biotechnol. 9 (2021) 1–11. https://doi.org/10.3389/fbioe.2021.608150.
135. L. An, Z. Li, L. Shi, L. Wang, Y. Wang, L. Jin, X. Shuai, J. Li, Inflammation-Targeted Celastrol Nanodrug Attenuates Collagen-Induced Arthritis through NF-κB and Notch1 Pathways, Nano Lett. 20 (2020) 7728–7736. https://doi.org/10.1021/acs.nanolett.0c03279.
136. J. Li, Y. Long, R. Guo, K. Ren, Z. Lu, M. Li, X. Wang, Y. Wang, Z. Zhang, Q. He, Shield and sword nano-soldiers ameliorate rheumatoid arthritis by multi-stage manipulation of neutrophils, J. Control. Release. 335 (2021) 38–48. https://doi.org/10.1016/j.jconrel.2021.05.008.
137. N. Zhang, C. Xu, N. Li, S. Zhang, L. Fu, X. Chu, H. Hua, X. Zeng, Y. Zhao, Folate receptor-targeted mixed polysialic acid micelles for combating rheumatoid arthritis: In vitro and in vivo evaluation, Drug Deliv. 25 (2018) 1182–1191. https://doi.org/10.1080/10717544.2018.1472677.

138. G. Seetharaman, A.R. Kallar, V.M. Vijayan, J. Muthu, S. Selvam, Design, preparation and characterization of pH-responsive prodrug micelles with hydrolyzable anhydride linkages for controlled drug delivery, J. Colloid Interface Sci. 492 (2017) 61–72. https://doi.org/10.1016/j.jcis.2016.12.070.
139. A. Chenxian Zhu, A. Zhongxing Zhang, Y. Wen, A. Xia Song, J. Zhu, Y. Yao, and J. Li, Cationic micelles as nanocarriers for enhancing intra-cartilage drug penetration and retention, J. Mater. Chem. B. 10 (2021) 4118. https://doi.org/10.1039/D2TB02050E.
140. M. Jaiswal, R. Dudhe, P.K. Sharma, Nanoemulsion: an advanced mode of drug delivery system, 3 Biotech. 5 (2015) 123–127. https://doi.org/10.1007/s13205-014-0214-0.
141. F.U. Rehman, K.U. Shah, S.U. Shah, I.U. Khan, G.M. Khan, A. Khan, From nanoemulsions to self-nanoemulsions, with recent advances in self-nanoemulsifying drug delivery systems (SNEDDS), Expert Opin. Drug Deliv. 14 (2017) 1325–1340. https://doi.org/10.1080/17425247.2016.1218462.
142. J. Tudela, M. Martínez, R. Valdivia, J. Romo, M. Portillo, R. Rangel, Enhanced Reader.pdf, Nature. 388 (2010) 539–547.
143. E.B. Souto, A. Cano, C. Martins-Gomes, T.E. Coutinho, A. Zielińska, A.M. Silva, Microemulsions and Nanoemulsions in Skin Drug Delivery, Bioengineering. 9 (2022) 1–22. https://doi.org/10.3390/bioengineering9040158.
144. J. Yoo, C. Park, G. Yi, D. Lee, H. Koo, Active targeting strategies using biological ligands for nanoparticle drug delivery systems, Cancers (Basel). 11 (2019). https://doi.org/10.3390/cancers11050640.
145. C. Li, X. Chen, X. Luo, H. Wang, Y. Zhu, G. Du, W. Chen, Z. Chen, X. Hao, Z. Zhang, X. Sun, Nanoemulsions Target to Ectopic Lymphoids in Inflamed Joints to Restore Immune Tolerance in Rheumatoid Arthritis, Nano Lett. 21 (2021) 2551–2561. https://doi.org/10.1021/acs.nanolett.0c05110.
146. J.P. Gokhale, H.S. Mahajan, S.S. Surana, Quercetin loaded nanoemulsion-based gel for rheumatoid arthritis: In vivo and in vitro studies, Biomed. Pharmacother. 112 (2019) 108622. https://doi.org/10.1016/j.biopha.2019.108622.
147. R. Hamed, M. Basil, T. AlBaraghthi, S. Sunoqrot, O. Tarawneh, Nanoemulsion-based gel formulation of diclofenac diethylamine: design, optimization, rheological behavior and in vitro diffusion studies, Pharm. Dev. Technol. 21 (2016) 980–989. https://doi.org/10.3109/10837450.2015.1086372.
148. J. Kousalová, T. Etrych, Polymeric nanogels as drug delivery systems, Physiol. Res. 67 (2018) s305–s317. https://doi.org/10.33549/physiolres.933979.
149. G. Soni, K.S. Yadav, Nanogels as potential nanomedicine carrier for treatment of cancer: A mini review of the state of the art, Saudi Pharm. J. 24 (2016) 133–139. https://doi.org/10.1016/j.jsps.2014.04.001.
150. Z. An, Q. Qiu, G. Liu, Synthesis of architecturally well-defined nanogels via RAFT polymerization for potential bioapplications, Chem. Commun. 47 (2011) 12424–12440. https://doi.org/10.1039/c1cc13955j.
151. M. Suhail, J.M. Rosenholm, M.U. Minhas, S.F. Badshah, A. Naeem, K.U. Khan, M. Fahad, Nanogels as drug-delivery systems: A comprehensive overview, Ther. Deliv. 10 (2019) 697–717. https://doi.org/10.4155/tde-2019-0010.
152. F. Sultana, Manirujjaman, Imran-Ul-Haque, M. Arafat, S. Sharmin, An overview of nanogel drug delivery system, J. Appl. Pharm. Sci. 3 (2013) 95–105. https://doi.org/10.7324/JAPS.2013.38.S15.
153. K. Mizuno, Y. Ikeuchi-Takahashi, Y. Hattori, H. Onishi, Preparation and evaluation of conjugate nanogels of glycyl-prednisolone with natural anionic polysaccharides as anti-arthritic delivery systems, Drug Deliv. 28 (2021) 144–152. https://doi.org/10.1080/10717544.2020.1865478.
154. T. Adachi, N. Miyamoto, H. Imamura, T. Yamamoto, E. Marin, W. Zhu, M. Kobara, Y. Sowa, Y. Tahara, N. Kanamura, K. Akiyoshi, O. Mazda, I. Nishimura, G. Pezzotti, Three-Dimensional Culture of Cartilage Tissue on Nanogel-Cross-Linked Porous Freeze-Dried Gel Scaffold for Regenerative Cartilage Therapy: A Vibrational Spectroscopy Evaluation, Int. J. Mol. Sci. 23 (2022). https://doi.org/10.3390/ijms23158099.

155. N. Feng, M. Yang, X. Feng, Y. Wang, F. Chang, J. Ding, Reduction-Responsive Polypeptide Nanogel for Intracellular Drug Delivery in Relieving Collagen-Induced Arthritis, ACS Biomater. Sci. Eng. 4 (2018) 4154–4162. https://doi.org/10.1021/acsbiomaterials.8b00738.
156. S. Bale, A. Khurana, A.S.S. Reddy, M. Singh, C. Godugu, Overview on therapeutic applications of microparticulate drug delivery systems, Crit. Rev. Ther. Drug Carrier Syst. 33 (2016) 309–361. https://doi.org/10.1615/CritRevTherDrugCarrierSyst.2016015798.
157. N.A. Ballesteros, M. Alonso, S.R. Saint-Jean, S.I. Perez-Prieto, An oral DNA vaccine against infectious haematopoietic necrosis virus (IHNV) encapsulated in alginate microspheres induces dose-dependent immune responses and significant protection in rainbow trout (Oncorrhynchus mykiss), Fish Shellfish Immunol. 45 (2015) 877–888. https://doi.org/10.1016/j.fsi.2015.05.045.
158. Y.M. Yoon, J.S. Lewis, M.R. Carstens, M. Campbell-Thompson, C.H. Wasserfall, M.A. Atkinson, B.G. Keselowsky, A combination hydrogel microparticle-based vaccine prevents type 1 diabetes in non-obese diabetic mice, Sci. Rep. 5 (2015) 1–13. https://doi.org/10.1038/srep13155.
159. A. Sheikh Hasan, A. Sapin, C. Damgé, P. Leroy, M. Socha, P. Maincent, Reduction of the in vivo burst release of insulin-loaded microparticles, J. Drug Deliv. Sci. Technol. 30 (2015) 486–493. https://doi.org/10.1016/j.jddst.2015.06.020.
160. J. Castro-Rosas, C.R. Ferreira-Grosso, C.A. Gómez-Aldapa, E. Rangel-Vargas, M.L. Rodríguez-Marín, F.A. Guzmán-Ortiz, R.N. Falfan-Cortes, Recent advances in microencapsulation of natural sources of antimicrobial compounds used in food – A review, Food Res. Int. 102 (2017) 575–587. https://doi.org/10.1016/j.foodres.2017.09.054.
161. N. Choudhury, M. Meghwal, K. Das, Microencapsulation: An overview on concepts, methods, properties and applications in foods, Food Front. 2 (2021) 426–442. https://doi.org/10.1002/fft2.94.
162. H. Yu, C. Huang, X. Kong, J. Ma, P. Ren, J. Chen, X. Zhang, H. Luo, G. Chen, Nanoarchitectonics of Cartilage-Targeting Hydrogel Microspheres with Reactive Oxygen Species Responsiveness for the Repair of Osteoarthritis, ACS Appl. Mater. Interfaces. 14 (2022) 40711–40723. https://doi.org/10.1021/acsami.2c12703.
163. R.J. Kulchar, B.R. Denzer, B.M. Chavre, M. Takegami, J. Patterson, A review of the use of microparticles for cartilage tissue engineering, Int. J. Mol. Sci. 22 (2021). https://doi.org/10.3390/ijms221910292.
164. S. Cosenza, M. Ruiz, K. Toupet, C. Jorgensen, D. Noël, Mesenchymal stem cells derived exosomes and microparticles protect cartilage and bone from degradation in osteoarthritis, Sci. Rep. 7 (2017) 1–12. https://doi.org/10.1038/s41598-017-15376-8.
165. J. Zhang, X. Zhang, Y. Hong, Q. Fu, Q. He, A. Mechakra, Q. Zhu, F. Zhou, R. Liang, C. Li, Y. Hu, Y. Zou, S. Zhang, H. Ouyang, Tissue-Adhesive Paint of Silk Microparticles for Articular Surface Cartilage Regeneration, ACS Appl. Mater. Interfaces. 12 (2020) 22467–22478. https://doi.org/10.1021/acsami.0c01776.
166. M. Li, M.J. Mondrinos, X. Chen, M.R. Gandhi, F.K. Ko, P.I. Lelkes, Elastin Blends for Tissue Engineering Scaffolds, J. Biomed. Mater. Res. Part A. 79 (2006) 963–73. https://doi.org/10.1002/jbm.a.
167. K. Peh, T. Khan, H. Ch'ng, Mechanical, bioadhesive strength and biological evaluations of chitosan films for wound dressing., J. Pharm. Pharm. Sci. 3 (2000) 303–311.
168. J. Di, S. Yao, Y. Ye, Z. Cui, J. Yu, T.K. Ghosh, Y. Zhu, Z. Gu, Stretch-Triggered Drug Delivery from Wearable Elastomer Films Containing Therapeutic Depots, ACS Nano. 9 (2015) 9407–9415. https://doi.org/10.1021/acsnano.5b03975.
169. J. Li, W.R.K. Illeperuma, Z. Suo, J.J. Vlassak, Hybrid hydrogels with extremely high stiffness and toughness, ACS Macro Lett. 3 (2014) 520–523. https://doi.org/10.1021/mz5002355.
170. J. Li, D.J. Mooney, Designing hydrogels for controlled drug delivery, Nat. Rev. Mater. 1 (2016) 1–18. https://doi.org/10.1038/natrevmats.2016.71.
171. W. Liu, M. Griffith, F. Li, Alginate microsphere-collagen composite hydrogel for ocular drug delivery and implantation, J. Mater. Sci. Mater. Med. 19 (2008) 3365–3371. https://doi.org/10.1007/s10856-008-3486-2.

172. L. Yu, J. Ding, Injectable hydrogels as unique biomedical materials, Chem. Soc. Rev. 37 (2008) 1473–1481. https://doi.org/10.1039/b713009k.
173. T. Nii, 38. Honary, S.; Zahir, F. Effect of zeta potential on the properties of nano-drug delivery systems, Molecules. 26 (2021) 1–45.
174. K.D. Ngadimin, A. Stokes, P. Gentile, A.M. Ferreira, Biomimetic hydrogels designed for cartilage tissue engineering, Biomater. Sci. 9 (2021) 4246–4259. https://doi.org/10.1039/d0bm01852j.
175. M.R. Prausnitz, R. Langer, Nihms121685, Nat Biotechnol. 26 (2009) 1261–1268. https://doi.org/10.1038/nbt.1504.Transdermal.
176. M.T.C. Mccrudden, B.M. Torrisi, M.J. Garland, T. Raghu, R. Singh, R.F. Donnelly, Microneedles for intradermal and transdermal delivery, Eur. J. Pharm. Sci. 50 (2013) 623–637. https://doi.org/10.1016/j.ejps.2013.05.005.Microneedles.
177. A. Tucak, M. Sirbubalo, L. Hindija, O. Rahić, J. Hadžiabdić, K. Muhamedagić, A. Čekić, E. Vranić, Microneedles: Characteristics, materials, production methods and commercial development, Micromachines. 11 (2020) 1–30. https://doi.org/10.3390/mi11110961.
178. P.S. Kumbhar, T.P. Jadhav, S.S. Chopade, T.T. Gavade, R.C. Sorate, T.U. Shinde, P.P. Maske, J.I. Disouza, A.S. Manjappa, Microneedles: An advanced approach for transdermal delivery of biologics, Asian J. Pharm. Res. 11 (2021) 46–54. https://doi.org/10.5958/2231-5691.2021.00010.1.
179. P. Zhou, C. Chen, X. Yue, J. Zhang, C. Huang, S. Zhao, A. Wu, X. Li, Y. Qu, C. Zhang, Strategy for osteoarthritis therapy: Improved the delivery of triptolide using liposome-loaded dissolving microneedle arrays, Int. J. Pharm. 609 (2021) 121211. https://doi.org/10.1016/j.ijpharm.2021.121211.
180. M. PISUTTANAWAT, S.S. Khuangsirikul, D. Heebthamai, O. Phruetthiphat, T. Chotanaphuti;, Effect of transdermal microneedle patch with NSAID in Osteoarthritis knee, Osteoarthr. Cartil. 28 (2020) S499. https://doi.org/10.1016/j.joca.2020.02.783.
181. D. Javelaud, A. Mauviel, Mammalian transforming growth factor-βs: Smad signaling and physio-pathological roles, Int. J. Biochem. Cell Biol. 36 (2004) 1161–1165. https://doi.org/10.1016/S1357-2725(03)00255-3.
182. G.S. Anusuya, M. Kandasamy, S.A.J. Raja, S. Sabarinathan, P. Ravishankar, B. Kandhasamy, Bone morphogenetic proteins: Signaling periodontal bone regeneration and repair, J. Pharm. Bioallied Sci. 8 (2016) S39–S41. https://doi.org/10.4103/0975-7406.191964.
183. C. Wen, L. Xu, X. Xu, D. Wang, Y. Liang, L. Duan, Insulin-like growth factor-1 in articular cartilage repair for osteoarthritis treatment, Arthritis Res. Ther. 23 (2021). https://doi.org/10.1186/S13075-021-02662-0.
184. M.J. Cross, J. Dixelius, T. Matsumoto, L. Claesson-Welsh, VEGF-receptor signal transduction, Trends Biochem. Sci. 28 (2003) 488–494. https://doi.org/10.1016/S0968-0004(03)00193-2.
185. O.K. Hwang, Y.W. Noh, J.T. Hong, J.W. Lee, Hypoxia Pretreatment Promotes Chondrocyte Differentiation of Human Adipose-Derived Stem Cells via Vascular Endothelial Growth Factor, Tissue Eng. Regen. Med. 2020 173. 17 (2020) 335–350. https://doi.org/10.1007/S13770-020-00265-5.
186. Z. Jia, S. Wang, Y. Liang, Q. Liu, Combination of kartogenin and transforming growth factor-β3 supports synovial fluid-derived mesenchymal stem cell-based cartilage regeneration, Am. J. Transl. Res. 11 (2019) 2056–2069.
187. Y. Zhao, B. Teng, X. Sun, Y. Dong, S. Wang, Y. Hu, Z. Wang, X. Ma, Q. Yang, Synergistic Effects of Kartogenin and Transforming Growth Factor-β3 on Chondrogenesis of Human Umbilical Cord Mesenchymal Stem Cells In Vitro, Orthop. Surg. 12 (2020) 938–945. https://doi.org/10.1111/OS.12691.
188. J. Lin, L. Wang, J. Lin, Q. Liu, W. Lin, L. Zhu, D. Raucher, Dual Delivery of TGF-β3 and Ghrelin in Microsphere/Hydrogel Systems for Cartilage Regeneration, Mol. 2021, Vol. 26, Page 5732. 26 (2021) 5732. https://doi.org/10.3390/MOLECULES26195732.
189. Y. Li, Y. Liu, Q. Guo, Silk fibroin hydrogel scaffolds incorporated with chitosan nanoparticles repair articular cartilage defects by regulating TGF-β1 and BMP-2, Arthritis Res. Ther. 23 (2021) 1–11. https://doi.org/10.1186/S13075-020-02382-X/FIGURES/6.

190. T. Zhou, X. Li, G. Li, T. Tian, S. Lin, S. Shi, J. Liao, X. Cai, Y. Lin, Injectable and thermosensitive TGF-β1-loaded PCEC hydrogel system for in vivo cartilage repair, Sci. Reports 2017 71. 7 (2017) 1–13. https://doi.org/10.1038/s41598-017-11322-w.
191. Y. Chen, T. Wu, S. Huang, C.W.W. Suen, X. Cheng, J. Li, H. Hou, G. She, H. Zhang, H. Wang, X. Zheng, Z. Zha, Sustained Release SDF-1α/TGF-β1-Loaded Silk Fibroin-Porous Gelatin Scaffold Promotes Cartilage Repair, ACS Appl. Mater. Interfaces. 11 (2019) 14608–14618. https://doi.org/10.1021/ACSAMI.9B01532/ASSET/IMAGES/LARGE/AM-2019-015322_0007.JPEG.
192. T. Wu, Y. Chen, W. Liu, K.L. Tong, C.W.W. Suen, S. Huang, H. Hou, G. She, H. Zhang, X. Zheng, J. Li, Z. Zha, Ginsenoside Rb1/TGF-β1 loaded biodegradable silk fibroin-gelatin porous scaffolds for inflammation inhibition and cartilage regeneration, Mater. Sci. Eng. C. 111 (2020) 110757. https://doi.org/10.1016/J.MSEC.2020.110757.
193. W. Ye, Z. Yang, F. Cao, H. Li, T. Zhao, H. Zhang, Z. Zhang, S. Yang, J. Zhu, Z. Liu, J. Zheng, H. Liu, G. Ma, Q. Guo, X. Wang, Articular cartilage reconstruction with TGF-β1-simulating self-assembling peptide hydrogel-based composite scaffold, Acta Biomater. 146 (2022) 94–106. https://doi.org/10.1016/J.ACTBIO.2022.05.012.
194. R. Vayas, R. Reyes, M.R. Arnau, C. Évora, A. Delgado, Injectable Scaffold for Bone Marrow Stem Cells and Bone Morphogenetic Protein-2 to Repair Cartilage, Cartilage. 12 (2021) 293–306. https://doi.org/10.1177/1947603519841682/ASSET/IMAGES/LARGE/10.1177_1947603519841682-FIG2.JPEG.
195. T. Taniyama, T. Masaoka, T. Yamada, X. Wei, H. Yasuda, T. Yoshii, Y. Kozaka, T. Takayama, M. Hirano, A. Okawa, S. Sotome, Repair of osteochondral defects in a rabbit model using a porous hydroxyapatite collagen composite impregnated with bone morphogenetic protein-2, Artif. Organs. 39 (2015) 529–535. https://doi.org/10.1111/AOR.12409.
196. A.C. Kuo, J.J. Rodrigo, A.H. Reddi, S. Curtiss, E. Grotkopp, M. Chiu, Microfracture and bone morphogenetic protein 7 (BMP-7) synergistically stimulate articular cartilage repair, Osteoarthr. Cartil. 14 (2006) 1126–1135. https://doi.org/10.1016/J.JOCA.2006.04.004.
197. S. Odabas, G.A. Feichtinger, P. Korkusuz, I. Inci, E. Bilgic, A.S. Yar, T. Cavusoglu, S. Menevse, I. Vargel, E. Piskin, Auricular cartilage repair using cryogel scaffolds loaded with BMP-7-expressing primary chondrocytes, J. Tissue Eng. Regen. Med. 7 (2013) 831–840. https://doi.org/10.1002/TERM.1634.
198. R. Kuroda, A. Usas, S. Kubo, K. Corsi, H. Peng, T. Rose, J. Cummins, F.H. Fu, J. Huard, Cartilage repair using bone morphogenetic protein 4 and muscle-derived stem cells, Arthritis Rheum. 54 (2006) 433–442. https://doi.org/10.1002/ART.21632.
199. H.J. Kim, M.A. Han, J.Y. Shin, J.H. Jeon, S.J. Lee, M.Y. Yoon, H.J. Kim, E.J. Choi, S.H. Do, V.C. Yang, H. He, Y. Il Yang, Intra-articular delivery of synovium-resident mesenchymal stem cells via BMP-7-loaded fibrous PLGA scaffolds for cartilage repair, J. Control. Release. 302 (2019) 169–180. https://doi.org/10.1016/J.JCONREL.2019.04.002.
200. K. Gavenis, N. Heussen, M. Hofman, S. Andereya, U. Schneider, B. Schmidt-Rohlfing, Cell-free repair of small cartilage defects in the Goettinger minipig: The effects of BMP-7 continuously released by poly(lactic-co-glycolid acid) microspheres, https://doi.org/10.1177/0885328213491440. 28 (2013) 1008–1015. https://doi.org/10.1177/0885328213491440.
201. H. Wu, Z. Peng, Y. Xu, Z. Sheng, Y. Liu, Y. Liao, Y. Wang, Y. Wen, J. Yi, C. Xie, X. Chen, J. Hu, B. Yan, H. Wang, X. Yao, W. Fu, H. Ouyang, Engineered adipose-derived stem cells with IGF-1-modified mRNA ameliorates osteoarthritis development, Stem Cell Res. Ther. 13 (2022) 1–15. https://doi.org/10.1186/S13287-021-02695-X/FIGURES/7.
202. H. Cho, J. Kim, S. Kim, Y.C. Jung, Y. Wang, B.J. Kang, K. Kim, Dual delivery of stem cells and insulin-like growth factor-1 in coacervate-embedded composite hydrogels for enhanced cartilage regeneration in osteochondral defects, J. Control. Release. 327 (2020) 284–295. https://doi.org/10.1016/J.JCONREL.2020.08.002.

203. S. Morscheid, J.K. Venkatesan, A. Rey-Rico, M. Cucchiarini, G. Schmitt, Remodeling of Human Osteochondral Defects via rAAV-Mediated Co-Overexpression of TGF-β and IGF-I from Implanted Human Bone Marrow-Derived Mesenchymal Stromal Cells, J. Clin. Med. 2019, Vol. 8, Page 1326. 8 (2019) 1326. https://doi.org/10.3390/JCM8091326.
204. X.B. Peng, Y. Zhang, Y.Q. Wang, Q. He, Q. Yu, IGF-1 and BMP-7 synergistically stimulate articular cartilage repairing in the rabbit knees by improving chondrogenic differentiation of bone-marrow mesenchymal stem cells, J. Cell. Biochem. 120 (2019) 5570–5582. https://doi.org/10.1002/JCB.27841.
205. S. Shi, B.J. Kelly, C. Wang, K. Klingler, A. Chan, G.J. Eckert, S.B. Trippel, Human IGF-I propeptide A promotes articular chondrocyte biosynthesis and employs glycosylation-dependent heparin binding, Biochim. Biophys. Acta – Gen. Subj. 1862 (2018) 567–575. https://doi.org/10.1016/J.BBAGEN.2017.11.017.
206. M.B. Gugjoo, Amarpal, A. Abdelbaset-Ismail, H.P. Aithal, P. Kinjavdekar, A.M. Pawde, G.S. Kumar, G.T. Sharma, Mesenchymal stem cells with IGF-1 and TGF- β1 in laminin gel for osteochondral defects in rabbits, Biomed. Pharmacother. 93 (2017) 1165–1174. https://doi.org/10.1016/J.BIOPHA.2017.07.032.
207. C. Alemdar, I. Yücel, B. Erbil, H. Erdem, R. Atiç, E. Özku, Effect of insulin-like growth factor-1 and hyaluronic acid in experimentally produced osteochondral defects in rats, Indian J. Orthop. 2016 504. 50 (2016) 414–420. https://doi.org/10.4103/0019-5413.185607.
208. Z. Zhou, W. Song, G. Zhang, S. Zhan, Z. Cai, W. Yu, Y. He, The recombinant human fibroblast growth factor-18 (sprifermin) improves tendon-to-bone healing by promoting chondrogenesis in a rat rotator cuff repair model, J. Shoulder Elb. Surg. 31 (2022) 1617–1627. https://doi.org/10.1016/J.JSE.2022.01.137.
209. L.T. Kuhn, T. Peng, G. Gronowicz, M.M. Hurley, Endogenous FGF-2 levels impact FGF-2/BMP-2 growth factor delivery dosing in aged murine calvarial bone defects, J. Biomed. Mater. Res. Part A. 109 (2021) 2545–2555. https://doi.org/10.1002/JBM.A.37249.
210. Y.P. Morscheid, J.K. Venkatesan, G. Schmitt, P. Orth, D. Zurakowski, S. Speicher-Mentges, M.D. Menger, M.W. Laschke, M. Cucchiarini, H. Madry, rAAV-Mediated Human FGF-2 Gene Therapy Enhances Osteochondral Repair in a Clinically Relevant Large Animal Model Over Time In Vivo, https://doi.org/10.1177/0363546521988941. 49 (2021) 958–969. https://doi.org/10.1177/0363546521988941.
211. S. Zhou, Z. Wang, J. Tang, W. Li, J. Huang, W. Xu, F. Luo, M. Xu, J. Wang, X. Wen, L. Chen, H. Chen, N. Su, Y. Shen, X. Du, Y. Xie, L. Chen, Exogenous fibroblast growth factor 9 attenuates cartilage degradation and aggravates osteophyte formation in post-traumatic osteoarthritis, Osteoarthr. Cartil. 24 (2016) 2181–2192. https://doi.org/10.1016/J.JOCA.2016.07.005.
212. P. Orth, G. Kaul, M. Cucchiarini, D. Zurakowski, M.D. Menger, D. Kohn, H. Madry, Transplanted articular chondrocytes co-overexpressing IGF-I and FGF-2 stimulate cartilage repair in vivo, Knee Surg. Sports Traumatol. Arthrosc. 19 (2011) 2119–2130. https://doi.org/10.1007/S00167-011-1448-6.
213. H. Madry, P. Orth, G. Kaul, D. Zurakowski, M.D. Menger, D. Kohn, M. Cucchiarini, Acceleration of articular cartilage repair by combined gene transfer of human insulin-like growth factor I and fibroblast growth factor-2 in vivo, Arch. Orthop. Trauma Surg. 130 (2010) 1311–1322. https://doi.org/10.1007/S00402-010-1130-3.
214. H. Maehara, S. Sotome, T. Yoshii, I. Torigoe, Y. Kawasaki, Y. Sugata, M. Yuasa, M. Hirano, N. Mochizuki, M. Kikuchi, K. Shinomiya, A. Okawa, Repair of large osteochondral defects in rabbits using porous hydroxyapatite/collagen (HAp/Col) and fibroblast growth factor-2 (FGF-2), J. Orthop. Res. 28 (2010) 677–686. https://doi.org/10.1002/JOR.21032.
215. J. Ide, K. Kikukawa, J. Hirose, K. ichi Iyama, H. Sakamoto, H. Mizuta, The Effects of Fibroblast Growth Factor-2 on Rotator Cuff Reconstruction With Acellular Dermal Matrix Grafts, Arthrosc. J. Arthrosc. Relat. Surg. 25 (2009) 608–616. https://doi.org/10.1016/J.ARTHRO.2008.11.011.

216. J.S. Lee, P. Guo, K. Klett, M. Hall, K. Sinha, S. Ravuri, J. Huard, W.L. Murphy, VEGF-attenuated platelet-rich plasma improves therapeutic effect on cartilage repair, Biomater. Sci. 10 (2022) 2172–2181. https://doi.org/10.1039/D1BM01873F.
217. H. Utsunomiya, X. Gao, H. Cheng, Z. Deng, G. Nakama, R. Mascarenhas, J.L. Goldman, S.K. Ravuri, J.W. Arner, J.J. Ruzbarsky, W.R. Lowe, M.J. Philippon, J. Huard, Intra-articular Injection of Bevacizumab Enhances Bone Marrow Stimulation–Mediated Cartilage Repair in a Rabbit Osteochondral Defect Model, https://doi.org/10.1177/03635465211005102. 49 (2021) 1871–1882. https://doi.org/10.1177/03635465211005102.
218. R. Chung, B.K. Foster, C.J. Xian, The potential role of VEGF-induced vascularisation in the bony repair of injured growth plate cartilage, J. Endocrinol. 221 (2014) 63–75. https://doi.org/10.1530/JOE-13-0539.
219. F. Geiger, H. Bertram, I. Berger, H. Lorenz, O. Wall, C. Eckhardt, H.G. Simank, W. Richter, Vascular endothelial growth factor gene-activated matrix (VEGF165-GAM) enhances osteogenesis and angiogenesis in large segmental bone defects, J. Bone Miner. Res. 20 (2005) 2028–2035. https://doi.org/10.1359/JBMR.050701.
220. R. Becker, T. Pufe, S. Kulow, N. Giessmann, W. Neumann, R. Mentlein, W. Petersen, Expression of vascular endothelial growth factor during healing of the meniscus in a rabbit model, J. Bone Joint Surg. Br. 86 (2004) 1082–1087. https://doi.org/10.1302/0301-620X.86B7.14349.
221. T. Tarkka, A. Sipola, T. Jämsä, Y. Soini, S. Ylä-Herttuala, J. Tuukkanen, T. Hautala, Adenoviral VEGF-A gene transfer induces angiogenesis and promotes bone formation in healing osseous tissues, J. Gene Med. 5 (2003) 560–566. https://doi.org/10.1002/JGM.392.
222. X. Li, G. Su, J. Wang, Z. Zhou, L. Li, L. Liu, M. Guan, Q. Zhang, H. Wang, Exogenous bFGF promotes articular cartilage repair via up-regulation of multiple growth factors, Osteoarthr. Cartil. 21 (2013) 1567–1575. https://doi.org/10.1016/J.JOCA.2013.06.006.
223. Y. Hong, N. Liu, R. Zhou, X. Zhao, Y. Han, F. Xia, J. Cheng, M. Duan, Q. Qian, X. Wang, W. Cai, H. Zreiqat, D. Feng, J. Xu, D. Cui, Combination Therapy Using Kartogenin-Based Chondrogenesis and Complex Polymer Scaffold for Cartilage Defect Regeneration, ACS Biomater. Sci. Eng. 6 (2020) 6276–6284. https://doi.org/10.1021/ACSBIOMATERIALS.0C00724.
224. N. Asgari, F. Bagheri, M.B. Eslaminejad, M.H. Ghanian, F.A. Sayahpour, A.M. Ghafari, Dual functional construct containing kartogenin releasing microtissues and curcumin for cartilage regeneration, Stem Cell Res. Ther. 11 (2020). https://doi.org/10.1186/S13287-020-01797-2.
225. W.N. Zeng, Y. Zhang, D. Wang, Y.P. Zeng, H. Yang, J. Li, C.P. Zhou, J.L. Liu, Q.J. Yang, Z.L. Deng, Z.K. Zhou, Intra-articular Injection of Kartogenin-Enhanced Bone Marrow-Derived Mesenchymal Stem Cells in the Treatment of Knee Osteoarthritis in a Rat Model, Am. J. Sports Med. 49 (2021) 2795–2809. https://doi.org/10.1177/03635465211023183.
226. B.C. Geiger, A.J. Grodzinsky, P. Hammond, Designing drug delivery systems for articular joints, Chem. Eng. Prog. 114 (2018).

Chapter 4
Drug Delivery Systems for Intervertebral Disc

Isma Liza Mohd Isa, Syafira Masri, Loai A. Elfawy, Safa Tahri, Ibrahim N. Amirrah, Revatyambigai Subramaniam, Raed H. Ogaili, Wan Chiew Ng, Nur Amalia Ra'oh, Mh. Busra Fauzi, Akshay Srivastava, and Sabarul Afian Mokhtar

Contents

I. L. Mohd Isa (✉)
CÚRAM, SFI Research Centre for Medical Devices, School of Medicine,
University of Galway, Galway, Ireland

Department of Anatomy, Faculty of Medicine, Universiti Kebangsaan Malaysia,
Kuala Lumpur, Malaysia
e-mail: ismaliza.mohd@universityofgalway.ie

S. Masri · L. A. Elfawy · I. N. Amirrah · R. Subramaniam · M. B. Fauzi
Centre for Tissue Engineering and Regenerative Medicine (CTERM), Faculty of Medicine,
University Kebangsaan Malaysia, Kuala Lumpur, Malaysia

S. Tahri
Centre for Tissue Engineering and Regenerative Medicine (CTERM), Faculty of Medicine,
University Kebangsaan Malaysia, Kuala Lumpur, Malaysia

Laboratory of Auto-immunity, Cancer and Immunogenetics, University Hospital Habib
Bourguiba, Sfax, Tunisia

R. H. Ogaili
CÚRAM, SFI Research Centre for Medical Devices, School of Medicine,
University of Galway, Galway, Ireland

W. C. Ng
Department of Otorhinolaryngology-Head and Neck Surgery, Faculty of Medicine,
Universiti Kebangsaan Malaysia, Kuala Lumpur, Malaysia

S. Murab (ed.), *Drug Delivery Systems for Musculoskeletal Tissues*,
https://doi.org/10.1007/978-3-031-55653-1_4

1 Introduction

The cause of low back pain (LBP) varies. However, intervertebral disc (IVD) degeneration is thought to be one of the main aetiologies affecting the spinal column between 26% and 42% of patients with LBP [1]. The IVD is classified as a secondary cartilaginous joint known as symphyses, located between adjacent vertebrae, which facilitate spinal movements, including axial bending, rotating, compressing and absorbing shocks [2]. It consists of the outer ring annulus fibrosus (AF) and inner region nucleus pulposus (NP). The NP is confined by a cartilaginous end plate (CEP) superiorly and inferiorly [3]. IVD degeneration begins in young adults and advances with age and ongoing degeneration in response to pathological insults [4]. Dysregulation of cellular phenotype and functions, imbalance of extracellular

N. A. Ra'oh
Department of Ophthalmology, Faculty Medicine, Universiti Kebangsaan Malaysia, Kuala Lumpur, Malaysia

A. Srivastava
Department of Medical Devices, National Institute of Pharmaceutical Education and Research-Ahmedabad (NIPER-A), Gandhinagar, Gujarat, India

S. A. Mokhtar
Department of Orthopaedics and Traumatology, Faculty of Medicine, Universiti Kebangsaan Malaysia, Kuala Lumpur, Malaysia

matrix (ECM) homeostasis and increased inflammation are hallmarks of IVD degeneration [5]. The development of discogenic LBP has been linked to increased inflammation, which induces nociceptive nerve ingrowth into the aneural disc and contributes to the symptoms of pain [6]. The current treatment of IVD degeneration, which includes analgesic medication, rehabilitation and spinal surgery to alleviate pain, does not directly address the underlying pathology of IVD degeneration. Therefore, designing an advanced delivery system for therapeutic agents tailored to the pathological disc microenvironment offers a definitive therapy to halt or at least delay the progress of IVD degeneration, which ultimately promotes tissue regeneration and alleviates pain [7].

2 Anatomy of Intervertebral Disc

The human spine comprises a total of 23 IVDs. Six cervical, twelve thoracic, and five lumbar discs are the total number of IVDs that form a quarter of the total height of the spine. Anatomically, an IVD consists of three different distinct components: The core is the gelatinous NP, the outer AF and the CEP that anchor the discs to adjacent vertebrae (Fig. 4.1a) [8]. Each IVD's thickness is approximately 7–10 mm, and the diameter is 4 cm. Generally, cells in both AF and NP can have several thin, long cytoplasmic projections more than 30 mm long. Such features are not seen in cells of other articular cartilage. It is suggested that they may play a role in the communications of mechanical strain within the tissue [9]. In healthy adults, most of the nutrients delivered to IVD depend mainly on the diffusion from the CEP due to the lack of blood vessels in this region. It has some nerve supplies mainly to the outer AF lamellae. Blood vessels can be found in the longitudinal ligaments next to the disc, but in young cartilage endplates (aiming infant), there are branches of the spinal artery [10].

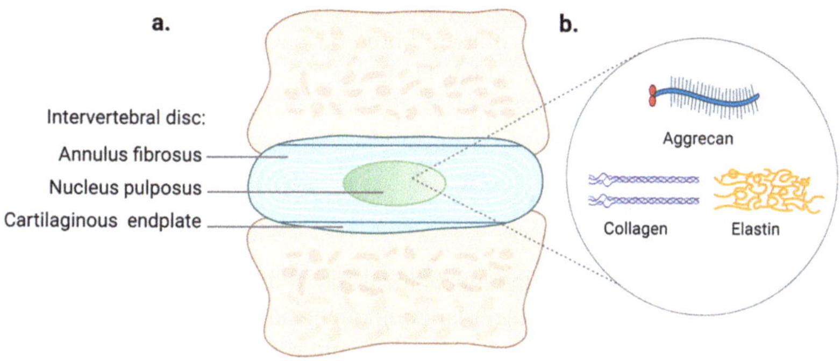

Fig. 4.1 Schematic of a healthy IVD. (**a**) Anatomy of the IVD consists of AF, NP and CEP. (**b**) The ECM of IVD is composed of macromolecules such as proteoglycan, collagen and elastin. (Created with BioRender.com)

2.1 Annulus Fibrosus

The AF contains a series of 15–25 concentric rings (lamellae) that lay parallel with collagen fibres. The angle of the fibres oriented to the vertical axis is approximately 60°. Elastin fibres lie between the lamellae, possibly helping the disc return to the following bending, whether it is flexion or extension. Elastin fibres are bound to the lamellae, which pass radically from one lamella to another. It plays a significant role in helping the disc return to its original arrangement. The cells in the outer region of the AF tend to be elongated, thin, fibroblast-like and aligned parallel to the collagen fibres. In the inner annulus, the cells could be more oval-shaped [11].

2.2 Nucleus Pulposus

The NP is the soft, gelatinous central region of the IVD. The NP supports disc mobility and its ability to absorb mechanical loads under compression. The ability of the NP to resist compression is associated with proteoglycan composition to retain water. This ability decreases with disc age and the decline of its aggrecan content [12]. The NP cells are chondrocyte-like, with a round cytoplasm occupying the ECM of the NP in the adult. Large, vacuolated notochordal cells are a second cell population within the NP that are more abundant in youth and sparse in adult IVD [13]. The proteoglycans are abundantly found in NP which comprises 50% of the dry weight of the IVD. Other major ECM components are collagen, including type II, III, V, VI, IX and XI [7].

2.3 Cartilaginous Endplate

The third morphological region is the cartilaginous endplate (CEP). It is a thin horizontal layer with less than 1 mm of hyaline cartilage, which interfaces the disc and the vertebral body. It consists of collagen fibres. These collagen fibres run horizontally and parallel to the vertebral bodies [14].

2.4 Extracellular Matrix

The ECM components support the mechanical functions of the IVD. Collagen and aggrecan are two major macromolecular components in IVD that play a mechanical role (Fig. 4.1b). Most type I and type II collagen fibrils form the collagen network, making up approximately 70% and 20% of the total dry weight of the AF and NP, respectively. This collagen network provides tensile strength to the disc [15].

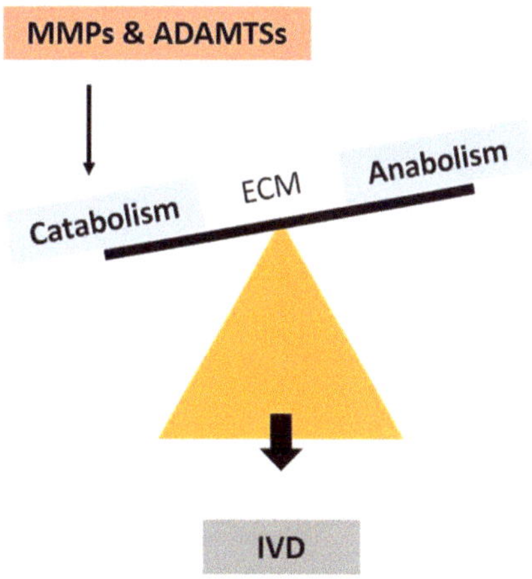

Fig. 4.2 Schematic figure of MMPs and ADAMTSs in IVD indicating regulation of catabolism and anabolism in ECM homeostasis. (Created with BioRender.com)

Aggrecan is the major proteoglycan of the disc. It maintains tissue hydrated due to the osmotic pressure provided by its constituent chains of chondroitin and keratan sulphate. The water content of the NP forms around 80% of the wet weight and 15% proteoglycan, which is greater than what it is in the annulus (approximately 70% and 5% of the wet weight, respectively). The matrix molecules are continually broken down by proteinases such as aggrecanases of a disintegrin and metalloprotease with thrombospondin motifs (ADAMTS) and matrix metalloproteinases (MMPs), which are synthesised by the cells in the disc. The degradation-synthesis balance of ECM macromolecules determines the integrity of the matrix and the mechanical properties of the disc. It is important for maintaining the healthy state of the disc (Fig. 4.2) [16].

3 Intervertebral Disc Degeneration

Degeneration of IVD is a widely known contributor to LBP, which is one of the most prevalent musculoskeletal disorders worldwide and results in a massive socioeconomic burden. Several factors contribute to IVD degeneration, such as heredity factors, obesity, ageing, heavy mechanical loading and even smoking. The ageing process and morphological damage of the IVD can lead to IVD degeneration caused by several complex molecular mechanisms leading to discogenic LBP [17, 18]. The senescent phenotype is associated with increased catabolism, implicating cellular senescence in the pathogenesis of IVD degeneration [19, 20]. Increased cellular apoptosis in IVD causes an imbalance of ECM homeostasis with an increase in ECM degradation and a decrease in ECM anabolism [21]. The loss of proteoglycan in the ECM can lead to severe dehydration, especially in the NP [22]. Increase production of pro-inflammatory cytokines and neurogenic mediators in the IVD, inducing inflammation [22]. TNF- and IL-1 have been linked to a catabolic phenotype,

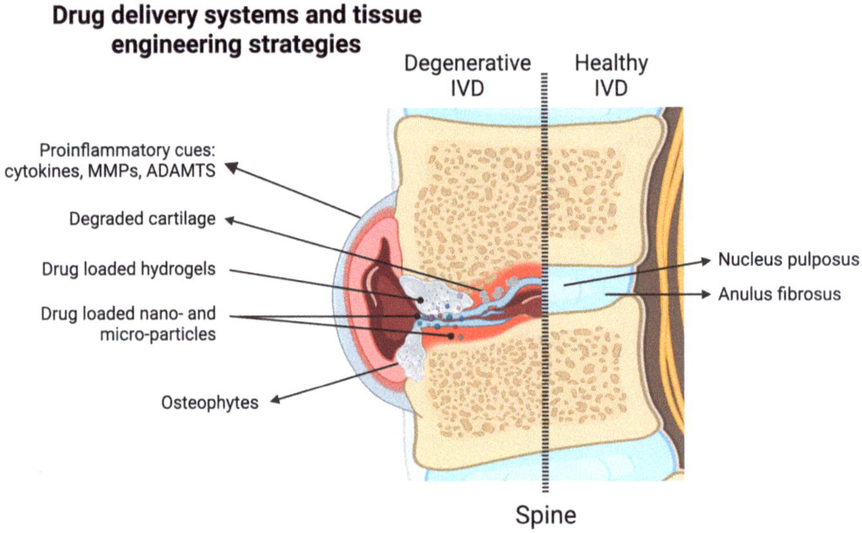

Fig. 4.3 Schematic representation of healthy IVD, the main morphological and pathological changes in IVD degeneration, and drug delivery strategies to target the pathological changes in IVD. (Created with BioRender.com)

which includes increased matrix-degrading enzymes MMP-3, MMP-13, ADAMTS-1 and ADAMTS-5 [23]. Degenerative IVDs show structural failure that is characterised by disc height collapse, loss of proteoglycans and water content in NP, osteophyte formation and calcification formation in CEP [22]. The degeneration of IVDs affects the biomechanics of IVD and the stability of the spinal column [24]. The AF fissures cause NP to extrude, resulting in nerve ingrowth and vascularisation in the IVD, which contribute to the development of a discogenic LBP [25] (Fig. 4.3).

4 Current Clinical Treatments

Current clinical treatments for IVD degeneration are divided into conservative treatments, including pharmacological and non-pharmacological therapies and surgical interventions. These treatments aim to reduce pain or relieve symptoms, hence improving the quality of life in patients. The main drawback of these treatments is that they do not halt the underlying pathogenesis of IVD degeneration. With the aggravation of IVD degeneration, symptoms cannot be controlled under the current treatments [17].

4.1 Pharmacological Therapy

Common drugs for relieving LBP include non-steroidal anti-inflammatory drugs (NSAIDs), muscle relaxants, opioids as painkillers and occasionally steroids. NSAIDs are the most used to treat LBP patients, including selective COX-2 NSAIDs and non-selective NSAIDs. A published Cochrane review suggested that NSAIDs are more effective than placebo for treating LBP with minimum risk of gastrointestinal side effects [26]. Opioids treat severe pain as a pain reliever but are frequently associated with constipation and sedation [27]. The muscle spasm around the spine can be relieved by taking muscle relaxants [28]. Taken together, these medications and others have their side effects and complications. In clinical practice, the severity of pain, the persistence of the symptoms, the degree of pain and the risk factors for complications should be considered when selecting the drug of choice for each case. It is about the risk-benefit balance while considering the patient's overall situation [29].

4.2 Non-Pharmacological Therapy

Non-drug therapy mainly includes drug-free guidelines to enhance general health and decrease LBP. Bed rest, traction, massage, exercise therapy, acupuncture, electromagnetic and electrothermal therapy and psychotherapy are the common non-pharmacological approaches prescribed to patients. Physical exercise improves the strength of muscles that support the spine. It also contributes to the proliferation of IVD cells [30]. Electrical nerve stimulation helps in relieving pain, according to two different trials. However, it has no advantages compared with exercise therapy. Although these treatments cannot reverse the degenerative changes in IVD, they can relieve LBP associated with intervertebral disc degeneration (IDD) patients by stimulating cells, preventing adhesion or promoting metabolite transport [31, 32].

4.3 Surgical Treatments

Intervertebral disc fusion (IDF) and total disc replacement (TDA) are surgical techniques in IDD treatment. IDF is the standard surgical treatment for symptomatic IDD when conservative treatment fails. IDF has many different approaches, such as the anterior approach, posterior approach and posterolateral approach [33]. It could be either minimally invasive or open surgery. Minimally invasive surgery has more advantages than open surgery in reducing post-surgery oedema, controlling the bleeding during the surgery and contributing to early post-operative functional recovery. In minimally invasive IDF, the postoperative pain will be less since the operation has been done by using a smaller skin incision and reducing the dissection of the paraspinal muscle. Therefore, the area of injury will be smaller than in open surgery, accelerating the postoperative functional recovery [34]. Unlike open surgery, IDF, the minimally invasive approach takes a longer learning curve for the surgeons. TDA was introduced as an alternative to fusion surgery and used a stainless-steel ball as a replacement for IVD. It was proposed to replicate the function of the NP, but it has severe complications due to the sinking and squeezing of the ball. Since then, more sophisticated designs of implants have been fabricated with unsettled clinical results.

Researchers are trying to innovate new approaches to make the treatment of IDD much safer and more accessible than current interventions. With the development of biomaterial science, cell therapy and drug delivery systems and nanomedicine, many studies produced evidence that this new era of using these multidisciplinary fields can change the prospective future of the treatment of IDD. This chapter focuses on reviewing the standard and new therapeutic approaches of IDD and puts forward the directions and prospects for the future.

Classes of Therapeutics
Alternative therapeutic strategies aim at increasing the effects of therapeutic molecules by tailoring the needs of the IVD microenvironment to regulate ECM homeostasis, promoting IVD regeneration and halting the progression of IVD degeneration [35, 36]. Here, we discuss the classes of therapeutic molecules, including small molecules, proteins and peptides, antibodies and inhibitors, enzyme inhibitors, nucleic acid and cells (Fig. 4.4).

4.4 Small Molecules

Small molecules are referred to as substances that attach to macromolecules and aid in the regulation of a particular biological process. In general, the molecular weight for a tiny molecule that needs to quickly diffuse through the cell membrane and be absorbed by the digestive system is 900 Daltons. Molecules greater than 550 Daltons have more difficulty absorbing, although some molecules up to 900 Daltons can successfully pass barriers [37, 38]. A few studies that evaluated the mechanism of action of small molecules in targeting different signalling pathways towards an effective therapy in vitro, in vivo and clinical studies show encouraging results [39, 40]. Small molecules were provided as drugs to modulate different cell processes [41]. Several of these small molecules effectively inhibit apoptosis, inflammation, oxidative stress and senescence, which can prevent the degeneration of the disc cells. They also showed anabolic and anti-catabolic effects, which are critical factors for IVD regeneration. According to previous literature, small molecules, such as icariin, resveratrol and epigallocatechin gallate EGCG, showed the ability to inhibit NF-kB and p38/MAPK signalling pathways and modulate the inflammatory response and prevent the development of a degenerative cascade [42, 43]. Also, they showed an impact on the downregulation of IL-1 and TNF-α levels in IVD cells,

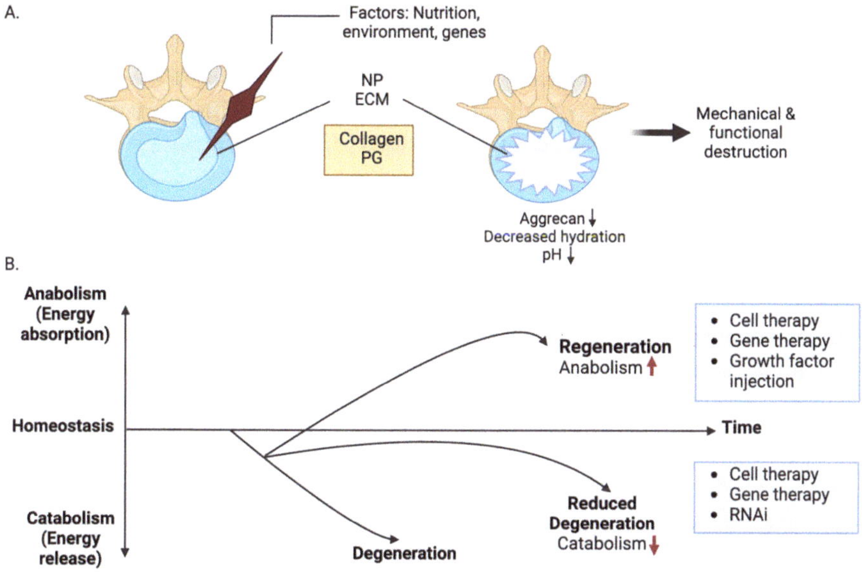

Fig. 4.4 An illustration showing the causes of intervertebral disc degeneration and strategies to prevent it and promote disc regeneration. (**a**) The extracellular matrix (ECM) in the nucleus pulposus (NP) has altered due to changes in collagen and proteoglycan (PG), as well as nutritional, environmental and genetic factors. These changes resulted in a decrease in aggrecan and pH and a loss of hydration, which led to the mechanical and functional destruction of NP. (**b**) Increased anabolism may promote the regeneration of degenerative discs, whereas a reduction in catabolism may promote the delaying of disc degeneration. (Created with BioRender.com)

including naringin, cannabidiol (CBD), epigallocatechin gallate (EGCG), curcumin, icariin, resveratrol, berberine and tofacitinib. Gefitinib, kaempferol and berberine are other small molecules that exclusively block the NF-kB signalling pathway [40, 44, 45]. On the other hand, intracellular p38/MAPK signals could be blocked by rhein and urolithin in vitro [46, 47].

4.5 Proteins and Peptides

Protein factors such as growth factors or proteinase inhibitors may effectively target the early stage of IVD degeneration (grades II–III) in controlling cell proliferation, differentiation and ECM deposition. Proteins may be introduced directly into the disks to promote anabolic responses or cell development that could stop the degeneration of the disks [48–51]. Among these proteins, a family of growth factors that have been found to induce bone and cartilage formation are bone morphogenetic proteins (BMPs) [51]. Another therapeutic option is the recombinant human BMP-2, which has been shown to promote IVD regeneration [51, 52]. One of the benefits of protein treatment is that recombinant human proteins can be produced without the threat of disease transmission. They can also be administered directly into the IVD without requiring a more invasive delivery method. On the other side, protein injections are only effective for a short period. Moreover, they show decreased efficacy in advanced stages of IVD degeneration and unfavourable side effects like inflammation and ossification when doses are misused [53].

4.6 Antibodies and Cytokine Inhibitors

One of the most tested therapeutic strategies is monoclonal antibodies and cytokine inhibitors. Tanezumab is a humanised monoclonal antibody that specifically inhibits nerve growth factor as a treatment for chronic pain [54]. In several chronic pain disorders, elevated NGF levels are linked to heightened pain perception [55]. To find new pain treatments, NGF inhibition is a promising target [56]. However, serious side effects were observed mainly when combined with NSAIDs [54, 57, 58]. Intra-articular administration of Adalimumab, an antibody against tumour necrosis factor-alpha (TNF-α), was also evaluated in the TNF-α inhibition [59]. Similarly, the use of anti-cytokine therapy for low back pain has been investigated [60]. Etanercept, a TNF-α selective inhibitor [61, 62] showed conflicting results. Several studies reported improvements in back and leg pain and function. Yet, in two other studies, Etanercept did not lead to improved primary outcomes, leg pain and disability index [63, 64]. Whether these treatments are also able to promote disc regeneration remains unclear. Interleukin-1 (IL-1) has also been targeted in joint degeneration. Anakinra, an IL-1 receptor antagonist (IL-1Ra) and AMG 108, an antibody against IL-1 receptor type 1, did not lead to symptomatic improvements for OA patients [61, 65].

Inhibition of cytokines has also been demonstrated to be a valid approach to reducing inflammation and disc degeneration. In this respect, interleukin-1β is a key cytokine [66] that IL-1Ra can block. Growth factors (GFs) are typically cytokines, defined by their ability to bind to specific transmembrane receptors resulting in the activation of an intercellular signalling cascade [67]. In general, the biological effect of cytokines can be various effects such as cell proliferation, differentiation, migration, apoptosis and regulation of ECM metabolism. However, there may be drawbacks to treating IVD degeneration, as IVD regeneration is much more challenging to achieve with GF therapy than it is with the regeneration of other tissues, like bone, due to the different cell populations and low cell density of the IVD [48, 68]. Additionally, the very short half-life of GF proteins may necessitate recurrent intradiscal GF injections. As a result, the degeneration process could be sped up [69, 70], which would increase the likelihood of toxic or unfavourable side effects, including ectopic bone formation after direct BMP injection [71].

4.7 Enzymes Inhibitors

Enzyme inhibitors are classes of drugs widely employed to treat inflammation and joint pain. Generally, enzyme inhibitors will attach to an enzyme and prevent it from actively functioning. The enzyme inhibitors, namely, MMP inhibitors, have also been suggested as potential treatments for degenerative joint conditions [72]. Few investigations have pinpointed the source of synthesis of these enzymes to the cells of the degenerate disc despite the suggestion by several researchers that MMPs and ADAMTS are involved in IVD degeneration [73]. MMPs are proteolytic enzymes that repair and deteriorate ECM proteins and their constituents. The MTMMPs, stromelysins, collagenases and gelatinases are the four categories of enzymes that make up the MMPs. The MMPs were considered necessary for the breakdown and reabsorption in the IVD [74].

MMPs play a key role in IVD degeneration, resulting in the loss of normal disc function and low-back discomfort by detecting the expression of MMPs. Increased levels of MMPs are associated with inflammation, in which 23 MMP proteins have been discovered in humans. MMPs have endogenous inhibitors known as tissue inhibitors of metalloproteinases (TIMPs) that provide strict control of their activity [75]. IVD may result from imbalances between proteinases and inhibitors, which may change the amounts of ECM components and interfere with the control of the ECM, as illustrated in Fig. 4.2. They can particularly disturb the equilibrium of IVD tissue anabolism and catabolism, stimulate extracellular matrix breakdown and induce intervertebral disc herniation [74].

Notably, MMPs are normally released inactively and need the regulatory activator proteins to become active. MMP expression is minimal or absent in normal IVD tissue [76]. However, increased expression of several MMPs has recently been discovered in degenerative IVD tissue. Since IVD is characterised by the breakdown of the ECM, treatments that can return discs to a state of normal homeostasis while causing the lowest side effects are required. TIMPs, of which there are now four varieties, are particular tissue inhibitors that prevent the action of MMPs [77]. TIMP-1 and TIMP-2 bind to active MMPs in a 1:1 stoichiometric way to form irreversible non-covalent complexes that block the enzyme activity [66]. A cysteine residue in TIMP may bind the zinc ion in the MMP active site to stop MMP from activating [78]. Therefore, TIMPs are considered promising therapeutic targets for IVD degeneration since they inhibit all MMPs and are produced by chondrocytes.

4.8 Nucleic Acids

The use of nucleic acid transfer, either RNA or DNA, to treat or prevent the progression of the disease is typically one of the gene therapies. Generally, significant genetic risk factors are present in IVD. Biological treatments aim to regenerate the disc and alleviate discomfort. It has been suggested that introducing genes to IVD cells that encode factors that govern matrix synthesis and catabolism can reduce an imbalance in the synthesis and catabolism of several significant ECM components [36]. For instance, the therapeutic potential of nucleic acid noncoding sequences exists. It is feasible to deliver oligonucleotides containing response elements in gene promoter regions [79]. These decoys attach to their corresponding transcription factors, rendering them inactive in driving gene expression. A technique for delivering the desired gene to the target disc cell is one of the technologies needed for gene therapy. A vector is necessary because it makes it easier for the transgene to be expressed and translocated into the nucleus [36]. Although non-viral-mediated gene transfer to disc cells has been created due to the potential risks associated with viral gene therapy, viral vectors have been commonly employed because they can easily rearrange their genetic material [80].

Briefly, Wehling et al. successfully transfected the target gene into the in vitro-cultured chondrocyte cells derived from bovine coccygeal vertebral endplates in 1997, introducing the concept of employing in vitro gene transfer to reverse IVDD for the first time [81]. However, there is also a treatment without involving a virus vector that utilises nucleic acid-based therapy to inhibit the gene expression at the transcriptional or post-transcriptional level. The therapies based on nucleic acids can be broadly divided into two categories: RNA-based therapeutics and DNA-based therapeutics [82].

4.9 Cell Therapies

Cell-based therapy appears to be the most promising method for treating IVD among the numerous biological approaches. Reducing the number of viable and functioning cells, with a significant part of cells being in a senescent condition, is an essential aspect of IVD degeneration. Thus, the injured disc might get a direct transplant of stem cells. Briefly, the mesenchymal stem cells (MSCs) derived from bone marrow, adipose tissue, Wharton's jelly of the umbilical cord and synovial membrane have been employed in this procedure because of their capability for self-renewal, anti-inflammatory qualities and aptitude to replace damaged disc cells [83]. According to reports, stem cell treatment can restore degenerative discs by regulating cellular development and have a paracrine impact (anti-inflammatory, anti-apoptotic) when inserted into NP [84].

Besides, the transplantation of mesenchymal stem cells (MSCs) in disc degeneration models has been the subject of several in vivo research such as rabbits and porcine model [85–88]. Overall, the outcomes of these in vivo investigations have shown that MSCs can survive and multiply after being implanted into the disc and develop the phenotypic traits of IVD cells. Furthermore, there is strong evidence that MSCs can reduce tissue inflammation. Figure 4.5 demonstrates cell transplantation approaches for cell therapy IVD treatment.

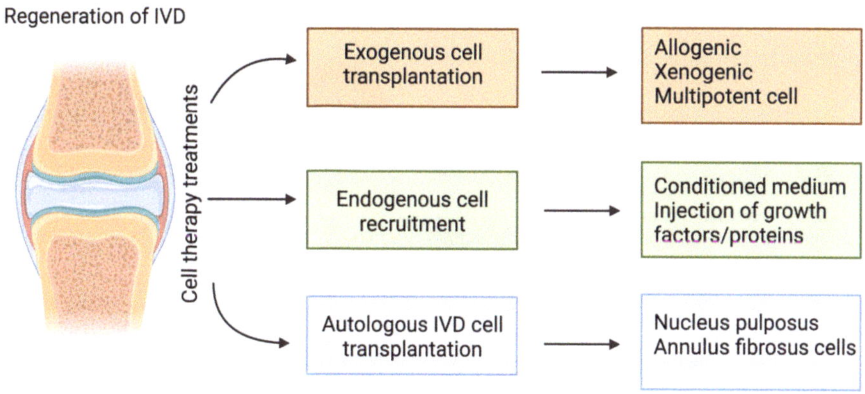

Fig. 4.5 Treatment options for IVD degeneration using cell therapy, including exogenous cell transplantation, endogenous cell recruitment and autologous IVD cell transplantation. (Created with BioRender.com)

5 Drug Delivery Nanosystems

IVD treatment mainly focuses on relieving the clinical symptom through a therapeutic approach [89]. However, the current drug administration effectiveness is limited due to poor distribution and indefinite target region [90]. Figure 4.3 summarises that the drug delivery system provides an advanced approach to target the pathological microenvironment of IVD for long-term efficacy. The development of drug delivery ranges from macro- to nanosystems based on their sizes (Fig. 4.6).

The drug delivery nanosystems held promise for promoting IVD regeneration and achieved excellent biocompatibility and biodegradability [91], attributing active delivery of the drugs directly to the targeted region. This system protects the drug degradation and allows sustained drug release targeting specific cells or tissues. They facilitate the administration of lower concentrations of required drugs and dosage precisely [92]. These systems provide improved pharmacodynamics and pharmacokinetics between the host and drug, promising positive therapeutic effects with lower toxicity [93] (Table 4.1).

5.1 Nanoparticles

Nanotechnology is one of the immense findings in pharmaceutical, therapeutic and diagnostic applications due to its target specificity and precise range. Nanoparticles can be particles, dendrimers, spheres, gold particles, micelles, liposomes and crystal [92, 94]. A nanoparticle's interference with the host is measured through its pharmacokinetics and pharmacodynamics properties. Nanoparticles range from 1 to 100 nm and the microscopic size has a unique physicochemical property size, thus

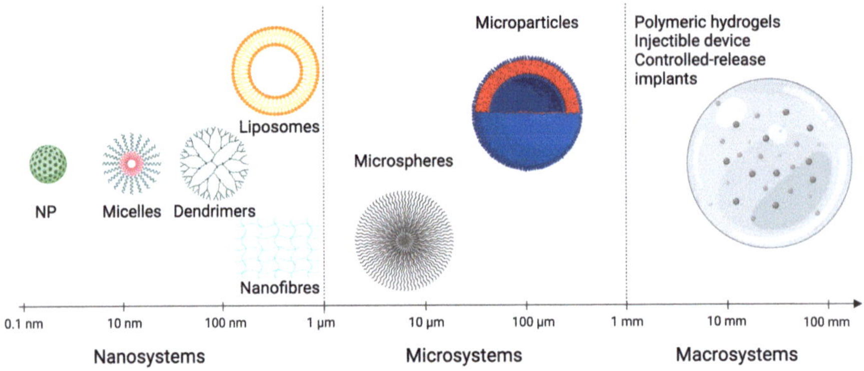

Fig. 4.6 Illustration of the different drug delivery platforms based on their size, from nano- to macrosystems. (Created with BioRender.com)

allowing its penetration and diffusion with the target readily [95]. The content, production, size and surface properties of nanoparticles can be modified according to their function. The function of these nanoparticles varies according to their core material and combination of biologics or drugs. Its release profile depends on the formulation, pH, temperature, binding capacity and degradation properties [96]. For instance, nanohydroxyapatite is widely used in bone implants due to its similar content to the bone composition of calcium phosphate. This forms a platform for new bone regeneration between the host and implant [97].

Nanoparticles can also be incorporated with various therapeutic agents to function as carriers. Nanoparticles such as liposomes, solid lipids, dendrimers, polymers, silicon or carbon materials and magnetic are widely used as nanocarriers to deliver drugs [92]. The efficacy relies on conjugating it with the drug by directing it to the target cell or tissue. It may be through a simple method such as adsorption, covalent attachment, encapsulation, or electroporation. Upon reaching their target, the nanocarriers release the therapeutic agents due to their physicochemical, temperature, pH, osmolality, or enzymatic reaction. It is crucial for these nanocarriers to integrate with the host without eliciting any immune response or causing toxicity [98]. The diversity of nanoparticles has been tested for IVD preclinical models, for example, polymeric, micelles and liposomes. The cells internalised these particles ranging from 1 to 100 nm encapsulated with antibodies or peptides. This resulted in the penetration of the biological membrane, allowing for targeted drug delivery and enhanced drug stability [99]. These particles offer rapid binding to the cartilage by penetrating the thick matrix. Besides, albumin and heparin nanoparticles are widely used for IVD treatment to release stromal-derived factor 1-alpha (SDF-1α) and the recruitment bone MSC [99]. In the results, enhanced disc regeneration was observed compared to the absence of a nanoparticle carrier. Overall, nanoparticle carrier has been a breakthrough for IVD treatment, as previously, systemic drug administration has limited bioavailability. These nanoparticles have been tailored to provide a locally targeted and sustained drug delivery [100].

Table 4.1 Description and advantages and disadvantages regarding nanosystems drug delivery

Category	Size	Mechanism	Possible modifications	Pros	Cons
Nanosystems (nanoparticles (NP), polymeric, micelles, liposomes)	1–1000 nm diameter in at least 1 dimension. NP < 10 nm; Micelles 10–100 nm; Dendrimers 100–500 nm; Liposomes 100 nm–1 μm.	Intracellular/intratissue protein target. Easily removed by synovial membrane foreign body reaction, cell internalisation, or enzymatic biodegradation in hours or days.	Resize, modify surfaces to control charges and interact with matrix or cells. Targeted antibody/peptide delivery. Permeation of cartilage matrix.	Targeted therapeutical delivery; penetrates biological membranes; improves hydrophobic drug solubility and stability.	Can agglomerate and cause injury; immunogenicity; particle solvent may be poisonous.

5.2 Nanofibres

Biological materials include composite hydrogels that are reinforced by fibres like the ECM of mammalian tissues—as such, developing bioinspired fibre-filled hydrogels is attractive in tissue engineering. The NP of IVD is a hydrated gelatinous core within the AF with an ECM consisting of proteoglycan hydrogel supported by collagen fibres. Treatments targeted to support the NP are generally used to deter mechanical failure, and integration with bioactive molecules may also promote cell growth, proteoglycan production and disc longevity. As such, nanofibres with a gel-like formulation that is biomimetic to the NP with improved modifications have a high potential to prevent height loss in the disc and overcome bio-mechanical and chemical changes related to degeneration. Nanofibres in gel formation can be administered without invasive surgery and as a tissue substitute. For example, it was found that a cellulose hydrogel formulated with chitosan/glycerophosphate/hydroxyethyl was suitable for cell-based supplementation to restore NP in the early stages of IVDD [101]. Other formulations, such as photo-crosslinked hyaluronic acid hydrogel, could also fill the NP and provide an in situ resistant hydrogel [102].

Injectable and gellable cellulose nanofibres and chitosan integrated within the disc nucleus and repaired and regenerated the NP of pig and rabbit spine models with disc degeneration [103]. Another material that could be used to replace the NP material is silicone gel. Electrospun fibre-anchored silicone discs were tested on rabbit models and showed restoration of NP post-implant with promising clinical potential [104]. The nanosurface modification also plays a vital role, whereby bone tissues and bone-forming cells are receptive to nanometre roughness. Traditional synthetic metal implants had roughness on the micrometre scale, resulting in improper adhesion and callus formation. Nanofibres with adjustable roughness can achieve the necessary nanoscale of materials such as alumina nanofibre and carbon nanofibre, which has shown success for osteoblast adhesion [105].

5.3 Liposomes

Liposomes are vesicles with an aqueous core surrounded by a phospholipid bilayer(s), arranged structurally as a cellular membrane and were the first nanoparticle to translate successfully to clinical applications, primarily for cancer treatment [106]. Figure 4.7 depicts the multifunctional components of liposomes.

The size of liposomes can vary between 50 nm and 50 μm depending on the number of bilayers in the structure, the buffer, lipids composition and the filtration strategy [106]. Liposomes have an aqueous and hydrophilic core with an external hydrophobic lipid bilayer. Both hydrophilic and hydrophobic drugs can be encapsulated [107]. Liposomes are attractive because of their high biocompatibility, ability

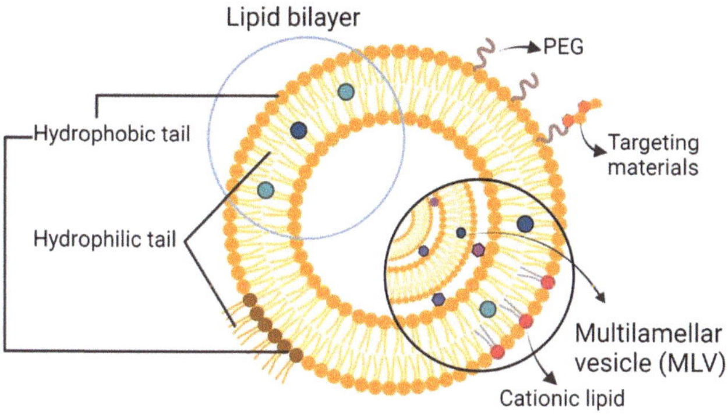

Fig. 4.7 A schematic illustration of liposomal structures depicting the ability of liposomes to carry both hydrophilic and hydrophobic drugs, solubilised in the core or embedded within the lipid bilayer. Variations of traditional liposomes are also represented as adding targeting ligands or poly(ethylene glycol) (PEG), as a multilamellar vesicle (MLV) or cationic or ionised lipid. (Created with BioRender.com)

to encapsulate soluble and insoluble drugs and immense possibilities for augmentation in physical and chemical properties for the intended use. In addition, multiple components can be integrated into the lipid bilayer for longer circulation times, controlled release profiles and targeting specific tissues.

Previously, liposomes were used to transfect human telomerase gene therapy and improve the cellular longevity of the nucleus pulposus in the IVD [108]. Small interfering RNAs (siRNAs)-mediated gene therapy for IVD degeneration also used liposomes in a Caspase 3-siRNA and dual siRNA formulations to downregulate Caspase 3 and ADAMT5 levels in disc degeneration rabbit models, which had beneficial synergistic effects [109]. After intravenous injection, liposomes are removed by mononuclear phagocyte systems, which can be improved by adding PEGylation, a biochemical modification of bioactive molecules with polyethylene glycol (PEG), which increases stealth and provides the tool to functionalise with targeting ligands [110].

Methods of formulation for drug-loaded liposomes can be either passive or active loading techniques [111]. Passive loading occurs during the formation of liposomes, which are techniques for solvent dispersion, mechanical dispersion and detergent removal. Mechanical dispersion includes freeze-thaw cycles, extrusion, sonication, hydration of lipid film, membrane extrusion and micro-emulsification. On the other hand, active loading techniques are conducted after the formation of liposomes, such as pH gradients, to encourage drugs into liposomes using protonatable amine functionalities of water-soluble drugs. Higher pH inside the liposome can precipitate the drug. Using pH response allows high loading efficiency in liposomes for drug deliveries such as doxorubicin (Doxil®), the first FDA-approved nanodrug in 1995 [112].

5.4 Dendrimers

Dendrimers consist of a nucleus or group of atoms known as the core with the projection of tree branch-like structures known as dendrons. Dendrimers are in the nanosize range in a radial symmetric shape, a globular macromolecule [100]. The branches signify its generation; the more branching units, the higher its generation. The core is designated as G0, and the subsequent layers add on G1, G2 and G5 [113]. Higher generation consists of multiple functional groups ranging from carbohydrate (glycodendrimers), peptides (peptide dendrimers), silicon-based dendrimers or polymers (polydendrimers) [114]. Dendrimers are polyionic in that they do not have consistent shapes and constantly change in size, shape and flexibility as their generation increases. A dendrimer is part of molecular chemistry due to the controlled synthesis and partly belongs to polymer due to their repetitive structure of monomers [113]. The chemical property of dendrimers allows their self-assembly by the precise association of chemical species and intramolecular forces. First, the core and the dendron are created either ditopic or polytopic, followed by the spontaneous formation of the dendrimer.

The polymer function of dendrimer aids its unique properties to be monodisperse forming multivalent interactions. The standard polymer fabrication of dendrimers is polyethyleneimine, polyamidoamine (PAMAM), poly-(N-isopropylacrylamide) and polypropyleneimine. The pharmacokinetic properties of the dendrimers are crucial for successful drug delivery [98]. The flexibility of dendrimer peripheral in modification allows the formation of antibody dendrimer or peptide dendrimer. Dendrimers' polyvalency properties widen their potency to form multiple interactions between the receptors [100]. Dendrimers are analogous to proteins, enzymes and viruses and are easily functionalised. Dendrimers can be conjugated or attached peripherally through covalent/non-covalent attachment or be encapsulated interiorly with other molecules [115]. This approach has been associated with PAMAM and anti-inflammatory drugs such as ibuprofen, indomethacin and ketoprofen. PAMAM dendrimers encapsulation data showed better internalisation and enhanced immune response due to their multivalency and higher generation. In IVD, PAMAM dendrimers were incorporated in the liposomes studied to modulate cytokine and chemokine release and exhibited anti-inflammatory reactions [116].

5.5 Micelles

Micelles describe nanoparticles that self-assemble spontaneously into a supramolecular form when amphiphiles are hydrated. Amphiphiles are molecules that contain both hydrophilic and hydrophobic parts. When amphiphiles are hydrated in an aqueous solution, they reorganise to form nanoparticles. Unlike liposomes, micelles have a hydrophobic core with an external shell that faces the water and is hydrophilic. Figure 4.8 provides a schematic illustration of the multifunctional

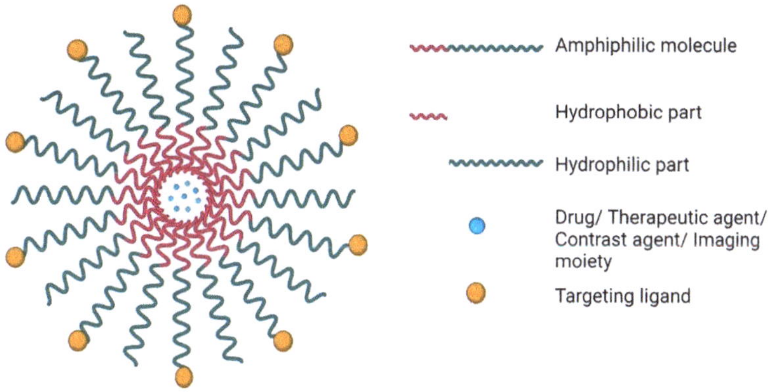

Fig. 4.8 Illustration on the multiple components that can be modulated in micelle formulations. (Created with BioRender.com)

components used in micelle formulations. When the amphiphilic molecule reaches a concentration threshold, called critical micelle concentration (CMC), vesicles are formed in the aqueous solution. Vesicles can encapsulate insoluble and hydrophobic drugs and other ligands, such as antibody fragments or growth factors within the hydrophobic core. Lowering CMC using block copolymers and advances in polymer chemistry in formulating polymeric micelles have resulted in vesicles with increased in vivo stability, making them attractive as a drug delivery system [117]. Similarly, to liposomes, micelles have versatile and tailorable physical and chemical properties for the intended use.

Although the micellar core is advantageous due to its loading capability for hydrophobic drugs or imaging agents, it cannot load hydrophilic drugs unless they are covalently linked to the micelle [118]. To successfully encapsulate hydrophobic drugs in the micelle core with colloidal stability of the less soluble drugs, methods for solvent evaporation, polymerisations, emulsions and nanoprecipitation have been utilised. The diameter of micellar vesicles usually ranges between 10 and 100 nm. It depends on the hydrophilic and hydrophobic ratio of the amphiphilic molecule, the polymer chemistry, the packing factor, concentration and the nature of the encapsulated agent [91, 118].

IVD therapy targeting direct regulation of the cells utilised micelle polyplexes consisting of PEG-polyamino acid block copolymers and integrated with an anabolic mRNA on rat model was performed to alter the homeostasis in IVD degeneration [119]. The strategy led to a significant increase in disc integrity maintenance, and the nanomicelles prevented inflammation induced by injected naked mRNA. MicroRNA-29 (miR29), with the ability to suppress fibrosis, loaded in micelle polyplexes and encapsulated in MMP-degradable hydrogels was also used to treat IVDD [120]. The hydrogel and particle polymers were both responsive to MMP, which led to the micelles being released from the gel. The following removal of the PEG shell from the particles enhances their cellular uptake and escape from endosomes. As a result, MMP-2 levels were reduced, and IVD fibrosis in vivo was mitigated [120].

6 Drug Delivery Microsystems

Traditional methods of drug administration are always associated with difficulties in controlling drug release. Micro-drug delivery system is one of the greatest scientific revolutions in delivering drugs to the IVD with its ability to sustain or control drug release. Due to their size, ranging from one to hundreds of microns, they can retain IVD for a few days and avoid the primary clearing mechanism: the lymphatic systems [121]. It is usually removed by foreign body response or biodegradation [91]. Micro-drug delivery systems also reduce the degradation rate of the drug, reduce local toxicity and increase patient compliance. It depends on its different sizes, surface coating for controlled drug release, tuneable drug loading and releases dynamics [91, 122].

6.1 Microspheres

A microsphere is a flowable spherical microparticle that ably delivers the drug to the interior of the targeted IVD tissue and controlled drug release [91, 123]. It releases the loaded drug slowly to employ a prolonged therapeutic effect and thus improve the IVD disease state at the source [124]. This further restores the anatomical structure and biological function of the IVD [123]. Generally, microspheres include both microcapsules and micromatrices. Microcapsule consists of a therapeutic drug as a core structure and is enclosed by the material as a shell. At the same time, micromatrices comprise a therapeutic drug that is embedded or dispersed in the entire material matrix [125].

Accurately controlled drug delivery depends on the encapsulated material [126, 127]. Organic polymers are currently used as microsphere material, including natural and synthetic polymers. Natural polymers such as collagen [128, 129], gelatine [130], alginate [131], chitosan [130] and starch [132] are selected as the microsphere material due to their low toxicity, biocompatible and biodegradability. Synthetic polymers such as poly(l-lactic acid) (PLLA) [35], poly(lactic-co-glycolic acid) (PLGA) [133] and poly(ester amide) (PEA) [134] have also become choices as microsphere material as it is mechanically stable, can be modified and has good degradability.

Various methods have been used to fabricate the microsphere, including microfluidic technology, spray-drying, phase separation, emulsion-solvent evaporation and electrospraying [123]. Different fabrication methods produce different structures of microspheres, such as porous, hollow, solid and bilayer structures, which give different medical uses to the microsphere [123]. According to Kim et al. [135], the rate of drug release depends on drug diffusion through the region [135]. Crystalline PLLA and semicrystalline PCL microspheres have a more sustained release profile of capsaicin (analgesic drug towards IVD) than amorphous PLGA. The rate of capsaicin released increases as the microsphere size decreases, and it is also correlated to drug diffusion through amorphous regions [126, 127].

The microsphere delivery system has low toxicities as it reduces the systemic drug dose, decreases the frequencies of administration and protects biological activity. Thus, it needs to adapt the release profile of the drug [136]. As the microsphere-drug delivery system can sustain the release of corticosteroids to reduce inflammation and the progression of IVDD, it can prevent the need for multiple injections of corticosteroids [22]. The sustained release of corticosteroids from the local injection of corticosterone-loaded resorbable and non-immunogenic tricalcium phosphate (TCP) microcapsules in traumatised IVD rat model was able to reduce the IVD degeneration [137]. TCP microcapsules containing corticosterone can prevent degeneration in an IVD trauma rat model [137]. Platelet-rich plasma (PRP) containing gelatine hydrogel microspheres can also delay the progression of IVD degeneration in the rabbit model [138, 139].

Interleukin-1 receptor antagonist (IL-1Ra) is encapsulated in the PLGA to overcome its short biological half-life (around 5 h) and to prolong its delivery by a mixed ratio of lactide to glycolide to increase the hydrophobicity of the microsphere. This study showed that the IL-1Ra-PLGA microsphere could inhibit nucleus pulposus (NP) degeneration within 20 days [140]. IL-1Ra- PLGA microspheres reduced the degradative effects of IL-1β on the NP for up to 20 days in vitro [141]. According to Gorth et al. [140], attenuation of NP degeneration in in vitro human IVD tissue culture is observed when it sustained received IL-1Ra-loaded PLGA microspheres [140]. Successfully delivered and retained IL-1ra-PLGA microsphere for 28 days in both NP and AF in vivo and effectively prevented IL-1β-induced NP glycosaminoglycan reduction. IL-1Ra- PLGA can be retained in IVD up to 28 days post-injection in the NP and AF [141].

Liang et al. delivered different combination agents, dexamethasone and coated with the polycation poly(ethylene imine) loaded in the PLGA microsphere to relieve pain and inhibit IVDD. This microsphere increases proteoglycan production and rebuilding the disc height by approximately 70% compared to healthy control animals [35]. Moreover, PLGA microparticles loaded with dexamethasone and FGF-2-embedded heparin/poly(l-lysine) nanoparticles also promote the proliferation of rat mesenchymal stem cells and differentiation into NP cells in vitro [142], and these induce partial tissue regeneration in an IVDD rat model [143]. Celecoxib-loaded PEA microspheres can inhibit the formation of osteophytes and sclerosis and prevent further degeneration of the IVD in a canine model [134].

6.2 *Microparticles*

One of the significant challenges in advancing pharmaceutical formulation is to sustain drug release. Microparticles can control and sustain drug release patterns. It is also biocompatible, stable, targets specific IVD sites, has uniform encapsulation, is less toxic and reduces the dosage frequency [35]. Microparticles are particle sizes ranging from 1 to 1000um that can be used as a drug carrier to the IVD. Microparticles can encapsulate many drugs, including small hydrophobic drugs such as corticosteroids and NSAIDs, to large macromolecules such as antibodies and proteins. A different formulation of biomaterials can tune their degradability and release profile [91]. Their big sizes limit their clearance via small capillary networks; microparticles can retain the drug for a long time compared to small delivery systems such as nanosystems [144].

To develop microparticles, many factors must be considered, such as physicochemical properties, biodegradation rate and degradation product [91]. Biocompatibility of the microparticle also depends on its degradation product. For example, the acidic breakdown product of the PLGA can reduce the surrounding pH and lead to inflammation [145]. Microparticles delivered the drug to the targeted IVD and sustained the release of the encapsulated drug [146]. Radiculopathy is always associated with increased levels of TNF-α in IVD locally. Systemic administration of anti-inflammatory drugs of TNF-α antagonists for radiculopathy has shown mixed results. Thus, many researchers are interested in studying the local delivery of anti-inflammatory drugs to treat this pathology. Chemically modified curcumin (TNF-α antagonist) conjugated with thermally responsive elastin-like polypeptide (ELP) (ELP-curcumin conjugate) developed into an injectable depot for sustained, local delivery of curcumin to treat radiculopathy. Radiculopathy is usually accompanied by IVD herniation [147]. Moreover, gellan gum (GG)-hydrogel disc is another exciting study involving microparticles for IVDD treatment. Its mechanical properties were improved by mixing the hydrogels with the microparticles. The GG-hydrogel was non-cytotoxic as lung fibroblast cell line (L929 cells) successfully encapsulated into it and remain viable after 72 h of culture. Thus, GG-hydrogel matrices with cell-loaded potential as an NP substitute [148].

Homeostasis of IVD degeneration can be restored by locally delivered anti-inflammatory compounds such as epigallocatechin-3-gallate (EGCG). Injectable and tissue adhesive-laden hydrogel depot (EGCG HYPOT) is fabricated by utilising the phenylborate ester reaction of EGCG and phenylboronic acid (PBA) of methacrylate hyaluronic acid (HAMA-PBA). EGCG HYPOT is smart delivery of EGCG, which can achieve the antioxidant and anti-inflammatory effect for IVDD treatment. EGCG HYPOT has good mechanical properties, tissue adhesion and sustained acid-responsive release of EGCG. EGCG HYPOT also scavenges oxygen and nitrogen-free radicals and suppresses the expression of pro-inflammatory factors [149]. The other study showed that encapsulation of EGCG into electrospraying glutaraldehyde-crosslinked gelatine microparticles could overcome the rapid degeneration/depletion properties of EGCG. This EGCG microparticle showed a trend towards modulating ECGC activity/release and successfully reducing the expression of the pro-inflammatory agent, which are catabolic mediators (MMP1, MMP3, MMP13) and inflammatory mediators (IL-6, IL-8, COX-2). Thus, the biocompatible microparticle downregulated the local inflammation in IVD efficiently by preserving ECGC activity and controlling the ECGC release [149].

7 Drug Delivery Macrosystems

Macroscopic biomaterials appear in many forms, such as ceramic, glass and polymer and range from millimetres to centimetres [150]. Macrosystem-based biomaterials aim to imitate the composition of native tissue and organ in the biological system [151]. Nevertheless, the mimicking effect will help to replenish moisture, stabilise disc height and improve extracellular matrix composition (ECM) in intervertebral disc disease (IVD) [152]. In IVD, the degenerated tissue causes inflammation and an acidic pH environment, causing ineffective treatment by drug alone. This is because the drug's efficacy is sensitive to environmental factors such as moisture, temperature and pH [153]. Macrosystem-based biomaterials incorporated with drugs protect and ensure the gradual release of drugs in the affected area. With the application, the efficacy of drug delivery can be improved by preventing re-injection of the drug over time and preventing toxicity due to high doses of the drug [154]. Macrosystem-based polymeric biomaterials are degradable and can be fabricated by natural or synthetic material [155]. Some tissue engineering applications might favour different polymeric biomaterial forms such as sponge, hydrogel and film.

7.1 Polymeric Hydrogels

In the IVD application, the hydrogel is the main application as it involves less invasive procedures than other polymeric biomaterials. Therefore, this chapter will discuss hydrogel application in IVD. Hydrogel is a three-dimensional (3D) polymeric biomaterial with high water content and can retain structure while absorbing water [156]. It can be prepared by monomers, prepolymers or exiting hydrophilic polymers [157]. Due to its characteristics, it provides a suitable environment for cell growth and is extensively used for cell immobilisation and drug delivery [158, 159]. Hydrogel is also applied in IVD to mimic NP tissue, as they have comparable mechanical characteristics [35]. To date, there are many types of crosslinking in forming good hydrogel properties. Interpenetrating networks (IPNs), stimuli-sensitive hydrogel, self-assemble hydrogel, physical crosslinking, chemical crosslinking and sliding crosslinking are among the common crosslinking techniques in hydrogel [156, 160, 161] (Table 4.2).

Polymeric hydrogel for IVD treatment needs to fulfil several requirements to suit the treatment. It has to be biocompatible, biodegradable and in situ curing to prevent leakage, have strong mechanical properties and sustain drug release over time [167]. Therefore, selecting a suitable crosslinking technique is vital to provide appropriate characteristics as injectable hydrogel in IVD. Researchers have frequently studied IPNs as they involve a combination of two polymeric biomaterials but possess entanglement with each other [168, 169]. It can improve the mechanical strength of hydrogel while retaining its biocompatibility and enriching the ECM composition

Table 4.2 Summary of the type of crosslinking technique and its characteristics for tissue engineering applications

Types of crosslinking techniques	Crosslinking properties	Characteristics
Interpenetrating networks	The interlinked molecules from two or more polymers' network form crosslinking effect [162].	It has multiple components which have a higher mimicking effect of the native ECM. It can enhance mechanical properties without adding crosslinker [163].
Stimuli-responsive hydrogel	The polymers are coated with the functional group, which is sensitive to environment factors such as temperature, light, magnetic field, electric field, pH, chemical and biological trigger [164].	It is widely used for controllable gelation time, degradation and flexible mechanical properties suiting physiological properties.
Self-assemble hydrogel	The crosslinking between polymers forms via non-covalent interaction such as hydrogen bonding, hydrophobic interaction, aromatic stacking and electrostatic interaction [165].	It is suitable for cell delivery. It maintains cell viability during injection by protecting the cell from mechanical strength.
Physical crosslinked	Crosslinking between polymers that involve the interaction of hydrophobic, electrostatic and hydrogen bonds [160].	The hydrogel is mostly reversible and less cytotoxic.
Chemical crosslinked	Crosslinking between polymers that involve chemical modification [160].	The mechanical properties of hydrogel can be enhanced and allow water absorption without dissolution. Drug release can be prolonged by diffusion or degradation.
Sliding crosslinked	The slidable structure between polymers [166].	This type of hydrogel can withstand higher stretch and compress, causing more elastic hydrogel.

of the hydrogel [170]. Stimuli-responsive hydrogels are widely investigated in IVD as the hydrogel can be engineered to respond to the specific environment in IVD (pH, reactive oxygen species and temperature) [164]. Self-assembly hydrogel is suitable for cell delivery as it possesses shear-thinning properties and can protect cells from damage during injection [165]. A future study is suggested to investigate suitable polymeric hydrogel through sliding crosslinking for IVD treatment as it owns a slidable structure, making the hydrogel have a higher flexibility in withstanding higher stretch and compression [166]. Physically crosslinked hydrogel might not fit as injectable hydrogel in IVD as it has low mechanical properties. In contrast, chemically crosslinked hydrogel requires a strong or high dose of crosslinkers to form satisfying gelation time and mechanical strength, which is inappropriate in IVD.

Selecting the proper type of hydrogel is also crucial because the intervertebral disc is rich with different types of ECM. Water, proteoglycans, collagen type II and elastin are the main ECM composition in IVD. Other minority ECM compositions such as collagen III, VI, IX and laminins also exist in forming suitable microenvironments [171]. IVD degeneration is associated with structural changes in ECM, such as reduced water content and collagen type III [172]. Moreover, it is also summarised that proteoglycan content decreases while collagen type I increases in a degenerated intervertebral disc [173], leading to increased MMP expression, which causes inflammation in the specific site. Continuous from that, the degenerated intervertebral disc is hardly regenerated as the specific site is commonly acidic and lacks nutrient and growth factors [174, 175]. This study suggested that hydrogel derived from laminin-containing ECM should be applied in IVD tissue engineering because laminin-containing ECM is important to regulate healthy, immature NP cells [171]. There are mainly two types of hydrogels, names derived from synthetic or natural polymers. Natural hydrogel is favoured in tissue engineering because it is non-cytotoxic as it makes up ECM. Collagen, elastin, fibrin, gelatine, silk fibroin, glycosaminoglycans, hyaluronic acid (HA), alginate and chitosan are among the components [176]. Implantation of HA hydrogel has been demonstrated to alter glycosylation, promote ECM repair and prevent pain phenotype in the disc injury model of rat tail [177]. However, to maintain the longevity of hydrogel in the targeted tissue, synthetic hydrogel is applied to enhance the mechanical properties [178].

Synthetic hydrogel (a mixture of natural and synthetic polymers) is commonly developed to obtain slower biodegradation and controlled release of drugs [179]. Drug release by hydrogel can undergo different pathways, such as diffusion, electrostatic interaction, hydrophobic association, cleavable covalent conjugation, degradation and hydrophobic association [180]. For instance, polysaccharide-based hydrogel applies physical entrapment and covalent bonding to encapsulate the drug [181]. The interaction between the drug and different hydrogel types had distinct outcomes. The previous study showed that protein-based hydrogel (sodium caseinate) had better niacin encapsulation efficacy than polysaccharide sodium carboxymethyl cellulose [182]. Moreover, the native tissue in IVD is mostly negatively charged due to the glycosaminoglycan and proteoglycan structures [183]. This structure reduces the diffusion efficiency of the drug. Thus, positive charge hydrogel can be applied as a drug delivery vehicle [184]. Therefore, different types of drugs should be tailored with different types of hydrogels, depending on their chemical properties, to ensure gradual drug release in IVD [180].

7.2 *Injectable Devices*

For severe IVD, conventional intervention surgery such as lumbar fusion and total disc replacement is performed. This procedure yields high risk for the patient, requires highly skilled surgeons and is time-consuming [185, 186]. Moreover, artificial material replacement in the NP resulted in the migration of foreign materials and incompatibility of the native tissue [187]. Continuous from that, the injectable hydrogel is favoured in IVD as it can improve the condition of IVD in goat models for up to 12 weeks because the hydrogel enriched native ECM and provided mechanical support [188].

Rapid gelation time is one of the most important characteristics of injectable hydrogel because it can prevent leakage after injection. The HA hydrogel has been shown to increase disc height and hydration and histological evidence to promote tissue repair in the lumbar disc of the rabbit model [189]. The stimuli-sensitive hydrogel enables hydrogel to form gelation faster. The thermo-responsive hydrogel can ensure the gelling of hydrogel within a specific temperature, specifically under physiological temperature. The thermo-responsive hydrogel can be prepared by blending the polymers with other materials. For instance, chitosan can be added with glycerol-phosphate, poly(ethylene glycol), poly(vinyl alcohol) or hexanoyl anhydride to achieve thermo-sensitive hydrogel [190]. This study proved that N-hexanoylation of glycol chitosan could achieve stability at 37 °C after injection in vitro and in vivo rat model [190]. Besides, photo-sensitive poly(ethylene glycol) dimethacrylate nanofibrillated cellulose hydrogel formed quick gelation using transdermal illumination in an ex vivo bovine IVD organ [191]. Chemically crosslinked hydrogel such as genipin-crosslinked fibrin hydrogel was proven effective in acting as filler in AF in the IVD bovine ex vivo model [192]. Chemically crosslinked hydrogel possesses a disadvantage in that the crosslinkers might cause cytotoxicity at a threshold of concentration [193]. Therefore, an interpenetrating network (IPN) could be the substitute to enhance the mechanical properties of hydrogel in the absence of crosslinkers. Fibronectin-conjugated fibrin and poly (ethylene glycol) diacrylate (PEGDA) is an IPN that effectively repaired disc herniation in ex vivo model [194]. An in vitro study showed that alginate with calcium carbonate and glucono-δ-lactone can form gelation without injection site leakage and can be applied as an NP substitute in an ex vivo IVD model [195].

Each hydrogel type caters to different needs, such as stimuli-responsive yield, faster gelation time and prevention leakage during injection [196]. At the same time, chemically crosslinked hydrogel yields slower biodegradation but higher cytotoxicity [193]. Moreover, natural hydrogel possesses biocompatibility but relatively low mechanical strength than synthetic hydrogel. On the other side, synthetic hydrogel might restrain cellular interaction as it lacks cell detection structure [167]. With that, researchers need to select the most suitable crosslinking mechanism and the type of biomaterials to achieve the most optimum characteristic of the injectable hydrogel as drug delivery in IVD. A clinical trial on hydrogel injection into IVD was performed through a discography procedure (ClinicalTrial.gov: NCT04727385). The procedure involved the injection of a fluorescence probe into the spine and reflected in an x-ray image [197]. This procedure increases UV exposure to the patient and surgeon, thus imposing the risk of radiation [198].

7.3 Controlled-Release Implants

The hydrogel can be administrated into IVD through injection and surgical incision. For surgical incisions, the hydrogel is prepared earlier according to the native tissue, and this fabrication method can produce a higher mechanical hydrogel [199]. However, this route of implantation requires a more invasive procedure causing more pain and time-consuming [152].

Multiple types of polymeric biomaterial have been studied, but limited is undergoing clinical trials, specifically HA (ClinicalTrials.gov: NCT04688281) and amniotic membrane (ClinicalTrials.gov: NCT04688281). Interestingly, Medtronic introduced an implanted bone graft made with a bovine collagen sponge incorporated with recombinant human bone morphogenetic protein-2 (rhBMP-2). This scaffold aims to provide a chemotaxis effect on the native cells, thus promoting proliferation by native cells and differentiation into osteoblasts (patent number: US8690957B2) [200]. Osteogenic proteins (OP) were delivered by implanted biomaterials such as hydroxyapatite, collagen and carboxymethylcellulose (patent number: US5916870A).

The hydrogel can accommodate many therapeutic molecules such as proteins, nucleic acids and small biomolecules [201]. The drug can be released from hydrogel via several mechanisms such as diffusion-controlled, swelling-controlled and chemical-controlled release [202]. Diffusion-controlled release occurs when the hydrogel pore size is larger than the drug size. The drug can diffuse out through the porosity of the hydrogel [203]. In swelling-controlled release hydrogel, the ability of water uptake by hydrogel is proportional to the drug release profile [204, 205]. Lastly, chemically controlled release involved drug release via enzymatic or hydrolytic reaction on the hydrogel [206, 207]. For instance, by modifying genipin concentration in fibrin hydrogel, the release of TGF-β3 can be controlled [193]. Recent studies focused on developing drug-encapsulated hydrogel with the regenerating effect and immunomodulation on IVD in the preliminary stage [208]. There are lacking clinical data for injectable hydrogel in IVD [209].

Quantities of study applied stimuli-sensitive hydrogel in delivering the anti-inflammatory drug. For instance, this study targeted to reduce the inflammation of IVD mice by injecting gefitinib-encapsulated thermo-sensitive hyaluronic acid hydrogel [40]. The hydrogel can be tailored based on the characteristics of the specific site. As such, this study developed an inflammation-responsive gelatine hydrogel to deliver curcumin and cholesterol-modified miRNA-21 inhibitor as an anti-inflammatory drug in IVD [210]. Celecoxib also acted as an anti-inflammatory drug and was delivered via thermo-sensitive N-isopropylacrylamide hydrogel that was able to reduce prostaglandin E_2 (PGE_2) in an inflamed intervertebral disc of mice model [211]. Direct injection of celecoxib might result in a shorter release profile and side effects due to drug-drug incompatibility [196]. Interestingly, this study inhibited Src homology-2 domain-containing protein tyrosine phosphatase-2 (SHP2) expression in IVD sample by applying thermo-sensitive N-isopropylacrylamide hydrogel incorporated with SHP099 (inhibitor of SHP2) and the regeneration of NP in mice was observed [212]. Thalidomide and ruxolitinib are also applied to regulate inflammation in IVD and were able to achieve a sustained drug profile by using chitosan with pluronic F127 hydrogel encapsulation [213]. Chen and co-researchers suggested semi-synthetic hydrogel, composed of hyaluronic acid and poly(amidoamine), to deliver siRNA to regulate the STING pathway in IVD rats. siRNA is susceptible to endogenous enzymes, and its anionic properties restrain it from entering the plasma membrane. While delivering through the hydrogel, the successful rate of delivery increased and reduced cost [214]. Uncommonly, reactive oxygen species (ROS)-sensitive hydrogel is developed by Bai and co-researchers to incorporate with rapamycin and effectively increase M2-like macrophage in IVD rat [215]. Besides stimuli-sensitive hydrogel, the chemical crosslinked hydrogel can also be utilised for anti-inflammatory drug delivery. In particular, genipin-crosslinked fibrin with collagen type I was developed to encapsulate infliximab (anti-TNF-α). Chemical crosslinks enhance the biodegradation profile of the hydrogel and sustain its anti-inflammatory properties for up to 20 days in vitro [216].

Other than the anti-inflammation effect of the injectable hydrogel in IVD, other studies also focused on regenerating native tissue in IVD. Decellularised ECM is applied as hydrogel to deliver adipose stem cell (ASC)-derived exosome and improve rats' degenerated NP structure [217]. The scaffold protects for the exosome to retain for a more extended period, and the hydrogel has similar characteristics to native tissue. Bone morphogenetic proteins 2 (BMP-2) and 7 (BMP-7) were encapsulated in fibrin/hyaluronic acid hydrogel in the IVD goat model, but no significant result was observed [218]. However, another study applying BMP-2 and BMP-7 in fibrin-hyaluronan hydrogel showed increased aggrecan, collagen type II and glycosaminoglycan production by NP cells [219]. More studies will need to prove the inconsistency of study outcomes to investigate the suitable dose and coupling effect of the growth factors. For instance, a hydrogel with BMP-2 and simvastatin was injected into a rat model and showed improved aggrecan and collagen type II regulation. Simvastatin was coupled to enhance the expression of the BMP-2 [220]. Injectable graphene oxide incorporating transforming growth factor beta-3 and anabolic growth factor can promote aggrecan and collagen type II expression by gradually releasing growth factor in bovine NP cells [221]. Insulin-like growth factor-1 (IGF-1) and BMP-2 encapsulated in polyvinyl alcohol (PVA) hydrogel can promote chondroadherin and hypoxia-inducible factor-1α in annulus fibrosus cell and nucleus pulposus cell in vitro [222]. In an in vivo setting, thiol-modified hyaluronic acid hydrogel loaded with PDGF-BB could reduce further degeneration of IVD in rabbit models for up to 8 weeks [223].

In short, immunomodulator biomolecules and growth factors are among the most studied to be encapsulated in hydrogel for IVD. The hydrogel can sustain the release of biomolecules and provide a positive result in a preliminary study. However, gene therapy in IVD through hydrogel should be conducted more extensively as some gene expressions, such as circ-FAM169S and circ-4099, were proven to be high in NP tissue of the IVD model, contributing to inflammation.

8 Conclusion and Future Perspective

The healthy IVD is rich in ECM components with a low cell number in the AF and NP. Changes in shape and size, loss of water, degradation of ECM and loss of cells are the hallmarks of IVD degeneration. The current surgical and conservation treatment does not reverse the disease process; hence, novel therapies have been investigated to delay or reverse IVD degeneration. The conservative treatment mainly involves pharmaceutical drugs as analgesics to reduce low back pain. Most importantly, the choice of drug is an important consideration for developing efficient drug-based therapeutics in IVD. Small molecules targeting NF-kB and p38/MAPK signalling pathways have been shown to help modulate inflammatory responses and prevent disease progression. Alternatively, the delivery of notable growth factors can control cell proliferation, differentiation and ECM deposition. It is also reported that delivery of therapeutic proteins such as growth factors and ECM-mimicking peptides may be introduced directly into the discs to promote anabolic responses or cell development that could stop the degeneration of the disks. Other therapeutic protein molecules like antibodies and cytokines have been shown to provide symptomatic relief by suppressing and blocking the activity of biomolecules and cell surface receptors. The higher activity of MMPs and ADAMTS was found to be involved in IVD degeneration. Thus, appropriate enzyme inhibitors are widely used to treat inflammation and pain in the lower back caused by IVD inflammation or degeneration. The potential of gene therapy has also been investigated for reversing the degeneration process of IVD. However, the delivery of all the molecules mentioned earlier, and biomolecules, has been a primary concern. Among the classical delivery platforms, the polymeric particles are the most attractive platform for the delivery molecules concerning achieving the required bioavailability with targeted and sustained delivery. These polymeric nanoparticle platforms were suitable for drugs, growth factors, peptides and other biologically relevant molecules.

Similarly, a liposome-based drug delivery platform can deliver hydrophobic and hydrophilic drugs with excellent control of the release profile. Moreover, liposome-based delivery platforms have been investigated for gene therapy applications in IVD. Recently, a multivalent dendrimer-based platform showed better internalisation than other conventional platforms. The PAMAM-based delivery platform showed excellent results in the delivery of drugs, biomolecules and genes. Overall, polymeric nanoparticles, liposomes or micelles demonstrated favourable outcomes in vitro and preclinical models of IVD. It was also observed that microparticles could be retained in IVD more extended than nanoparticles. PCL, PLLA, PLGA, PLA, PEA, alginate and chitosan are widely explored polymers for developing microparticle-based delivery systems to treat IVD degeneration. Other drug delivery platforms, such as nanofibres and hydrogels, are gaining popularity for the controlled release of drugs and biomolecules and their ability to act as drug reservoir systems and support cell growth and proliferation. There are multifactorial causes for IVD degeneration, which may require multidrug delivery approaches for the treatment of IVD degeneration. It is well understood that targeting more than one pathophysiological pathway could provide a more effective solution for treating diseases, which supports the idea of developing multidrug delivery platforms. However, a careful selection of delivery platforms is required for their cytotoxicity and pro-inflammatory properties.

In the future, studies may also focus on developing intelligent drug delivery and reservoir systems capable of delivering more than one type of drug in a dose-dependent manner.

Funding We received funding from the Health Research Board, Ireland for the Emerging Investigator Award for Health 2022 (grant number EIA-2022-010), Ministry of Higher Education, Malaysia for the Fundamental Research Grant Scheme (grant number FRGS/1/2022/SKK10/UKM/02/12) and Universiti Kebangsaan Malaysia for the Geran Galakan Penyelidik Muda (grant number GGPM-2022-029).

References

1. B.-G. Peng, "Pathophysiology, Diagnosis, and Treatment of Discogenic Low Back Pain," *World J. Orthop.*, vol. 4, no. 2, pp. 42–52, 2013, https://doi.org/10.5312/wjo.v4.i2.42.
2. K. L. Moore, A. F. Dalley I, and A. M. R. Agur, *Clinially Oriented Anatomy*. 2014.
3. I. M. Shapiro and M. V. Risbud, "Introduction to the Structure, Function, and Comparative Anatomy of the Vertebrae and the Intervertebral Disc," in *The Intervertebral Disc*, 2014, pp. 3–429.
4. D. Sakai *et al.*, "Exhaustion of Nucleus Pulposus Progenitor Cells with Ageing and Degeneration of the Intervertebral Disc," *Nat. Commun.*, vol. 3, no. May, p. 1264, 2012, https://doi.org/10.1038/ncomms2226.
5. P. Bermudez-Lekerika *et al.*, "Immuno-Modulatory Effects of Intervertebral Disc Cells," *Front. Cell Dev. Biol.*, vol. 10, no. June, pp. 1–32, 2022, https://doi.org/10.3389/fcell.2022.924692.
6. F. J. Lyu *et al.*, "Painful Intervertebral Disc Degeneration and Inflammation: From Laboratory Evidence to Clinical Interventions," *Bone Res.*, vol. 9, no. 1, 2021, https://doi.org/10.1038/s41413-020-00125-x.
7. I. L. Mohd Isa, S. A. Mokhtar, S. A. Abbah, M. B. Fauzi, A. Devitt, and A. Pandit, "Intervertebral Disc Degeneration: Biomaterials and Tissue Engineering Strategies Towards Precision Medicine," *Adv. Healthc. Mater.*, vol. 2102530, 2022, https://doi.org/10.1002/adhm.202102530.
8. S. Grassel and A. Aszodi, *Cartilage Pathophysiology*, vol. 2. 2017.
9. C. Liu, L. Xiao, Y. Zhang, Q. Zhao, and H. Xu, "Regeneration of annulus fibrosus tissue using a DAFM/PECUU-blended electrospun scaffold," *J. Biomater. Sci. Polym. Ed.*, vol. 31, no. 18, pp. 2347–2361, 2020, https://doi.org/10.1080/09205063.2020.1812038.
10. R. P. Prithvi, "Intervertebral Disc: Anatomy-Physiology-Pathophysiology-Treatment," *Pain Pract.*, vol. 8, no. 1, pp. 18–44, 2008, https://doi.org/10.1111/j.1533-2500.2007.00171.x.
11. P. J. Roughley, "Biology of Intervertebral Disc Aging and Degeneration Involvement of the Extracellular Matrix," *Spine (Phila. Pa. 1976).*, vol. 29, no. 23, pp. 2691–2699, 2004.
12. J. D. Placzek and D. A. Boyce, *Orthopaedic Physical Therapy Secrets – E-Book*. 2016.
13. C. A. Séguin, D. Chan, C. L. Dahia, and Z. Gazit, "Latest Advances in Intervertebral Disc Development and Progenitor Cells," *JOR Spine*, vol. 1, no. 3, pp. 1–11, 2018, https://doi.org/10.1002/jsp2.1030.
14. Y. Wu, J. Loaiza, R. Banerji, O. Blouin, and E. Morgan, "Structure-function relationships of the human vertebral endplate," *JOR Spine*, vol. 4, no. 3, pp. 1–13, 2021, https://doi.org/10.1002/jsp2.1170.

15. D. R. Eyre and H. Muir, "Quantitative analysis of types I and II collagens in human intervertebral discs at various ages," *BBA – Protein Struct.*, vol. 492, no. 1, pp. 29–42, 1977, https://doi.org/10.1016/0005-2795(77)90211-2.
16. Q. Wei, X. Zhang, C. Zhou, Q. Ren, and Y. Zhang, "Roles of large aggregating proteoglycans in human intervertebral disc degeneration," *Connect. Tissue Res.*, vol. 60, no. 3, pp. 209–218, 2019, https://doi.org/10.1080/03008207.2018.1499731.
17. J. Xin, Y. Wang, Z. Zheng, S. Wang, S. Na, and S. Zhang, "Treatment of Intervertebral Disc Degeneration," *Orthop. Surg.*, vol. 14, no. 7, pp. 1271–1280, 2022, https://doi.org/10.1111/os.13254.
18. S. Elmasry, S. Asfour, J. Pablo, D. R. Vaccari, and F. Travascio, "Effects of Tobacco Smoking on the Degeneration of the Intervertebral Disc : A Finite Element Study," pp. 1–22, 2015, https://doi.org/10.1371/journal.pone.0136137.
19. C. L. Le Maitre, A. J. Freemont, and J. A. Hoyland, "Accelerated Cellular Senescence in Degenerate Intervertebral Discs: A Possible Role in the Pathogenesis of Intervertebral Disc Degeneration," *Arthritis Res. Ther.*, vol. 9, no. 3, pp. 1–12, 2007, https://doi.org/10.1186/ar2198.
20. K. Joyce, I. L. Mohd Isa, A. Krouwels, L. B. Creemers, A. Devitt, and A. Pandit, "The Role of Altered Glycosylation in Human Nucleus Pulposus Cells in Inflammation and Degeneration," *Eur. Cells Mater.*, vol. 40, pp. 401–420, 2021, https://doi.org/10.22203/eCM.v041a26.
21. H. Morris, C. F. Gonçalves, M. Dudek, J. Hoyland, and Q. J. Meng, "Tissue Physiology Revolving Around the Clock: Circadian Rhythms as Exemplified by the Intervertebral Disc," *Ann. Rheum. Dis.*, vol. 80, no. 7, pp. 828–839, 2021, https://doi.org/10.1136/annrheumdis-2020-219515.
22. S. E. Navone *et al.*, "Inflammatory Mediators and Signalling Pathways Controlling Intervertebral Disc Degeneration," *Histol. Histopathol.*, vol. 32, no. 6, pp. 523–542, 2017, https://doi.org/10.14670/HH-11-846.
23. Y. Wang, M. Che, J. Xin, Z. Zheng, J. Li, and S. Zhang, "The Role of IL-1β and TNF-α in Intervertebral Disc Degeneration," *Biomed. Pharmacother.*, vol. 131, 2020, https://doi.org/10.1016/j.biopha.2020.110660.
24. X. B. Zhang *et al.*, "Targeted therapy for intervertebral disc degeneration: Inhibiting apoptosis is a promising treatment strategy," *Int. J. Med. Sci.*, vol. 18, no. 13, pp. 2799–2813, 2021, https://doi.org/10.7150/ijms.59171.
25. M. Stefanakis *et al.*, "Annulus Fissures are Mechanically and Chemically Conducive to the Ingrowth of Nerves and Blood Vessels," *Spine (Phila. Pa. 1976).*, vol. 37, no. 22, pp. 1883–1891, 2012, https://doi.org/10.1097/BRS.0b013e318263ba59.
26. N. Vaudreuil *et al.*, "乳鼠心肌提取 HHS Public Access," *Physiol. Behav.*, vol. 176, no. 1, pp. 139–148, 2016, https://doi.org/10.1016/j.spinee.2017.04.012.NSAID.
27. M. E. Hale, C. Dvergsten, and J. Gimbel, "Efficacy and safety of oxymorphone extended release in chronic low back pain: Results of a randomized, double-blind, placebo- and active-controlled phase III study," *J. Pain*, vol. 6, no. 1, pp. 21–28, 2005, https://doi.org/10.1016/j.jpain.2004.09.005.
28. M. W. van Tulder, T. Touray, A. D. Furlan, S. Solway, and L. M. Bouter, "Muscle relaxants for non-specific low-back pain," *Cochrane Database Syst. Rev.*, vol. 2017, no. 3, 2003, https://doi.org/10.1002/14651858.CD004252.
29. D. Brötz *et al.*, "Is there a role for benzodiazepines in the management of lumbar disc prolapse with acute sciatica?," *Pain*, vol. 149, no. 3, pp. 470–475, 2010, https://doi.org/10.1016/j.pain.2010.02.015.
30. O. Buyukturan, B. Buyukturan, S. Sas, C. Kararti, and I. Ceylan, "The effect of mulligan mobilization technique in older adults with neck pain: A randomized controlled, double-blind study," *Pain Res. Manag.*, vol. 2018, 2018, https://doi.org/10.1155/2018/2856375.
31. F. J. R. França *et al.*, "Motor Control Training Compared with Transcutaneous Electrical Nerve Stimulation in Patients with Disc Herniation with Associated Radiculopathy,"

Am. J. Phys. Med. Rehabil., vol. 98, no. 3, pp. 207–214, 2019, https://doi.org/10.1097/PHM.0000000000001048.
32. L. A. V. Ramos *et al.*, "Comparison Between Transcutaneous Electrical Nerve Stimulation and Stabilization Exercises in Fatigue and Transversus Abdominis Activation in Patients With Lumbar Disk Herniation: A Randomized Study," *J. Manipulative Physiol. Ther.*, vol. 41, no. 4, pp. 323–331, 2018, https://doi.org/10.1016/j.jmpt.2017.10.010.
33. F. Taher *et al.*, "Lumbar degenerative disc disease: current and future concepts of diagnosis and management," *Adv. Orthop.*, vol. 2012, p. 970752, Jan. 2012, https://doi.org/10.1155/2012/970752.
34. O. Rabau *et al.*, "Lateral Lumbar Interbody Fusion (LLIF): An Update," *Glob. Spine J.*, vol. 10, no. 2_suppl, pp. 17S–21S, 2020, https://doi.org/10.1177/2192568220910707.
35. S. B. G. Blanquer, D. W. Grijpma, and A. A. Poot, "Delivery Systems for the Treatment of Degenerated Intervertebral Discs," *Adv. Drug Deliv. Rev.*, vol. 84, pp. 172–187, 2015, https://doi.org/10.1016/j.addr.2014.10.024.
36. E. J. Roh *et al.*, "Genetic Therapy for Intervertebral Disc Degeneration," *International Journal of Molecular Sciences*, vol. 22, no. 4. pp. 1–14, 2021, https://doi.org/10.3390/ijms22041579.
37. W. J. Sandborn, "New targets for small molecules in inflammatory bowel disease," *Gastroenterol. Hepatol.*, vol. 11, no. 5, pp. 338–340, 2015.
38. M. Hojjat-Farsangi, "Small-molecule inhibitors of the receptor tyrosine kinases: Promising tools for targeted cancer therapies," *Int. J. Mol. Sci.*, vol. 15, no. 8, pp. 13768–13801, 2014, https://doi.org/10.3390/ijms150813768.
39. M. Molinos, C. R. Almeida, J. Caldeira, C. Cunha, R. M. Gonçalves, and M. A. Barbosa, "Inflammation in intervertebral disc degeneration and regeneration," *J. R. Soc. Interface*, vol. 12, no. 104, 2015, https://doi.org/10.1098/rsif.2014.1191.
40. Z. Pan *et al.*, "Therapeutic effects of gefitinib-encapsulated thermosensitive injectable hydrogel in intervertebral disc degeneration," *Biomaterials*, vol. 160, pp. 56–68, 2018, https://doi.org/10.1016/J.BIOMATERIALS.2018.01.016.
41. J. A. Efe and S. Ding, "The evolving biology of small molecules: Controlling cell fate and identity," *Philos. Trans. R. Soc. B Biol. Sci.*, vol. 366, no. 1575, pp. 2208–2221, 2011, https://doi.org/10.1098/rstb.2011.0006.
42. L. Cao, X. Chen, X. Xiao, Q. Ma, and W. Li, "Resveratrol inhibits hyperglycemia-driven ROS-induced invasion and migration of pancreatic cancer cells via suppression of the ERK and p38 MAPK signaling pathways," *Int. J. Oncol.*, vol. 49, no. 2, pp. 735–743, 2016, https://doi.org/10.3892/ijo.2016.3559.
43. W. Hua *et al.*, "Icariin Attenuates Interleukin-1β-Induced Inflammatory Response in Human Nucleus Pulposus Cells," *Curr. Pharm. Des.*, vol. 23, no. 39, pp. 6071–6078, 2017, https://doi.org/10.2174/1381612823666170615112158.
44. J. Zhu *et al.*, "Kaempferol slows intervertebral disc degeneration by modifying LPS-induced osteogenesis/adipogenesis imbalance and inflammation response in BMSCs," *Int. Immunopharmacol.*, vol. 43, pp. 236–242, 2017, https://doi.org/10.1016/j.intimp.2016.12.020.
45. L. Lu *et al.*, "Berberine prevents human nucleus pulposus cells from IL-1β-induced extracellular matrix degradation and apoptosis by inhibiting the NF-κB pathway," *Int. J. Mol. Med.*, vol. 43, no. 4, pp. 1679–1686, 2019, https://doi.org/10.3892/ijmm.2019.4105.
46. H. Liu *et al.*, "Urolithin a inhibits the catabolic effect of TNFα on nucleus pulposus cell and alleviates intervertebral disc degeneration in vivo," *Front. Pharmacol.*, vol. 9, no. SEP, pp. 1–11, 2018, https://doi.org/10.3389/fphar.2018.01043.
47. H. Li, C. Liang, Q. Chen, and Z. Yang, "Rhein: A potential biological therapeutic drug for intervertebral disc degeneration," *Med. Hypotheses*, vol. 77, no. 6, pp. 1105–1107, 2011, https://doi.org/10.1016/j.mehy.2011.09.013.
48. K. Masuda, T. R. Oegema, and H. S. An, "Growth factors and treatment of intervertebral disc degeneration.," *Spine (Phila. Pa. 1976).*, vol. 29, no. 23, pp. 2757–69, Dec. 2004, [Online]. Available: http://www.ncbi.nlm.nih.gov/pubmed/15564925.
49. S. T. Yoon, "Molecular therapy of the intervertebral disc," *Spine J.*, vol. 5, no. 6 SUPPL., pp. S280–S286, 2005, https://doi.org/10.1016/j.spinee.2005.02.017.

50. C. J. Wallach *et al.*, "Gene transfer of the catabolic inhibitor TIMP-1 increases mesured proteoglycans in cells from degenerated human intervertebral discs," *Spine (Phila. Pa. 1976).*, vol. 28, no. 20, pp. 2331–2337, 2003, https://doi.org/10.1097/01.BRS.0000085303.67942.94.
51. Y. Zhang, H. S. An, E. J. M. A. Thonar, S. Chubinskaya, T. C. He, and F. M. Phillips, "Comparative effects of bone morphogenetic proteins and Sox9 overexpression on extracellular matrix metabolism of bovine nucleus pulposus cells," *Spine (Phila. Pa. 1976).*, vol. 31, no. 19, pp. 2173–2179, 2006, https://doi.org/10.1097/01.brs.0000232792.66632.d8.
52. S. T. Yoon *et al.*, "The effect of bone morphogenetic protein-2 on rat intervertebral disc cells in vitro," *Spine (Phila. Pa. 1976).*, vol. 28, no. 16, pp. 1773–1780, 2003, https://doi.org/10.1097/01.BRS.0000083204.44190.34.
53. Y. Zhang, A. Chee, E. J.-M. a Thonar, and H. S. An, "Intervertebral disk repair by protein, gene, or cell injection: a framework for rehabilitation-focused biologics in the spine.," *PM R*, vol. 3, no. 6 Suppl 1, pp. S88–94, Jun. 2011, https://doi.org/10.1016/j.pmrj.2011.04.020.
54. N. E. Lane *et al.*, "Tanezumab for the Treatment of Pain from Osteoarthritis of the Knee," *N. Engl. J. Med.*, vol. 363, no. 16, pp. 1521–1531, 2010, https://doi.org/10.1056/nejmoa0901510.
55. R. A. Deyo, S. K. Mirza, and B. I. Martin, "Back pain prevalence and visit rates: Estimates from U.S. national surveys, 2002," *Spine (Phila. Pa. 1976).*, vol. 31, no. 23, pp. 2724–2727, 2006, https://doi.org/10.1097/01.brs.0000244618.06877.cd.
56. M. Zorbas, S. Hurst, D. Shelton, M. Evans, D. Finco, and M. Butt, "A multiple-dose toxicity study of tanezumab in cynomolgus monkeys," *Regul. Toxicol. Pharmacol.*, vol. 59, no. 2, pp. 334–342, 2011, https://doi.org/10.1016/j.yrtph.2010.11.005.
57. T. J. Schnitzer, N. E. Lane, C. Birbara, M. D. Smith, S. L. Simpson, and M. T. Brown, "Long-term open-label study of tanezumab for moderate to severe osteoarthritic knee pain," *Osteoarthr. Cartil.*, vol. 19, no. 6, pp. 639–646, 2011, https://doi.org/10.1016/j.joca.2011.01.009.
58. C. Birbara *et al.*, "Safety and efficacy of subcutaneous tanezumab in patients with knee or hip osteoarthritis," *J. Pain Res.*, vol. 11, pp. 151–164, 2018, https://doi.org/10.2147/JPR.S135257.
59. D. Aitken *et al.*, "A randomised double-blind placebo-controlled crossover trial of HUMira (adalimumab) for erosive hand OsteoaRthritis – the HUMOR trial," *Osteoarthr. Cartil.*, vol. 26, no. 7, pp. 880–887, 2018, https://doi.org/10.1016/j.joca.2018.02.899.
60. M. V Risbud and I. M. Shapiro, "Role of Cytokines in Intervertebral Disc Degeneration: Pain and Disc-content," *Nat Rev Rheumatol.*, vol. 10, no. 1, pp. 44–56, 2016, https://doi.org/10.1038/nrrheum.2013.160.Role.
61. S. B. Cohen *et al.*, "A randomized, double-blind study of AMG 108 (a fully human monoclonal antibody to IL-1R1) in patients with osteoarthritis of the knee," *Arthritis Res. Ther.*, vol. 13, no. 4, p. R125, 2011, https://doi.org/10.1186/ar3430.
62. S. Genevay, S. Stingelin, and C. Gabay, "Efficacy of etanercept in the treatment of acute, severe sciatica: A pilot study," *Ann. Rheum. Dis.*, vol. 63, no. 9, pp. 1120–1123, 2004, https://doi.org/10.1136/ard.2003.016451.
63. I. J. Dahabreh, M. Chung, E. M. Balk, W. W. Yu, and P. Mathew, "Annals of Internal Medicine Active Surveillance in Men With Localized Prostate Cancer," *Ann. Intern. Med.*, 2011.
64. T. Okoro, S. I. Tafazal, S. Longworth, and P. J. Sell, "Tumor necrosis α-blocking agent (Etanercept): A triple blind randomized controlled trial of its use in treatment of sciatica," *J. Spinal Disord. Tech.*, vol. 23, no. 1, pp. 74–77, 2010, https://doi.org/10.1097/BSD.0b013e31819afdc4.
65. X. Chevalier *et al.*, "Intraarticular injection of anakinra in osteoarthritis of the knee: A multicenter, randomized, double-blind, placebo-controlled study," *Arthritis Care Res.*, vol. 61, no. 3, pp. 344–352, 2009, https://doi.org/10.1002/art.24096.
66. C. L. Le Maitre, A. Pockert, D. J. Buttle, A. J. Freemont, and J. A. Hoyland, "Matrix synthesis and degradation in human intervertebral disc degeneration.," *Biochem. Soc. Trans.*, vol. 35, no. Part 4, pp. 652–5, Aug. 2007, https://doi.org/10.1042/BST0350652.

67. S. T. Yoon and N. M. Patel, "Molecular therapy of the intervertebral disc," *Eur. Spine J.*, vol. 15, no. SUPPL. 3, 2006, https://doi.org/10.1007/s00586-006-0155-3.
68. K. Masuda, "Biological repair of the degenerated intervertebral disc by the injection of growth factors," *Eur. Spine J.*, vol. 17, no. SUPPL. 4, 2008, https://doi.org/10.1007/s00586-008-0749-z.
69. A. J. Michalek, M. R. Buckley, L. J. Bonassar, I. Cohen, and J. C. Iatridis, "The effects of needle puncture injury on microscale shear strain in the intervertebral disc annulus fibrosus," *Spine J.*, vol. 10, no. 12, pp. 1098–1105, 2010, https://doi.org/10.1016/j.spinee.2010.09.015.
70. J. D. Kang, "Does a needle puncture into the annulus fibrosus cause disc degeneration?," *Spine J.*, vol. 10, no. 12, pp. 1106–1107, 2010, https://doi.org/10.1016/j.spinee.2010.10.014.
71. A. R. Poynton and J. M. Lane, "Safety profile for the clinical use of bone morphogenetic proteins in the spine," *Spine (Phila. Pa. 1976).*, vol. 27, no. 16 SUPPL., pp. 40–48, 2002, https://doi.org/10.1097/00007632-200208151-00010.
72. T. Jayakumar and P. S. Bhavan, "Molecular Targets of Natural Products for Chondroprotection in Destructive Joint Diseases," 2020.
73. C. L. Le Maitre, A. J. Freemont, and J. A. Hoyland, "Localization of degradative enzymes and their inhibitors in the degenerate human intervertebral disc," *J. Pathol.*, vol. 204, no. 1, pp. 47–54, 2004, https://doi.org/10.1002/path.1608.
74. S. Guo *et al.*, "The Mechanisms and Functions of GDF-5 in Intervertebral Disc Degeneration," *Orthop. Surg.*, vol. 13, no. 3, pp. 734–741, 2021, https://doi.org/10.1111/os.12942.
75. N. V. Vo, R. A. Hartman, T. Yurube, L. J. Jacobs, G. A. Sowa, and J. D. Kang, "Expression and regulation of metalloproteinases and their inhibitors in intervertebral disc aging and degeneration," *Spine J.*, vol. 13, no. 3, pp. 331–341, 2013, https://doi.org/10.1016/j.spinee.2012.02.027.
76. W. J. Wang *et al.*, "MMPs and ADAMTSs in intervertebral disc degeneration," *Clin. Chim. Acta*, vol. 448, pp. 238–246, 2015, https://doi.org/10.1016/j.cca.2015.06.023.
77. T. Cawston *et al.*, "The regulation of MMPs and TIMPs in cartilage turnover," *Ann. N. Y. Acad. Sci.*, vol. 878, pp. 120–129, 1999, https://doi.org/10.1111/j.1749-6632.1999.tb07678.x.
78. J. Feng Zhang, G. Liang Wang, Z. Jie Zhou, X. Qian Fang, S. Chen, and S. Wu Fan, "Expression of Matrix Metalloproteinases, Tissue Inhibitors of Metalloproteinases, and Interleukins in Vertebral Cartilage Endplate," *Orthop. Surg.*, vol. 10, no. 4, pp. 306–311, 2018, https://doi.org/10.1111/os.12409.
79. C. Evans, "Potential biologic therapies for the intervertebral disc," *J. Bone Jt. Surg.*, vol. 88, no. SUPPL. 2, pp. 95–98, 2006, https://doi.org/10.2106/JBJS.E.01328.
80. K. Nishida *et al.*, "Gene therapy approach for disc degeneration and associated spinal disorders," *Eur. Spine J.*, vol. 17, no. SUPPL. 4, pp. 459–466, 2008, https://doi.org/10.1007/s00586-008-0751-5.
81. P. Wehling, K.-P. Schulitz, P. D. Robbins, C. H. Evans, and J. A. Reinecke, "00007632-199705150-00008.pdf." 1997.
82. P. Sampara, R. R. Banala, S. K. Vemuri, G. R. Av, and S. Gpv, "Understanding the molecular biology of intervertebral disc degeneration and potential gene therapy strategies for regeneration: A review," *Gene Ther.*, vol. 25, no. 2, pp. 67–82, 2018, https://doi.org/10.1038/s41434-018-0004-0.
83. L. M. Benneker *et al.*, "Cell therapy for intervertebral disc repair: Advancing cell therapy from bench to clinics," *Eur. Cells Mater.*, vol. 27, no. SUPPL, pp. 5–11, 2014, https://doi.org/10.22203/eCM.v027sa02.
84. I. B. Han, "Moving forward: Gene therapy for intervertebral disc degeneration," *Neurospine*, vol. 17, no. 1, pp. 17–18, 2020, https://doi.org/10.14245/ns.2040108.054.
85. H. B. Henriksson *et al.*, "Transplantation of human mesenchymal stems cells into intervertebral discs in a xenogeneic porcine model," *Spine (Phila. Pa. 1976).*, vol. 34, no. 2, pp. 141–148, 2009, https://doi.org/10.1097/BRS.0b013e31818f8c20.
86. T. Ganey, W. C. Hutton, T. Moseley, M. Hedrick, and H. Meisel, "Intervertebral Disc Repair Using Adipose Tissue-Derived Stem and Regenerative Cells," vol. 34, no. 21, pp. 1–8, 2009.

87. D. Sakai *et al.*, "Transplantation of mesenchymal stem cells embedded in Atelocollagen® gel to the intervertebral disc: A potential therapeutic model for disc degeneration," *Biomaterials*, vol. 24, no. 20, pp. 3531–3541, 2003, https://doi.org/10.1016/S0142-9612(03)00222-9.
88. D. Sakai *et al.*, "Differentiation of mesenchymal stem cells transplanted to a rabbit degenerative disc model: Potential and limitations for stem cell therapy in disc regeneration," *Spine (Phila. Pa. 1976).*, vol. 30, no. 21, pp. 2379–2387, 2005, https://doi.org/10.1097/01.brs.0000184365.28481.e3.
89. M. A. Adams and P. J. Roughley, "What is intervertebral disc degeneration, and what causes it?," *Spine (Phila. Pa. 1976).*, vol. 31, no. 18, pp. 2151–61, Aug. 2006, https://doi.org/10.1097/01.brs.0000231761.73859.2c.
90. T. Oichi, Y. Taniguchi, Y. Oshima, S. Tanaka, and T. Saito, "Pathomechanism of intervertebral disc degeneration," *JOR Spine*, vol. 3, no. 1, pp. 1–9, 2020, https://doi.org/10.1002/jsp2.1076.
91. F. Colella *et al.*, "Drug delivery in intervertebral disc degeneration and osteoarthritis: Selecting the optimal platform for the delivery of disease-modifying agents," *J. Control. Release*, vol. 328, no. May, pp. 985–999, 2020, https://doi.org/10.1016/j.jconrel.2020.08.041.
92. M. J. Mitchell, M. M. Billingsley, R. M. Haley, M. E. Wechsler, N. A. Peppas, and R. Langer, "Engineering Precision Nanoparticles for Drug Delivery," *Nat. Rev. Drug Discov.*, pp. 1–24, 2020, https://doi.org/10.1038/s41573-020-0090-8.
93. K. Dzobo *et al.*, "Review Article Advances in Regenerative Medicine and Tissue Engineering : Innovation and Transformation of Medicine," vol. 2018, 2018.
94. N. I. Md Fadilah, I. L. Mohd Isa, W. S. Wan Kamarul Zaman, Y. Tabata, and M. B. Fauzi, "The Effect of Nanoparticle-Incorporated Natural-Based Biomaterials towards Cells on Activated Pathways: A Systematic Review," *Polymers (Basel).*, vol. 14, p. 476, 2022.
95. M. Cunniffe, F. J. O. Brien, S. Partap, T. J. Levingstone, K. T. Stanton, and G. R. Dickson, "The synthesis and characterization of nanophase hydroxyapatite using a novel dispersant-aided precipitation method," pp. 1142–1149, 2010, https://doi.org/10.1002/jbm.a.32931.
96. S. Vieira, S. Vial, R. L. Reis, and J. M. Oliveira, "Nanoparticles for bone tissue engineering," *Biotechnol. Prog.*, vol. 33, no. 3, pp. 590–611, 2017, https://doi.org/10.1002/btpr.2469.
97. G. Lewis, "Nanostructured Hydroxyapatite Coating on Bioalloy Substrates : Current Status and Future Directions," vol. 2, no. 1, pp. 65–82, 2017.
98. V. Patravale, P. Dandekar, and R. Jain, "3 – Characterization techniques for nanoparticulate carriers," in *Woodhead Publishing Series in Biomedicine*, V. Patravale, P. Dandekar, and R. B. T.-N. D. D. Jain, Eds. Woodhead Publishing, 2012, pp. 87–121.
99. I. Venugopal, A. I. Mehta, and A. A. Linninger, "Chapter 5 Drug Delivery Applications of Nanoparticles in the Spine," vol. 2059.
100. A. Z. Wilczewska, K. Niemirowicz, K. H. Markiewicz, and H. Car, "Nanoparticles as drug delivery systems," *Pharmacol. Reports*, vol. 64, no. 5, pp. 1020–1037, 2012, https://doi.org/10.1016/S1734-1140(12)70901-5.
101. P. Roughley, C. Hoemann, E. DesRosiers, F. Mwale, J. Antoniou, and M. Alini, "The potential of chitosan-based gels containing intervertebral disc cells for nucleus pulposus supplementation.," *Biomaterials*, vol. 27, no. 3, pp. 388–96, Jan. 2006, https://doi.org/10.1016/j.biomaterials.2005.06.037.
102. A. I. Chou and S. B. Nicoll, "Characterization of photocrosslinked alginate hydrogels for nucleus pulposus cell encapsulation," *J. Biomed. Mater. Res. A*, vol. 91, no. 1, pp. 187–194, 2009, https://doi.org/10.1002/JBM.A.32191.
103. I. Doench *et al.*, "Injectable and Gellable Chitosan Formulations Filled with Cellulose Nanofibers for Intervertebral Disc Tissue Engineering," *Polymers (Basel).*, vol. 10, no. 11, 2018, https://doi.org/10.3390/polym10111202.
104. M. Khandaker and S. Riahanizad, "Evaluation of electrospun nanofiber-anchored silicone for the degenerative intervertebral disc," *J. Healthc. Eng.*, 2017, https://doi.org/10.1155/2017/5283846.

105. A. Yau, I. Sands, and Y. Chen, "Nano-Scale Surface Modifications to Advance Current Treatment Options for Cervical Degenerative Disc Disease (CDDD)," *J. Orthop. Res. Ther.*, vol. 4, no. 9, 2019.
106. V. P. Torchilin, "Recent advances with liposomes as pharmaceutical carriers," *Nat. Rev. Drug Discov. 2005 42*, vol. 4, no. 2, pp. 145–160, Feb. 2005, https://doi.org/10.1038/nrd1632.
107. S. Hua and S. Y. Wu, "The use of lipid-based nanocarriers for targeted pain therapies," *Front. Pharmacol.*, vol. 4 NOV, p. 143, 2013, https://doi.org/10.3389/FPHAR.2013.00143/BIBTEX.
108. S. A. Chung *et al.*, "Nucleus pulposus cellular longevity by telomerase gene therapy," *Spine (Phila. Pa. 1976).*, vol. 32, no. 11, pp. 1188–1196, May 2007, https://doi.org/10.1097/BRS.0B013E31805471A3.
109. R. R. Banala *et al.*, "Efficiency of dual siRNA-mediated gene therapy for intervertebral disc degeneration (IVDD)," *Spine J.*, vol. 19, no. 5, pp. 896–904, May 2019, https://doi.org/10.1016/J.SPINEE.2018.10.016.
110. T. M. Allen and P. R. Cullis, "Liposomal drug delivery systems: from concept to clinical applications," *Adv. Drug Deliv. Rev.*, vol. 65, no. 1, pp. 36–48, Jan. 2013, https://doi.org/10.1016/J.ADDR.2012.09.037.
111. J. Gubernator, "Active methods of drug loading into liposomes: recent strategies for stable drug entrapment and increased in vivo activity," *Expert Opin. Drug Deliv.*, vol. 8, no. 5, pp. 565–580, May 2011, https://doi.org/10.1517/17425247.2011.566552.
112. Y. Barenholz, "Doxil® — The first FDA-approved nano-drug: Lessons learned," *J. Control. Release*, vol. 160, no. 2, pp. 117–134, Jun. 2012, https://doi.org/10.1016/J.JCONREL.2012.03.020.
113. E. Abbasi, S. F. Aval, A. Akbarzadeh, M. Milani, and H. T. Nasrabadi, "Dendrimers: synthesis, applications, and properties," vol. 9, no. 1, pp. 1–10, 2014, https://doi.org/10.1186/1556-276X-9-247.
114. V. Patravale, P. Dandekar, and R. Jain, "1 – Nanoparticulate systems as drug carriers: the need," in *Woodhead Publishing Series in Biomedicine*, V. Patravale, P. Dandekar, and R. B. T.-N. D. D. Jain, Eds. Woodhead Publishing, 2012, pp. 1–28.
115. N. Phogat, M. Kohl, I. Uddin, and A. Jahan, "Chapter 11 – Interaction of Nanoparticles With Biomolecules, Protein, Enzymes, and Its Applications," H.-P. Deigner and M. B. T.-P. M. Kohl, Eds. Academic Press, 2018, pp. 253–276.
116. K. Gardikis *et al.*, "New Drug Delivery Nanosystem Combining Liposomal and Dendrimeric Technology (Liposomal Locked-In Dendrimers) for Cancer Therapy," vol. 99, no. 8, pp. 3561–3571, 2010, https://doi.org/10.1002/jps.
117. U. Kedar, P. Phutane, S. Shidhaye, and V. Kadam, "Advances in polymeric micelles for drug delivery and tumor targeting," *Nanomedicine Nanotechnology, Biol. Med.*, vol. 6, no. 6, pp. 714–729, 2010, https://doi.org/10.1016/j.nano.2010.05.005.
118. T. E. Kavanaugh, T. A. Werfel, H. Cho, K. A. Hasty, and C. L. Duvall, "Particle-based technologies for osteoarthritis detection and therapy," *Drug Deliv. Transl. Res.*, vol. 6, no. 2, pp. 132–147, May 2016, https://doi.org/10.1007/s13346-015-0234-2.
119. C. Y. Lin, S. T. Crowley, S. Uchida, Y. Komaki, K. Kataoka, and K. Itaka, "Treatment of Intervertebral Disk Disease by the Administration of mRNA Encoding a Cartilage-Anabolic Transcription Factor," *Mol. Ther. – Nucleic Acids*, vol. 16, pp. 162–171, Jun. 2019, https://doi.org/10.1016/J.OMTN.2019.02.012.
120. G. Feng *et al.*, "Sustained and Bioresponsive Two-Stage Delivery of Therapeutic miRNA via Polyplex Micelle-Loaded Injectable Hydrogels for Inhibition of Intervertebral Disc Fibrosis," *Adv. Healthc. Mater.*, vol. 7, no. 21, p. 1800623, Nov. 2018, https://doi.org/10.1002/ADHM.201800623.
121. C. Evans, V. B. Kraus, and L. A. Setton, "Progress in intra-articular therapy," vol. 10, no. 1, pp. 11–22, 2015, https://doi.org/10.1038/nrrheum.2013.159.Progress.
122. J. Becerra, J. A. Andrades, E. Guerado, P. Zamora-Navas, J. M. López-Puertas, and A. H. Reddi, "Articular cartilage: Structure and regeneration," *Tissue Eng. – Part B Rev.*, vol. 16, no. 6, pp. 617–627, 2010, https://doi.org/10.1089/ten.teb.2010.0191.

123. T. Guo *et al.*, "New Hope for Treating Intervertebral Disc Degeneration: Microsphere-Based Delivery System," *Front. Bioeng. Biotechnol.*, vol. 10, no. July, pp. 1–22, 2022, https://doi.org/10.3389/fbioe.2022.933901.
124. R. J. Kulchar, B. R. Denzer, B. M. Chavre, M. Takegami, and J. Patterson, "A review of the use of microparticles for cartilage tissue engineering," *Int. J. Mol. Sci.*, vol. 22, no. 19, 2021, https://doi.org/10.3390/ijms221910292.
125. T. Virmani and J. Gupta, "PHARMACEUTICAL APPLICATION OF MICROSPHERES: AN APPROACH FOR THE TREATMENT OF VARIOUS DISEASES Tarun Virmani * and Jyoti Gupta School of Pharmaceutical Sciences, M. V. N. University, Palwal – 121105, Haryana, India.," *Int. J. Pharm. Sci. Res.*, vol. 8, no. 8, pp. 3252–3260, 2017, https://doi.org/10.13040/IJPSR.0975-8232.8(8).3252-60.
126. S. Freiberg and X. X. Zhu, "Polymer microspheres for controlled drug release," *Int. J. Pharm.*, vol. 282, no. 1–2, pp. 1–18, 2004, https://doi.org/10.1016/j.ijpharm.2004.04.013.
127. E. Mathiowitz *et al.*, "Biologically erodable microspheres as potential oral drug delivery systems," *Nature*, vol. 386, no. 6623. pp. 410–414, 1997, https://doi.org/10.1038/386410a0.
128. Y. Y. Li *et al.*, "Delivering mesenchymal stem cells in collagen microsphere carriers to rabbit degenerative disc: reduced risk of osteophyte formation.," *Tissue Eng. Part A*, vol. 20, no. 9–10, pp. 1379–91, 2014, https://doi.org/10.1089/ten.TEA.2013.0498.
129. M. Yuan, K. W. Leong, and B. P. Chan, "Three-dimensional Culture of Rabbit Nucleus Pulposus Cells in Collagen Microspheres," *Spine J.*, vol. 11, no. 10, pp. 947–960, 2011, https://doi.org/10.1016/j.spinee.2011.07.004.
130. Y. Xu *et al.*, "Metabolism Balance Regulation via Antagonist-Functionalized Injectable Microsphere for Nucleus Pulposus Regeneration," *Adv. Funct. Mater.*, vol. 30, no. 52, pp. 1–14, 2020, https://doi.org/10.1002/adfm.202006333.
131. S. S. Sivan, I. Bonstein, Y. N. Marmor, G. Pelled, Z. Gazit, and M. Amit, "Encapsulation of Human-Bone-Marrow-Derived Mesenchymal Stem Cells in Small Alginate Beads Using One-Step Emulsification by Internal Gelation: In Vitro, and In Vivo Evaluation in Degenerate Intervertebral Disc Model," *Pharmaceutics*, vol. 14, no. 6, 2022, https://doi.org/10.3390/pharmaceutics14061179.
132. S. O. Adeosun, M. O. Ilomuanya, O. P. Gbenebor, M. O. Dada, and C. C. Odili, "Biomaterials for Drug Delivery: Sources, Classification, Synthesis, Processing, and Applications," *Adv. Funct. Mater.*, 2020, https://doi.org/10.5772/intechopen.93368.
133. K. A. I. Zhu, F. Zhao, Y. Yang, and W. Mu, "Effects of simvastatin – loaded PLGA microspheres on treatment of rats with intervertebral disk degeneration and on 6 – K – PGF1 α and HIF – 1 α," pp. 579–584, 2020, https://doi.org/10.3892/etm.2019.8267.
134. A. R. Tellegen *et al.*, "PT," *J. Control. Release*, p. #pagerange#, 2018, https://doi.org/10.1016/j.jconrel.2018.08.019.
135. S. Kim *et al.*, "Nanoparticle Formulation for Controlled Release of Capsaicin," 2011, https://doi.org/10.1166/jnn.2011.3636.
136. N. Agnihotri, G. Soni, D. K. Chanchal, A. Khan, and S. Tiwari, "A REVIEW ON MICROSPHERES A NOVEL DRUG DELIVERY SYSTEM FOR MULTI- PARTICULATE DRUG RELEASE Neha Agnihotri *, G. C. Soni, Dilip Kumar Chanchal, Afrin Khan and Sakshi Tiwari Institute of Pharmacy, Bundelkhand University Jhansi – 284128, Uttar Pradesh, India," vol. 5, no. 1, pp. 6–15, 2019, https://doi.org/10.13040/IJPSR.0975-8232.IJLSR.5(1).6-15.
137. A. A. Ragab *et al.*, "O RIGINAL A RTICLE A Preliminary Report on the Effects of Sustained Administration of Corticosteroid on Traumatized Disc Using the Adult Male Rat Model," vol. 22, no. 7, pp. 473–478, 2009.
138. M. Nagae *et al.*, "Intervertebral Disc Regeneration Using Platelet-Rich Plasma and Biodegradable Gelatin Hydrogel Microspheres," vol. 13, no. 1, 2007, https://doi.org/10.1089/ten.2006.0042.
139. K. Sawamura, T. Ikeda, D. Ph, M. Nagae, and D. Ph, "Characterization of In Vivo Effects of Platelet-Rich Plasma and Biodegradable Gelatin Hydrogel Microspheres on Degenerated Intervertebral Discs," vol. 15, no. 12, 2009.

140. D. J. Gorth *et al.*, "IL-1ra delivered from poly (lactic- co -glycolic acid) microspheres attenuates IL-1 b -mediated degradation of nucleus pulposus in vitro," *Arthritis Res. Ther.*, vol. 14, no. 4, p. R179, 2012, https://doi.org/10.1186/ar3932.
141. D. J. Gorth *et al.*, "In vivo retention and bioactivity of IL-1ra microspheres in the rat intervertebral disc: a preliminary investigation," pp. 1–5, 2014, https://doi.org/10.1186/s40634-014-0015-8.
142. C. Z. Liang *et al.*, "Dual delivery for stem cell differentiation using dexamethasone and bFGF in / on polymeric microspheres as a cell carrier for nucleus pulposus regeneration," pp. 1097–1107, 2012, https://doi.org/10.1007/s10856-012-4563-0.
143. C. Liang *et al.*, "Acta Biomaterialia Dual release of dexamethasone and TGF- b 3 from polymeric microspheres for stem cell matrix accumulation in a rat disc degeneration model," *Acta Biomater.*, vol. 9, no. 12, pp. 9423–9433, 2013, https://doi.org/10.1016/j.actbio.2013.08.019.
144. J. M. Ho, H. Sung, R. Kim, Y. Wook, C. Myung, and J. Kang, "Recent advances in intra-articular drug delivery systems to extend drug retention in joint," *J. Pharm. Investig.*, vol. 0, no. 0, p. 0, 2018, https://doi.org/10.1007/s40005-018-0383-7.
145. B. S. Grund, M. Bauer, and D. Fischer, "Polymers in Drug Delivery---State of the Art and Future Trends **," no. 3, pp. 61–87, 2011, https://doi.org/10.1002/adem.201080088.
146. W. R. Gombotz and S. F. Wee, "Protein release from alginate matrices," vol. 31, pp. 267–285, 1998.
147. S. M. Sinclair *et al.*, "A genetically engineered thermally responsive sustained release curcumin depot to treat neuroin fl ammation," *J. Control. Release*, vol. 171, no. 1, pp. 38–47, 2013, https://doi.org/10.1016/j.jconrel.2013.06.032.
148. D. R. Pereira *et al.*, "Development of gellan gum-based microparticles/hydrogel matrices for application in the intervertebral disc regeneration," *Tissue Eng. – Part C Methods*, vol. 17, no. 10, pp. 961–972, 2011, https://doi.org/10.1089/ten.tec.2011.0115.
149. Z. He *et al.*, "Pr ep rin t n ot pe er re v Pr ep rin t n ot pe er," 2022.
150. I. Sheikh and Y. Dahman, "Chapter 2 – Applications of nanobiomaterials in hard tissue engineering," A. M. B. T.-N. in H. T. E. Grumezescu, Ed. William Andrew Publishing, 2016, pp. 33–62.
151. J. Leijten, J. Rouwkema, Y. S. Zhang, A. Nasajpour, M. R. Dokmeci, and A. Khademhosseini, "Advancing Tissue Engineering: A Tale of Nano-, Micro-, and Macroscale Integration.," *Small*, vol. 12, no. 16, pp. 2130–2145, Apr. 2016, https://doi.org/10.1002/smll.201501798.
152. Y.-C. Huang, Y. Hu, Z. Li, and K. D. K. Luk, "Biomaterials for intervertebral disc regeneration: Current status and looming challenges," *J. Tissue Eng. Regen. Med.*, vol. 12, no. 11, pp. 2188–2202, Nov. 2018, https://doi.org/10.1002/term.2750.
153. O. S. Fenton, K. N. Olafson, P. S. Pillai, M. J. Mitchell, and R. Langer, "Advances in Biomaterials for Drug Delivery," *Adv. Mater.*, vol. 30, no. 29, p. 1705328, Jul. 2018, https://doi.org/10.1002/adma.201705328.
154. T. R. Hoare and D. S. Kohane, "Hydrogels in drug delivery : Progress and challenges," *Polym. with aligned carbon Nanotub. Act. Compos. Mater.*, vol. 49, no. 8, pp. 1993–2007, 2008, https://doi.org/10.1016/j.polymer.2008.01.027.
155. B. Love, "Chapter 9 – Polymeric Biomaterials," B. B. T.-B. Love, Ed. Academic Press, 2017, pp. 205–238.
156. J. Kopecek, "Hydrogel biomaterials: a smart future?," *Biomaterials*, vol. 28, no. 34, pp. 5185–5192, Dec. 2007, https://doi.org/10.1016/j.biomaterials.2007.07.044.
157. E. H. Schacht, "Polymer chemistry and hydrogel systems," *J. Phys. Conf. Ser.*, vol. 3, pp. 22–28, 2004, https://doi.org/10.1088/1742-6596/3/1/004.
158. S. Mantha *et al.*, "Smart Hydrogels in Tissue Engineering and Regenerative Medicine.," *Mater. (Basel, Switzerland)*, vol. 12, no. 20, Oct. 2019, https://doi.org/10.3390/ma12203323.
159. Y. S. Zhang and A. Khademhosseini, "Advances in engineering hydrogels.," *Science*, vol. 356, no. 6337, May 2017, https://doi.org/10.1126/science.aaf3627.
160. R. Parhi, "Cross-Linked Hydrogel for Pharmaceutical Applications: A Review.," *Adv. Pharm. Bull.*, vol. 7, no. 4, pp. 515–530, Dec. 2017, https://doi.org/10.15171/apb.2017.064.

161. G. Malucelli *et al.*, "Sliding crosslinked thermoresponsive materials: Polypseudorotaxanes made of poly(N-Isopropylacrylamide) and acrylamide-γ-cyclodextrin," *Front. Chem.*, vol. 6, no. NOV, pp. 1–8, 2018, https://doi.org/10.3389/fchem.2018.00585.
162. N. Karak, "1 – Fundamentals of polymers," N. B. T.-V. O.-B. P. Karak, Ed. Woodhead Publishing, 2012, pp. 1–30.
163. S. Ishikawa *et al.*, "Interpenetrating Polymer Network Hydrogels via a One-Pot and in Situ Gelation System Based on Peptide Self-Assembly and Orthogonal Cross-Linking for Tissue Regeneration," *Chem. Mater.*, vol. 32, no. 6, pp. 2353–2364, Mar. 2020, https://doi.org/10.1021/acs.chemmater.9b04725.
164. Z. Li, Y. Zhou, T. Li, J. Zhang, and H. Tian, "Stimuli-responsive hydrogels: Fabrication and biomedical applications," *VIEW*, vol. 3, no. 2, p. 20200112, Mar. 2022, https://doi.org/10.1002/VIW.20200112.
165. L. Li, L. Xie, R. Zheng, and R. Sun, "Self-Assembly Dipeptide Hydrogel: The Structures and Properties," *Front. Chem.*, vol. 9, no. September, pp. 1–15, 2021, https://doi.org/10.3389/fchem.2021.739791.
166. Z. Wu, J. Zhang, Q. Lin, Y. Zhu, L. Wang, and Y. Li, "Movable-crosslinking tough hydrogels with lithium ion as sensitive and durable compressive sensor," *Polymer (Guildf).*, vol. 214, p. 123257, 2021, https://doi.org/10.1016/j.polymer.2020.123257.
167. C. Yan *et al.*, "Applications of Functionalized Hydrogels in the Regeneration of the Intervertebral Disc," *Biomed Res. Int.*, vol. 2021, p. 2818624, 2021, https://doi.org/10.1155/2021/2818624.
168. A. Gloria, R. De Santis, L. Ambrosio, F. Causa, and K. E. Tanner, "A Multi-component Fiber-reinforced PHEMA-based Hydrogel/HAPEXTM Device for Customized Intervertebral Disc Prosthesis," *J. Biomater. Appl.*, vol. 25, no. 8, pp. 795–810, May 2010, https://doi.org/10.1177/0885328209360933.
169. Y. Gan *et al.*, "An interpenetrating network-strengthened and toughened hydrogel that supports cell-based nucleus pulposus regeneration," *Biomaterials*, vol. 136, pp. 12–28, 2017, https://doi.org/10.1016/j.biomaterials.2017.05.017.
170. C. O. Crosby, B. Stern, N. Kalkunte, S. Pedahzur, S. Ramesh, and J. Zoldan, "Interpenetrating polymer network hydrogels as bioactive scaffolds for tissue engineering," vol. 38, no. 3, pp. 347–361, 2022, https://doi.org/10.1515/revce-2020-0039.
171. P. Y. Hwang, J. Chen, L. Jing, B. D. Hoffman, and L. A. Setton, "The role of extracellular matrix elasticity and composition in regulating the nucleus pulposus cell phenotype in the intervertebral disc: a narrative review.," *J. Biomech. Eng.*, vol. 136, no. 2, p. 21010, Feb. 2014, https://doi.org/10.1115/1.4026360.
172. J. Caldeira *et al.*, "Matrisome Profiling during Intervertebral Disc Development and Ageing," *Sci. Rep.*, vol. 7, no. 1, pp. 1–15, 2017, https://doi.org/10.1038/s41598-017-11960-0.
173. H. Liang, R. Luo, G. Li, W. Zhang, Y. Song, and C. Yang, "The Proteolysis of ECM in Intervertebral Disc Degeneration.," *Int. J. Mol. Sci.*, vol. 23, no. 3, Feb. 2022, https://doi.org/10.3390/ijms23031715.
174. H. E. Gruber and E. N. Hanley Jr, "Analysis of aging and degeneration of the human intervertebral disc: comparison of surgical specimens with normal controls," *Spine (Phila. Pa. 1976).*, vol. 23, no. 7, pp. 751–757, 1998.
175. A. Maroudas, R. A. Stockwell, A. Nachemson, and J. Urban, "Factors involved in the nutrition of the human lumbar intervertebral disc: cellularity and diffusion of glucose in vitro.," *J. Anat.*, vol. 120, no. Pt 1, p. 113, 1975.
176. M. C. Catoira, L. Fusaro, D. Di Francesco, M. Ramella, and F. Boccafoschi, "Overview of natural hydrogels for regenerative medicine applications.," *J. Mater. Sci. Mater. Med.*, vol. 30, no. 10, p. 115, Oct. 2019, https://doi.org/10.1007/s10856-019-6318-7.
177. I. L. Mohd Isa *et al.*, "Implantation of Hyaluronic Acid Hydrogel Prevents the Pain Phenotype in a Rat Model of Intervertebral Disc Injury," *Sci. Adv.*, vol. 4, no. eaaq0597, pp. 1–19, 2018, https://doi.org/10.1126/sciadv.aaq0597.

178. U. S. K. Madduma-Bandarage and S. V Madihally, “Synthetic hydrogels: Synthesis, novel trends, and applications,” *J. Appl. Polym. Sci.*, vol. 138, no. 19, p. 50376, May 2021, https://doi.org/10.1002/app.50376.
179. Y. Berkovitch and D. Seliktar, “Semi-synthetic hydrogel composition and stiffness regulate neuronal morphogenesis.,” *Int. J. Pharm.*, vol. 523, no. 2, pp. 545–555, May 2017, https://doi.org/10.1016/j.ijpharm.2016.11.032.
180. J. Li and D. J. Mooney, “Designing hydrogels for controlled drug delivery,” *Nat. Rev. Mater.*, vol. 1, no. 12, p. 16071, 2016, https://doi.org/10.1038/natrevmats.2016.71.
181. A. Sood, A. Gupta, and G. Agrawal, “Recent advances in polysaccharides based biomaterials for drug delivery and tissue engineering applications,” *Carbohydr. Polym. Technol. Appl.*, vol. 2, p. 100067, 2021, https://doi.org/10.1016/j.carpta.2021.100067.
182. D. H. Hanna, V. F. Lotfy, A. H. Basta, and G. R. Saad, “Comparative evaluation for controlling release of niacin from protein- and cellulose-chitosan based hydrogels,” *Int. J. Biol. Macromol.*, vol. 150, pp. 228–237, 2020, https://doi.org/10.1016/j.ijbiomac.2020.02.056.
183. A. R. Jackson, T.-Y. Yuan, C.-Y. Huang, and W. Y. Gu, “A Conductivity Approach to Measuring Fixed Charge Density in Intervertebral Disc Tissue,” *Ann. Biomed. Eng.*, vol. 37, no. 12, pp. 2566–2573, 2009, https://doi.org/10.1007/s10439-009-9792-0.
184. A. Vedadghavami, C. Zhang, and A. G. Bajpayee, “Overcoming negatively charged tissue barriers: Drug delivery using cationic peptides and proteins,” *Nano Today*, vol. 34, p. 100898, 2020, https://doi.org/10.1016/j.nantod.2020.100898.
185. H. Yoshihara and D. Yoneoka, “National trends in the surgical treatment for lumbar degenerative disc disease: United States, 2000 to 2009,” *Spine J.*, vol. 15, no. 2, pp. 265–271, 2015, https://doi.org/10.1016/j.spinee.2014.09.026.
186. R. A. Deyo, D. T. Gray, W. Kreuter, S. Mirza, and B. I. Martin, “United States Trends in Lumbar Fusion Surgery for Degenerative Conditions,” *Spine (Phila. Pa. 1976).*, vol. 30, no. 12, 2005.
187. R. D. Bowles and L. A. Setton, “Biomaterials for Intervertebral Disc Regeneration and Repair,” *Biomaterials*, vol. 129, pp. 54–67, 2017, https://doi.org/10.1016/j.biomaterials.2017.03.013.
188. A. Fusco *et al.*, “168. Minimally invasive hydrogel nucleoplasty in a goat model of moderate severity disc degeneration,” *Spine J.*, vol. 21, no. 9, Supplement, p. S84, 2021, https://doi.org/10.1016/j.spinee.2021.05.196.
189. M. Inoue *et al.*, “An Injectable Hyaluronic Acid Hydrogel Promotes Intervertebral Disc Repair in a Rabbit Model,” *Spine (Phila. Pa. 1976).*, vol. 46, no. 15, pp. E810–E816, 2021, https://doi.org/10.1097/brs.0000000000003921.
190. Z. Li *et al.*, “Thermo-sensitive injectable glycol chitosan-based hydrogel for treatment of degenerative disc disease,” *Carbohydr. Polym.*, vol. 184, pp. 342–353, 2018, https://doi.org/10.1016/j.carbpol.2018.01.006.
191. A. Schmocker *et al.*, “A photopolymerized composite hydrogel and surgical implanting tool for a nucleus pulposus replacement,” *Biomaterials*, vol. 88, pp. 110–119, 2016, https://doi.org/10.1016/j.biomaterials.2016.02.015.
192. D. A. Frauchiger Rahel *et al.*, “Genipinp-enhanced fibrin hydrogel and novel silk for intervertebral disc repair in a loaded bovine organ culture model,” *Journal of Functional Biomaterials*, vol. 9, no. 3. 2018, https://doi.org/10.3390/jfb9030040.
193. C. J. Panebianco, T. J. DiStefano, B. Mui, W. W. Hom, and J. C. Iatridis, “Crosslinker concentration controls TGFβ-3 release and annulus fibrosus cell apoptosis in genipin-crosslinked fibrin hydrogels.,” *Eur. Cell. Mater.*, vol. 39, pp. 211–226, May 2020, https://doi.org/10.22203/eCM.v039a14.
194. T. J. DiStefano *et al.*, “Development of a two-part biomaterial adhesive strategy for annulus fibrosus repair and ex vivo evaluation of implant herniation risk,” *Biomaterials*, vol. 258, p. 120309, 2020, https://doi.org/10.1016/j.biomaterials.2020.120309.
195. E. A. Growney Kalaf, M. Pendyala, J. G. Bledsoe, and S. A. Sell, “Characterization and restoration of degenerated IVD function with an injectable, in situ gelling alginate hydrogel: An in vitro and ex vivo study,” *J. Mech. Behav. Biomed. Mater.*, vol. 72, pp. 229–240, 2017, https://doi.org/10.1016/j.jmbbm.2017.05.014.

196. A. R. Tellegen *et al.*, "Intradiscal Application of A PCLA–PEG–PCLA Hydrogel Loaded with Celecoxib for the Treatment of Back Pain in Canines: What's In It for Humans?," *J. Tissue Eng. Regen. Med.*, vol. 12, no. 3, pp. 642–652, 2018, https://doi.org/10.1002/term.2483.
197. C. Gruver and K. B. Guthmiller, "Provocative Discography.," Treasure Island (FL), 2022.
198. M. G. Lykissas and D. Giannoulis, "Minimally invasive spine surgery for degenerative spine disease and deformity correction: a literature review.," *Ann. Transl. Med.*, vol. 6, no. 6, p. 99, Mar. 2018, https://doi.org/10.21037/atm.2018.03.18.
199. Ø. Øvrebø *et al.*, "Design and clinical application of injectable hydrogels for musculoskeletal therapy," *Bioeng. Transl. Med.*, vol. 7, no. 2, p. e10295, May 2022, https://doi.org/10.1002/btm2.10295.
200. Medtronic, "Infuse Bone Graft," *Medtronic*.
201. M. Vigata, C. Meinert, D. W. Hutmacher, and N. Bock, "Hydrogels as Drug Delivery Systems: A Review of Current Characterization and Evaluation Techniques.," *Pharmaceutics*, vol. 12, no. 12, Dec. 2020, https://doi.org/10.3390/pharmaceutics12121188.
202. P. Ghasemiyeh and S. Mohammadi-Samani, "Hydrogels as Drug Delivery Systems; Pros and Cons," *Trends Pharm. Sci.*, vol. 5, no. 1, pp. 7–24, 2019, https://doi.org/10.30476/tips.2019.81604.1002.
203. C.-C. Lin and A. T. Metters, "Hydrogels in controlled release formulations: network design and mathematical modeling.," *Adv. Drug Deliv. Rev.*, vol. 58, no. 12–13, pp. 1379–1408, Nov. 2006, https://doi.org/10.1016/j.addr.2006.09.004.
204. R. Bettini, P. Colombo, G. Massimo, P. L. Catellani, and T. Vitali, "Swelling and drug release in hydrogel matrices: polymer viscosity and matrix porosity effects," *Eur. J. Pharm. Sci.*, vol. 2, pp. 213–219, 1994.
205. J. Siepmann and N. A. Peppas, "Modeling of drug release from delivery systems based on hydroxypropyl methylcellulose (HPMC).," *Adv. Drug Deliv. Rev.*, vol. 48, no. 2–3, pp. 139–157, Jun. 2001, https://doi.org/10.1016/s0169-409x(01)00112-0.
206. C.-C. Lin and K. S. Anseth, "PEG hydrogels for the controlled release of biomolecules in regenerative medicine.," *Pharm. Res.*, vol. 26, no. 3, pp. 631–643, Mar. 2009, https://doi.org/10.1007/s11095-008-9801-2.
207. V. Ravaine, C. Ancla, and B. Catargi, "Chemically controlled closed-loop insulin delivery.," *J. Control. release Off. J. Control. Release Soc.*, vol. 132, no. 1, pp. 2–11, Nov. 2008, https://doi.org/10.1016/j.jconrel.2008.08.009.
208. S. Singh, A. A. Patel, and J. R. Singh, "Intervertebral Disc Degeneration: The Role and Evidence for Non-Stem-Cell-Based Regenerative Therapies.," *Int. J. spine Surg.*, vol. 15, no. s1, pp. 54–67, Apr. 2021, https://doi.org/10.14444/8055.
209. H. Jia *et al.*, "Injectable hydrogel with nucleus pulposus-matched viscoelastic property prevents intervertebral disc degeneration," *J. Orthop. Transl.*, vol. 33, pp. 162–173, 2022, https://doi.org/10.1016/j.jot.2022.03.006.
210. Y. Wang *et al.*, "Sustained gene delivery from inflammation-responsive anti-inflammatory hydrogels promotes extracellular matrix metabolism balance in degenerative nucleus pulposus," *Compos. Part B Eng.*, vol. 236, p. 109806, 2022, https://doi.org/10.1016/j.compositesb.2022.109806.
211. N. Willems *et al.*, "Biocompatibility and intradiscal application of a thermoreversible celecoxib-loaded poly-N-isopropylacrylamide MgFe-layered double hydroxide hydrogel in a canine model," *Arthritis Res. Ther.*, vol. 17, no. 1, p. 214, 2015, https://doi.org/10.1186/s13075-015-0727-x.
212. J. Wang *et al.*, "Therapeutic effect of the injectable thermosensitive hydrogel loaded with SHP099 on intervertebral disc degeneration," *Life Sci.*, vol. 266, p. 118891, 2021, https://doi.org/10.1016/j.lfs.2020.118891.
213. F. Hu *et al.*, "Identification of inflammatory regulation roles of thalidomide/ruxolitinib in nucleus pulposus and construction of polyelectrolyte nanocomplexes-impregnated injectable hydrogels for synergistic intervertebral disk degeneration treatment," *Nano Today*, vol. 44, p. 101462, 2022, https://doi.org/10.1016/j.nantod.2022.101462.

214. J. Chen *et al.*, "Injectable self-healing hydrogel with siRNA delivery property for sustained STING silencing and enhanced therapy of intervertebral disc degeneration," *Bioact. Mater.*, vol. 9, pp. 29–43, 2022, https://doi.org/10.1016/j.bioactmat.2021.08.003.
215. J. Bai *et al.*, "Reactive Oxygen Species-Scavenging Scaffold with Rapamycin for Treatment of Intervertebral Disk Degeneration," *Adv. Healthc. Mater.*, vol. 9, no. 3, pp. 1–10, 2020, https://doi.org/10.1002/adhm.201901186.
216. M. Likhitpanichkul *et al.*, "Fibrin-genipin Annulus Fibrosus Sealant as a Delivery System for Anti-TNFα Drug," *Spine J.*, vol. 15, no. 2015, pp. 2045–2054, 2015, https://doi.org/10.1016/j.spinee.2015.04.026.
217. H. Xing *et al.*, "Injectable exosome-functionalized extracellular matrix hydrogel for metabolism balance and pyroptosis regulation in intervertebral disc degeneration," *J. Nanobiotechnology*, vol. 19, no. 1, p. 264, 2021, https://doi.org/10.1186/s12951-021-00991-5.
218. M. Peeters *et al.*, "BMP-2 and BMP-2/7 Heterodimers Conjugated to a Fibrin/Hyaluronic Acid Hydrogel in a Large Animal Model of Mild Intervertebral Disc Degeneration," *Biores. Open Access*, vol. 4, no. 1, pp. 398–406, Oct. 2015, https://doi.org/10.1089/biores.2015.0025.
219. Z. Li *et al.*, "Heterodimeric BMP-2/7 for nucleus pulposus regeneration—In vitro and ex vivo studies," *J. Orthop. Res.*, vol. 35, no. 1, pp. 51–60, Jan. 2017, https://doi.org/10.1002/jor.23351.
220. K. D. Than *et al.*, "Intradiscal injection of simvastatin results in radiologic, histologic, and genetic evidence of disc regeneration in a rat model of degenerative disc disease," *Spine J.*, vol. 14, no. 6, pp. 1017–1028, 2014, https://doi.org/10.1016/j.spinee.2013.11.034.
221. C. Ligorio *et al.*, "TGF-β3-loaded graphene oxide – self-assembling peptide hybrid hydrogels as functional 3D scaffolds for the regeneration of the nucleus pulposus," *Acta Biomater.*, vol. 127, pp. 116–130, 2021, https://doi.org/10.1016/j.actbio.2021.03.077.
222. Y. Akyuva *et al.*, "Delivering growth factors through a polymeric scaffold to cell cultures containing both nucleus pulposus and annulus fibrosus," *Turk. Neurosurg.*, vol. 29, no. 2, pp. 180–193, 2019, https://doi.org/10.5137/1019-5149.JTN.22672-18.1.
223. D. N. Paglia, H. Singh, T. Karukonda, H. Drissi, and I. L. Moss, "PDGF-BB delays degeneration of the intervertebral discs in a rabbit preclinical model," *Spine (Phila. Pa. 1976).*, vol. 41, no. 8, pp. E449–E458, 2016, https://doi.org/10.1097/BRS.0000000000001336.

Chapter 5
Drug Delivery Systems for Tendon and Ligament

Aswin Sundarakrishnan

Contents

1 Introduction

1.1 Tendon and Ligament Injuries and the Need for New Drug Delivery Systems

Tendon and ligament injuries constitute around 50% of all musculoskeletal injuries [1, 2]. Every year, there are over 30 million tendon and ligament injuries in the United States, accounting for about US$40 billion in medical treatment costs [1, 3].

A. Sundarakrishnan (✉)
Northeastern University, Toronto, ON, Canada

S. Murab (ed.), *Drug Delivery Systems for Musculoskeletal Tissues*,
https://doi.org/10.1007/978-3-031-55653-1_5

Tendon and ligament injury treatment costs are exorbitant because these are bradytrophic tissues with poor vascularization, innervation, and inferior regenerative capacity [4].

Tendons transmit forces from muscles to the skeleton, and ligaments transmit forces from bone to bone. The structure of these tissues enables the storage of elastic energy and the ability to withstand tensile forces that will allow locomotion [1, 5]. When tensile forces exceed limits, it results in injuries. Acute injuries occur due to sudden trauma and are often observed in athletes and accident-prone children [6]. Conversely, chronic injuries are frequently observed in seniors due to age-related degeneration and persistent inflammation [6].

Tendon acute and chronic overuse injuries are collectively called tendinopathies [5, 6], caused by intrinsic (e.g., age, gender, and body weight) or extrinsic factors (e.g., sporting activities and physical loading) [5]. Certain tendons and ligaments are more prone to tendinopathies than others; these include rotator cuff tendons, forearm extensor tendons, Achilles tendons, tibialis posterior and patellar tendons, and anterior and posterior cruciate ligaments (ACL and PCL). The Achilles tendon is the most frequently injured, and it has found its way to Greek mythology, where the demigod hero Achilles is invincible except for his Achilles heel [7].

Depending on their severity, tendinopathies are treated using conventional physiotherapy (e.g., minor sprains of the wrist tendons) or surgery (e.g., tearing of extensor and flexor tendons). Besides these treatment options, drugs are routinely used to treat tendinopathies. For example, salicin crystals, the earlier forms of aspirin, have been used for treating rheumatism since the mid-eighteenth century [8]. Since then, many drug innovations have ensued, including entirely newer classes of biological drugs (e.g., growth factor proteins, monoclonal antibodies, nucleic acids, and live cells) that can elicit targeted effects on tissues. These contrast with the systemic non-targeted impact of small-molecule compounds. For example, platelet-rich plasma (PRP) is a biological injected directly into the tendon/ligament injury site, compared to the small-molecule drug aspirin delivered orally in tablet form. Rivaled to small-molecule compounds, working with biologics drugs poses additional challenges, such as maintaining stability (e.g., protein stability), successful intracellular delivery (e.g., delivery of nucleic acid-based drugs), and viability (e.g., live cells) [9]. There is a definite need for the development of newer classes of drug delivery systems.

Integrated drug delivery systems combining biologics such as biomaterials, drugs, proteins, and cells are the new frontier in preclinical tendon and ligament treatments [10]. Integrated drug delivery systems mimic the extracellular matrix (ECM) biophysical cues and modulate immune cells to resolve inflammation [10]. They can serve as sustained release depots owing to their physical interactions with drugs for controlled release. They provide the benefit of less frequent invasive dosing, thereby minimizing toxicity and fewer chances for infections [11–13]. Further, these systems offer synergistic mechanical and structural reinforcement through biomaterials and delivery of immunomodulatory growth factors and matrix proteins via cells. Several new integrated biomaterial drug delivery systems are currently in different stages of development in the clinic [10]. However, breakthrough therapies have yet to result. Currently, no treatment modality can fully return the function of tendon/ligaments back to their pre-injured state. There is a certain need for new research in developing integrated treatment modalities. In the subsequent sections, we will discuss current clinical treatments in detail with their advantages and disadvantages, followed by a discussion of integrated drug delivery systems research.

1.2 Tendon Structure-Informed Drug Delivery Systems

Reviewing what we know about tendon/ligament structure, assembly, and composition is essential, as these attributes inform the development of new drug delivery systems. The smallest unit of the tendon and ligament is the collagen fibril which is made of cross-linked tropocollagen molecules (Fig. 5.1) [15]. Cross-linked collagen molecules aggregate progressively into microfibrils, fibrils, fibers, fascicles, tertiary bundles, and the whole tendon [16]. Fibers are bundled and sheathed by a layer of connective tissue called the endotenon, containing the vascular, lymphatic, and nerve supply [16]. The epitenon is the connective tissue sheath for the whole tendon. The epitenon and endotenon support the smooth, frictionless gliding of collagen fibers during extension [5, 7].

Because tendons and ligaments are poorly vascularized, systemic drug delivery through the blood supply is inefficient. Instead, most drugs are delivered directly to the injury site via injection, ensuring faster, targeted delivery while minimizing off-target side effects. Most current drug delivery systems for tendon/ligament treatment utilize biomaterials as vehicles. Biomaterials drug carriers have been chosen based on the natural ECM composition of tendons and ligaments. Collagens (e.g., type I, III, V, VI, XII, XIV, and XV) constitute most of the extracellular matrix in tendons and ligaments. Collagen type I is the most widely used natural biomaterial

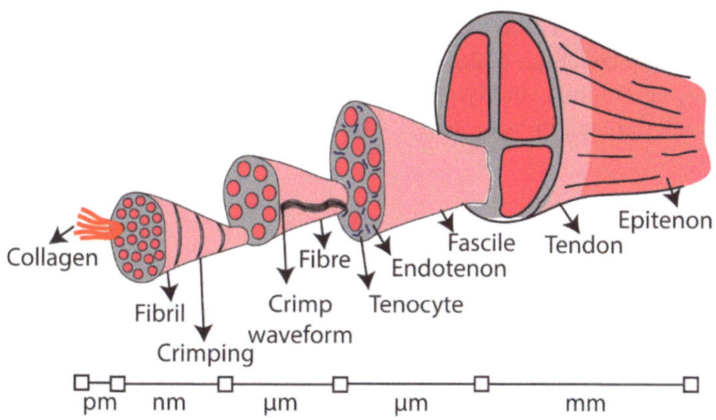

Fig. 5.1 Schematic of the hierarchical structure of a tendon or ligament unit. (The figure is redrawn from [14])

for delivering drugs in tendinopathies, as it includes 60% of the tendon dry mass and 95% of the total collagen [17–19]. Reconstituted collagen type I is readily available in large quantities. The next most frequently used biomaterial drug delivery system is fibrin, a provisional matrix secreted during the tissue healing response after tendon/ligament injury [20, 21]. Unlike collagen, fibrin makes an excellent reservoir for growth factor proteins owing to its binding sites [20, 22]. Elastin constitutes 1–2% of ECM mass in tendons and is also used as a drug delivery vehicle [7]. However, only engineered recombinant elastin-like peptides can deliver drugs, as natural elastin ECM is both hydrophobic and insoluble. Recombinant elastin-like peptides have been tested as treatments for Achilles injuries in preclinical animal models [23, 24]. The ground substance in tendons contains hyaluronan, elastin, tenascin, and proteoglycans such as decorin, biglycan, fibromodulin, and lumican [5, 7, 20]. Together, these proteins form a highly viscous, hydrophilic matrix supporting collagen fibers [5]. Among the ground substance proteins, hyaluronic acid is the most widely tested drug delivery system for tendinopathy-related pain, inflammation, and regeneration [20].

1.3 Tendon Healing-Informed Drug Delivery Systems

Collagen fibers in tendons exist in a crimp-like waveform in their natural state (Fig. 5.1). During locomotion, collagen fibers can be stretched up to 4%, losing their crimp-like pattern [25]. Injuries occur when stretched beyond this limit; excessive stretching beyond 4% strain results in microscopic tearing of collagen fibers, and further stretching beyond 8–10% causes macroscopic rupture [25]. Owing to their hypovascular nature, the healing processes of tendons and ligaments are extremely slow in the order of years (Fig. 5.2).

The natural healing response of the tendon involves three main phases: (i) inflammatory, (ii) proliferation/repair, and (iii) remodeling (Fig. 5.2) [6]. During the inflammatory phase, a fibrin-based blood clot forms releasing chemoattractants from activated platelets [5, 6, 26]. Immune cells migrate to the injury site, including monocytes, neutrophils, and lymphocytes. The immune cells work to phagocytize necrotic material and, in turn, secrete vasoactive, chemotactic, and pro-angiogenic

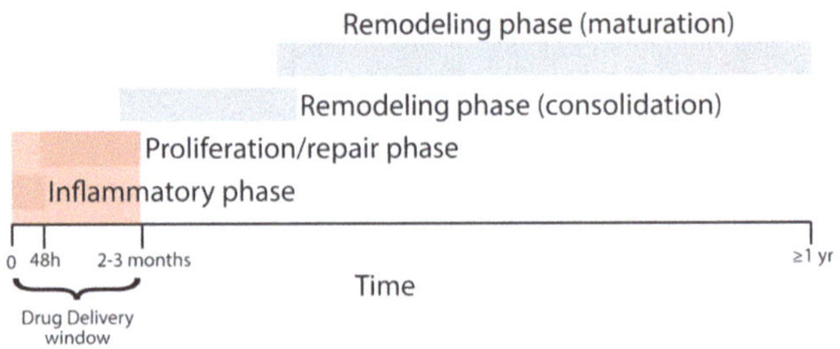

Fig. 5.2 Drug delivery window during phasic tendon healing. (The figure is redrawn from [5])

factors, thereby recruiting tenocytes and sprouting new blood vessels. After 2–3 days of injury, the second proliferative phase occurs, where fibroblasts migrate to the injury site along with tenocytes and proliferate. At this stage, neutrophils decline, and macrophages continuously produce growth factors. The ECM in this stage is unorganized, predominantly collagen type III, fibronectin, and proteoglycans [5, 6]. Following 2–3 months of injury, the remodeling phase starts when collagen type I replaces collagen type III and glycosaminoglycans. Cellularity decreases during this period, and collagen fibers begin to align along the longitudinal axis of the tendon. The final maturation stage includes increased collagen fibril cross-linking and the formation of mature tendon tissue [5, 27].

Many effective tendinopathy therapeutics tend to replicate or augment cellular and physiologic events during the inflammatory and proliferative phases [5]. For example, increasing the delivery of exogenous growth factor proteins is a favored therapeutic strategy, as these growth factor proteins have known roles during the inflammatory and proliferative stages of regeneration, including basic fibroblast growth factor (bFGF), insulin growth factor (IGF), platelet-derived growth factor (PDGF), transforming growth factor (TGF-β1), and vascular endothelial growth factor (VEGF) [5]. Along the same lines, cell-based therapeutics tend to replicate inflammatory and proliferative healing events such as tenocyte recruitment, tendon stem cell differentiation, and fibroblast-based ECM production. Insights from embryonic tendon development indicate that tendon regeneration can be promoted by tipping the balance toward intrinsic healing via tendon cells instead of extrinsic recovery provided by migrating inflammatory cells [26, 28]. Based on this information, drug delivery treatments focus on reducing the inflammatory burden during the early stages of injury in the hope of recapitulating healing events in the developing embryonic tendon.

2 Drug Delivery Systems and Treatments for Tendon and Ligament Injuries and Disorders

2.1 Surgery

The advantage of surgery is the ability to fix acute and macroscopic injuries such as tears and complete resections. It is the most widely used therapeutic intervention, with more than 300,000 rotator cuff tendon surgeries performed yearly [10]. Surgery can also treat chronic injuries but is associated with more significant complications [29].

The disadvantages of surgery include (i) the inability to restore tendon/ligament mechanical function back to its natural state and (ii) the development of postoperative complications including re-rupture, elongation, and muscle atrophy [30]. Recently, the use of surgery for treating tendinopathies has become controversial. Certain research studies have found surgery comparable to nonsurgical options for reducing re-rupture rates or improving tendon tissue properties [31, 32].

2.2 Physical Therapy

The advantage of physical therapy is the ability to ameliorate injuries and regenerate tendons with mechanical loading [25]. Physical therapy must be tailored for the specific tendon/ligament tissue based on anatomy, and no single treatment protocol fits all tissues [3, 10]. For example, the flexor tendon requires mobilization, while the rotator cuff tendon needs to remain immobilized for successful healing [10, 28]. Other therapeutic modalities under this category include cryotherapy, the application of magnetic fields, and ultrasound [33]. The obvious disadvantage is that physical therapy is not a stand-alone treatment for tendon and ligament injuries. Instead, it is a procedure used in conjunction with other treatment modalities.

2.3 Small-Molecule Drug Delivery

The main advantage of small-molecule drug delivery is their size, enabling them to cross the cell membrane passively after systemic administration. Nonsteroidal inflammatory drugs (NSAIDs) are the most frequently used small-molecule drugs for treating tendinopathy pain. NSAIDs work by nonselectively inhibiting cyclooxygenase (COX) enzymes and thereby indirectly prevent the production of prostaglandins, reducing inflammation and associated pain [10, 34]. Small molecules such as synthetic corticosteroids also work by reducing inflammation.

Small-molecule drugs can be administered orally, topically, or parenterally [10]. Oral tablet formulations are made with sugars (e.g., lactose, cellulose) or animal proteins such as gelatin [35]. Polypropylene glycol is the primary drug delivery vehicle in topical formulations [36]. Both tablet and topical formulations contain other binding agents, coatings, plasticizers, and preservatives. Some disadvantages of small-molecule drug delivery systems include cytotoxicity due to their ability to diffuse freely across the cell membrane when administered systemically. For example, certain classes of small-molecule drugs negatively affect tenocyte viability, fibroblastic presence, and collagen expression [37]. For these reasons, small-molecule treatment should be used only during the early stages of tendon healing [34].

2.4 Growth Factor Protein Delivery

Tendon and ligament injuries produce growth factors at different stages of the healing process, so exogenous growth factor delivery is an accepted treatment strategy [7, 38]. Exogenous growth factor delivery has been used for (i) recruiting immune cells, tenocytes, and blood vessels to the site of injury, (ii) supporting the proliferation of tenocytes and vascular cells, (iii) promoting secretion of growth factors in a positive feedback loop, and (iv) laying down new extracellular matrix proteins at the site of injury [3, 6]. Growth factor proteins, including basic fibroblast growth factor (bFGF), bone morphogenic protein (BMP-12, BMP-13, BMP-14), connective tissue growth factor (CTGF), insulin-like growth factor (IGF), platelet-derived growth factor (PDGF), transforming growth factor (TGF-β), and vascular endothelial growth factor (VEGF), have been tested in preclinical animal models [1, 39]. The benefits of these growth factors in tendon repair have been reviewed in detail previously [5, 7].

Growth factor proteins are biological drugs with the same limitations, including (i) reduced in vivo stability due to susceptibility to proteolytic cleavage within inflammatory environments, varied pH, temperature, and other conditions, (ii) off-target effects when not delivered directly to the injury site, such as triggering the immune system or promoting the growth of unwanted cell types, and (iii) beneficial only when delivery is performed during early stages of tendon and ligament healing [40]. Delivery of the growth factors in preclinical studies has been through injections directly to the injury site or drug delivery systems such as overexpression plasmid systems and biomaterial reservoirs. It is worth noting that there are currently no growth factor therapies approved by the FDA for treating tendinopathies.

2.5 Gene Delivery

Gene delivery systems aim to provide controlled and longer-lasting delivery of growth factor proteins to the injury site. Like growth factor systems, they aim to recruit cells to the injury site to augment tendon regeneration through cell proliferation and ECM production. Some other advantages of gene delivery systems include (i) the production of proteins through natural post-translational host cell mechanisms leading to greater biological activity, (ii) reduced risk of immune reactions to growth factor proteins produced by host cells, and (iii) improved efficacy owing to better protein stability and bioactivity [3, 5, 6, 13].

Genetically engineered viral genomes avoid replication and virulent components while maintaining their natural capacity to transfer genetic material to mammalian cells. The most efficient way to deliver growth factor genes is using viral vectors. The most common recombinant viruses for gene delivery include adenoviruses, lentiviruses, retroviruses, and adeno-associated viruses (AAVs) [5]. BMP-14 was delivered successfully using adenoviruses to the Achilles tendon in a rat model of laceration injury [41]. This study showed that transduced tendons exhibited 70% greater tensile strength than control tendons [41]. Further, transduced tendons also contained more neo-tenocytes at the healing site [41]. The use of viral vectors for gene delivery has been reviewed in detail previously [42, 43].

Over the past decade, more research studies have focused on nonviral vector-based gene delivery over viral vector-based delivery strategies [44]. The disadvantage of viral vectors is that they are cumbersome for clinical translation due to potential off-target effects and their capacity to trigger the immune system. Nonviral vectors can include the naked plasmid, but they often involve a polymer carrier such as a liposome or biomaterial. Poly(lactic-co-glycolic acid) (PLGA)-based nanoparticles have been used for delivering TGF-β1-miRNA for preventing flexor tendon adhesion formation in a chicken model of tendon injury [45]. PLGA has also been tested for co-delivering bFGF and VEGF genes, and FGF and VEGF transfected tendons showed increased strength and improved gliding function [46].

Disadvantages of gene delivery include (i) risk of in vivo degradation when injected without a carrier, (ii) safety concerns of using plasmids, as they can be associated with mutagenicity, (iii) carriers such as viral vectors often elicit life-threatening immune responses, and (iv) carriers such as nonviral vectors do not elicit an immune response, but they have diminished transfecting capacity.

2.6 Platelet-Rich Plasma (PRP) Delivery

Tendon and ligament healing requires a complex nonoverlapping network of different growth factor proteins. While delivering single growth factor proteins has yielded promising results, co-delivery of multiple growth factors could be a promising treatment strategy. Platelet-rich plasma (PRP) is an autologous blood product currently used in the clinic for treating tendinopathy and tendon injury patients [47]. Currently, PRP delivery is a treatment for various tendinopathies associated with rotator cuff tendon, patellar tendon, Achilles tendons, ulnar collateral ligament, and anterior cruciate ligament [1]. PRP is obtained by centrifuging peripheral venous blood, leaving behind concentrated growth factor proteins, platelets, fibrin, and leukocytes. Besides being a cocktail of different growth factor proteins, PRP also contains platelets and fibrin ECM [47]. Growth factor proteins found in PRP include platelet-derived growth factor (PDGF), vascular endothelial growth factor (VEGF), transforming growth factor-β (TGF-β), epidermal growth factor (EGF), fibroblast growth factor (FGF), and insulin-like growth factor (IGF) [47].

The advantages of PRP are as follows: (i) PRP has the ability to deliver multiple growth factor proteins in high concentrations along with platelets and ECM to promote tendon regeneration, (ii) PRP therapies are multifunctional as they can modulate inflammation, promote proliferation of cells, prevent apoptosis, induce differentiation, and also promote matrix secretion, (iii) as an autologous product, there are no immune-related issues with PRP delivery, (iv) PRP preparation involves simple laboratory instruments, and therefore, the therapy is feasible in clinical settings with minimal equipment [48].

There are many inconsistencies regarding the benefits of PRP therapy based on their studies in preclinical animal models and humans. While some studies have reported beneficial effects on tendon regeneration, others have found no significant difference [49, 50]. Further, studies have shown more significant biomechanical benefits with single growth factor delivery than PRP therapy [51]. The inconsistencies and disadvantages of PRP therapy are due to (i) donor-based variability, as PRP from different donors varies based on age, sex, pathology, etc., (ii) variability due to different preparation protocols leading to some PRP formulations containing more platelets or leukocytes, and (iii) variations due to delivery of the product to the patient in terms of composition, quantity, and frequency [1, 49]. Because of the lack of conclusive benefits of PRP treatments, they are not covered by medical insurance companies, and this is one of the primary drawbacks of this therapy [1, 10].

2.7 Cellular Drug Delivery Systems

During the natural healing process, different cell types arrive at the injury site for repair, and repair is mediated by cellular grafting, cell secretion of growth factors, and ECM proteins. Cellular drug delivery systems aim to recapitulate this process. The main advantage of cellular drug delivery systems is their ability to (i) integrate with the host tissue and (ii) sustain long-term delivery of favorable growth factors and ECM proteins. Cell-based drug delivery systems tested in preclinical and clinical models of therapy include bone marrow-derived mesenchymal stem cells (BMSCs), adipose-derived stem cells (ASCs), endogenous ligament-derived stem cells (LDSCs), tendon-derived stem cells (TDSCs), and MSCs from other sources (e.g., synovial fluid) [1, 5, 6].

Research studies have shown that autologous tenocytes harvested from a non-injured patellar tendon can be expanded in vitro and used as cell drug delivery systems for treating chronic lateral epicondylitis (LE) [52, 53]. Sixteen patients reported improved pain, function, and structural repair at 1-year post-treatment using this treatment strategy [52]. Follow-up at 4.5 years showed continued clinical benefit confirming the durability of autologous tenocyte therapy for treating tendinopathies [53]. Autologous tendon injection (ATI) is currently offered in Australia by a company called Orthocell for treating rotator cuff tendinopathy. Treatment benefits purport improved function and pain resolution 6 months post-treatment [54]. While autologous therapies do not elicit an immune response, they are limited in their in vitro expansion potential. Autologous tenocyte therapies are also associated with donor site morbidity.

Stem cells overcome the limitations of autologous therapies due to their unlimited expansion potential. Multipotent mesenchymal stem cells (MSCs) have been successfully tested in preclinical and clinical settings for treating tendinopathies [55]. In a preclinical equine model of tendinopathy, bone marrow-derived MSC treatment resulted in quicker recovery and reduced re-injury rates [13]. Further, in a human clinical trial of 14 patients with complete rotator cuff tears, iliac-derived MSC treatment resulted in improved healing and tissue integrity [5].

One of the significant advantages of using MSCs is their immunomodulatory capacity and ability to exert anti-inflammatory effects at the site of tendon injury [5]. MSCs have also been known to exert regenerative effects by secreting trophic mediators, growth factors, and angiogenic and anti-apoptotic factors [54]. It is also believed that MSCs could exert their anti-inflammatory and regenerative effects through the secretion of extracellular vesicles [56]. The disadvantages of using

MSCs and cell-based delivery systems, in general, are (i) the need for invasive extraction procedures, (ii) donor and extraction site-based variabilities, and (iii) MSCs may never fully integrate into the site of injury. MSCs delivered via biomaterial hydrogels (e.g., collagen or fibrin) have better tissue integration [3]. Combinatory systems possess better mechanical properties, histological scores, tissue integration, and functionality. Even with all these improvements, controlled human trials have not been unequivocally promising, with specific clinical trials showing positive results and others showing no improvement over control injections [57].

2.8 *Biomaterial-Based Drug Delivery Systems*

2.8.1 Tissue Grafts and Decellularized Matrices

The current standard for treating large-scale tendon defects is autologous, allogenic, or xenogeneic tissue grafts [3]. The source for autograft donor tissue is determined based on the size of the injury and the required mechanical properties. The drawback, however, is the donor morbidity associated with harvesting tissue. Allografts are suitable alternates, comparable in functional outcome, but contribute to less mature neo-tissue, are challenging to obtain, are more expensive, and are 6.7 times more prone to failure than autografts [3]. Other disadvantages of allografts include potential mechanical property mismatches due to age, body weight, and other characteristics.

Xenografts are suggested as a lower-cost alternative to allografts, prepared by removing cells through decellularization techniques and cross-linking and sterilization processes. Xenografts derived from the bovine dermis (Bio-Blanket), equine pericardium (OrthADAPT), and small intestinal mucosa (Restore) are FDA-approved scaffolds currently on the market for tendon tissue repair [6]. The major drawback of xenografts is their potential ability to trigger an immune response resulting in graft rejection. Revivicor, a subsidiary of United Therapeutics, has developed genetically modified GalSafe pigs lacking alpha-gal sugar, which causes immune rejection in patients [58]. Tendons from these pigs could be superior to current treatments, functioning almost like allografts, and will likely be tested in the clinic in the coming years.

Several research studies have attempted to improve the drawbacks of allografts and xenografts. Hyaluronic acid (HA), a natural component of the tendon's extracellular matrix, can improve allograft biocompatibility via pretreatment, significantly reducing adhesion formation [59]. Significantly, pretreatment with HA resulted in reduced adhesion formation [59]. Autologous cell seeding can also enhance allograft biocompatibility. Research has shown that adipose-derived mesenchymal stem cells (AD-MSCs) cultured on allografts increase tensile strength and collagen density [60]. Further, human donor tenocyte re-cellularized rabbit flexor tendons showed increased ultimate tensile strength and elastic modulus after 4 weeks of in vivo treatment [61].

2.8.2 Natural Biomaterial Drug Delivery Systems

As collagen is the major component of the tendon ECM, collagen biomaterials in different forms, such as gels, fibers, scaffolds, and composites, have been investigated as delivery vehicles for tendinopathies. Collagen gels make excellent drug delivery vehicles as they can promote the viability of cells and prevent growth factor proteins from protease degradation [17]. Recombinant BMP-2 released from a collagen gel resulted in more significant neo-tissue formation compared to growth factor delivery alone in a rabbit model of extensor tendon injury [62]. Compared to gel formats, collagen scaffolds provide relatively better biomechanical support at the injury site, mainly when they include aligned fibers [63]. Aligned collagen fiber substrates have been shown to induce tenogenic phenotype in fibroblasts and possess better regenerate capacity for treating tendons in vivo [64]. Even with different formats, pure collagen-based biomaterials are disadvantaged and unable to match the native tendon's mechanical properties. Collagen scaffolds can be fabricated as a composite with other biomaterials, such as silk, chitosan, and PLGA, to strengthen and better match native tissue mechanical properties [65, 66]. Another drawback with collagen-based materials is that they are xenogeneic sourced, with a constant risk of infection and immune reaction. Recent advances in biotechnology have enabled the production of standardized formulations of human collagen using plants [67, 68]. These could ease clinical application and regulatory approval of collagen-based biomaterials for treating tendinopathies.

Fibrin is needed for clotting and wound healing, and it is the second most widely used biomaterial drug delivery platform for treating tendinopathies. Fibrin is mainly used in a gel format to deliver growth factor proteins and cells [69, 70]. Like collagen, fibrin has inferior mechanical properties and primarily serves as a provisional scaffold at the injury site. Silk fibroin is a natural biomaterial drug delivery carrier used primarily for its superior mechanical properties [65, 71]. Because silk fibroin proteins do not have cellular binding sites, they are used as bulking agents or composite scaffolds. Knitted silk combined with collagen resulted in better tendon healing than collagen alone control scaffolds in a rabbit tendon healing model [72]. While collagen and silk reproduce the ECM and mechanical properties of the native tendon, respectively, chitosan and alginate are used to recapitulate the tendon's ground substance, which is rich in glycosaminoglycans (GAGs) [6, 73, 74]. Silk fibroin, chitosan, and alginate do not have any cellular binding sites. Therefore, they are often used as biocomposite scaffolds [74, 75], and previous studies have used them to deliver growth factors such as FGF, PDGF, VEGF, and EGF [76–79]. Hyaluronic acid is a non-sulfated glycosaminoglycan (GAG) and a natural component of the tendon extracellular matrix; it is also widely used for treating tendinopathies and drug delivery [80, 81].

2.8.3 Synthetic Biomaterial Drug Delivery Systems

Synthetic biomaterials have been used mainly because of their improved mechanical properties compared to natural biomaterials. Further, synthetic grafts can be customized to be biocompatible, degradable, and non-immunogenic. Several polyesters such as polylactic acid (PLA), polyglycolic acid (PGA), and poly(lactic-co-glycolic acid) (PLGA) have been used for treating tendinopathies, as their degradation by-products of lactic acid and glycolic acid are natural secondary metabolites of the body.

A composite PGA and PLA scaffold was used to deliver adipose-derived mesenchymal stem cells in a rabbit Achilles tendon injury model. Scaffolds contained an inner layer of PGA fibers and an outer layer of knitted PGA + PLA fibers forming a net-like structure. Compared to controls, Achilles injuries treated with these composite scaffolds displayed increased collagen alignment, fibril diameters, and tensile strength [82]. Most studies utilizing PLA, PGA, and PLGA have utilized these polymers in fiber format, often through an electrospinning technique. These polymers have also been woven into nets or mats and have been used for delivering growth factor proteins and cells [83]. The major drawback of PLA, PGA, and PLGA polymers is the acidic nature of their degradation products, which can trigger inflammation [84].

Alternatives such as poly(ε-caprolactone) (PCL), polydioxanone (PDS), and poly(trimethylene carbonate) (PTMC) have been used due to their nonacidic degradation products. Poly(ε-caprolactone) (PCL) composite scaffolds in a rabbit model of tendon healing showed good tissue regeneration capacity [83]. PCL scaffolds are also prepared by electrospinning or weaving to produce aligned structures. When adipose-derived mesenchymal stem cells (ADMSCs) were seeded on aligned woven scaffolds, it resulted in the expression of tendon markers, including collagen type I, type III, tenascin-C, and scleraxis (Scx) [85]. Further, PCL-based composite scaffolds have also been used for delivering growth factor proteins such as bFGF [86].

Although there has been a lot of progress in using synthetic biomaterial scaffolds as cell and protein carriers, these materials still possess low biocompatibility, degradation problems, and issues with mechanical compatibility with the native tendon [83]. Current strategies to improve biocompatibility have included surface functionalization with fibronectin or RGD peptides and reducing the size of the scaffolds to limit the production of secondary metabolites. Mechanical compatibility issues have been overcome mainly through the adoption of weaving techniques and by developing new composite scaffolds with other biomaterials. There are excellent detailed reviews on the use of biomaterial systems for tendon and ligament healing, and you can refer to these for additional information [86–91].

2.9 Other Drug Delivery Formats

Microspheres encapsulate protein drugs for sustained and localized delivery and protect drugs from degradation. Both natural and synthetic biomaterials have been used to fabricate microspheres. PLGA synthetic polymer-derived microspheres loaded with bFGF growth factor have been tested in vitro on tenocytes [92]. Treated cells displayed increased proliferation and viability compared to control groups [92]. Natural biomaterial chitosan has also been used to produce microspheres for tendon healing. Chitosan microspheres were loaded with curcumin and delivered in a rat tendon calcification model [93]. The released curcumin in this study reduced inflammation, rescued calcification, and enhanced regeneration [94]. The significant drawbacks of using microspheres for controlled drug delivery are related to the standardization of fabrication protocols, biocompatibility, and relatively high production costs when using expensive growth factor proteins [95].

Nanoparticles are particles less than 100 nm [96], used as drug carriers because they provide targeted and sustained release. Nanoparticles can also prolong circulation time, and the most significant benefit of these systems is their ability to enter cells via endocytic processes such as pinocytosis, phagocytosis, and endocytosis [97]. Nanoparticles developed using poly(vanillyl alcohol-co-oxalate) reduced inflammation and oxidation in a rodent model of Achilles tendon contusion [98]. Further, dextran-based nanoparticles fared better than emulsion-based systems for delivering FGF in a rodent model of Achilles tendon injury [99]. In this study, the nanoparticle-containing growth factor was encapsulated within an electrospun PLLA membrane [99], adding another layer of polymeric drug delivery to gain further control over drug release [90, 100]. Several other studies have gotten promising results using nanoparticle-based delivery systems [45, 101]. As nanoparticles are relatively newer drug delivery systems, rigorous safety assessments must be carried out before using them in the clinic, which is their major drawback.

Sutures are used during surgical therapy of tendons and ligaments to impart mechanical stability, and they have become ideal vehicles for drug delivery. Growth factors such as bFGF and VEGF loaded into PLGA nanoparticles have been coated on synthetic polyester biomaterial sutures. Tendons treated with modified sutures had higher strength and lower adhesion scores than controls [102]. Among sutures manufactured from natural biomaterials, aligned collagen sutures have been heparinized for prolonged PDGF delivery in a chicken flexor tendon injury model [103]. This study also showed that tendons treated with PDGF-releasing sutures have superior mechanical strength and vascularization compared to those treated with control sutures [103]. Several other studies have reported the beneficial effects of drug-releasing sutures for treating tendinopathies [90]. The drawback of sutures is their low drug-loading capacity, and improvements to loading capacity will open up more applications in the future.

Microneedles are arrays of tiny needles that penetrate the skin superficially. They are used for pain-free, non-invasive delivery of drugs and proteins to the human body. These systems are often fabricated using many polymers, although polyvinyl alcohol (PVA) is preferred due to the ease of manufacturing based on a simple freeze-drying process [104]. Microneedles manufactured using this technique have been used to deliver PDGF and parathyroid hormone (PTH) in an animal Achilles tendon injury model [104, 105]. PDGF and parathyroid hormone-treated tendons displayed improved healing compared to controls [104, 105]. Because microneedle drug delivery systems can only provide localized delivery to the dermal region, their application might be limited to treating only certain kinds of tendon and ligament injuries. For example, most preclinical studies have tested the Achilles tendon due to its proximity to the skin surface.

3 Integrated Drug Delivery Systems

The future of tendinopathy treatment is using integrated drug delivery systems. Novel integrated therapies target tendon tissue regeneration by providing ECM-derived biophysical cues and modulating the immune response for adequate resolution of inflammation [10]. These systems are formulated as stand-alone injectable formulations or combinatory biomaterial delivery systems. The novelty of these products is using the combination of drug and biological product therapies with biomaterials for additional synergistic effects [10]. Examples of novel integrated delivery systems include (i) biomaterial porous microspheres loaded with dexamethasone showing extended release in vitro compared to drug alone [106], (ii) recombinant BMP-2 released from a collagen gel resulted in more significant neo-tissue formation compared to growth factor delivery alone in a rabbit model of extensor tendon injury [62], and (iii) knitted silk scaffolds combined with collagen resulting in better tendon healing than collagen alone in a rabbit model of tendon injury [72]. While integrated drug delivery systems show promise, the application of these treatments will likely be limited to preclinical models in the near term, as significant regulatory submission-related challenges need to be overcome for these therapies to be used as human treatments.

4 Conclusion and Future Outlook

This book chapter provided a comprehensive summary of different therapies for treating tendon and ligament injuries while discussing their advantages and disadvantages in detail. Despite many technological advancements in drug delivery systems, we still do not have a therapeutic and drug delivery procedure for promoting tendon regeneration to the point where it resembles the pre-injured state. Research is currently headed toward developing novel integrated therapies, combining the beneficial effects of drugs, biologics, and biomaterials. As our understanding of embryonic tendon development and inflammatory mechanisms improve and technological advancements continue in fields such as 3D printing, integrated therapies will evolve. The focus in the coming years will be on discovering integrated treatments that are nonimmunogenic, nontoxic, bioresorbable, and scalable and figuring out the regulatory strategy for translating these therapies from the bench to the clinic.

References

1. Leong, N.L., et al., *Tendon and Ligament Healing and Current Approaches to Tendon and Ligament Regeneration.* Journal of Orthopaedic Research, 2020. **38**(1): p. 7–12.
2. Wu, F., M. Nerlich, and D. Docheva, *Tendon injuries: Basic science and new repair proposals.* EFORT Open Rev, 2017. **2**(7): p. 332–342.
3. Lomas, A.J., et al., *The past, present and future in scaffold-based tendon treatments.* Advanced Drug Delivery Reviews, 2015. **84**: p. 257–277.
4. Korntner, S., et al., *Limiting angiogenesis to modulate scar formation.* Advanced Drug Delivery Reviews, 2019. **146**: p. 170–189.
5. Docheva, D., et al., *Biologics for tendon repair.* Adv Drug Deliv Rev, 2015. **84**: p. 222–39.
6. Hafeez, M.N., et al., *Tendon tissue repair in prospective of drug delivery, regenerative medicines, and innovative bioscaffolds.* Stem cells international, 2021. **2021**.
7. Schneider, M., et al., *Rescue plan for Achilles: Therapeutics steering the fate and functions of stem cells in tendon wound healing.* Advanced Drug Delivery Reviews, 2018. **129**: p. 352–375.
8. Rainsford, K.D., *Aspirin and related drugs*. 2004: CRC Press.
9. Vargason, A.M., A.C. Anselmo, and S. Mitragotri, *The evolution of commercial drug delivery technologies.* Nat Biomed Eng, 2021. **5**(9): p. 951–967.
10. Freedman, B.R., D.J. Mooney, and E. Weber, *Advances toward transformative therapies for tendon diseases.* Science Translational Medicine, 2022. **14**(661): p. eabl8814.
11. Manning, C.N., et al., *Sustained delivery of transforming growth factor beta three enhances tendon-to-bone healing in a rat model.* J Orthop Res, 2011. **29**(7): p. 1099–105.
12. Li, L., et al., *Release of celecoxib from a bi-layer biomimetic tendon sheath to prevent tissue adhesion.* Materials Science and Engineering: C, 2016. **61**: p. 220–226.
13. Walden, G., et al., *A Clinical, Biological, and Biomaterials Perspective into Tendon Injuries and Regeneration.* Tissue Eng Part B Rev, 2017. **23**(1): p. 44–58.
14. Thorpe, C.T., et al., *Chapter 1 – Tendon Physiology and Mechanical Behavior: Structure–Function Relationships*, in *Tendon Regeneration*, M.E. Gomes, R.L. Reis, and M.T. Rodrigues, Editors. 2015, Academic Press: Boston. p. 3–39.
15. Laurent, C., *13 – Micromechanics of Ligaments and Tendons*, in *Multiscale Biomechanics*, J.-F. Ganghoffer, Editor. 2018, Elsevier. p. 489–509.
16. Tempfer, H., et al., *Biological augmentation for tendon repair: lessons to be learned from development, disease, and tendon stem cell research.* Cell Engineering and Regeneration, 2020: p. 735–765.
17. Amit Kumar, V., *Collagen-Based Biomaterial as Drug Delivery Module*, in *Collagen Biomaterials*, M. Nirmal and C. Sanjiban, Editors. 2022, IntechOpen: Rijeka. p. Ch. 5.
18. Juncosa-Melvin, N., et al., *Effects of mechanical stimulation on the biomechanics and histology of stem cell–collagen sponge constructs for rabbit patellar tendon repair.* Tissue engineering, 2006. **12**(8): p. 2291–2300.
19. Mozafari, M., et al., *Synthesis and characterisation of highly interconnected porous poly(ε-caprolactone)-collagen scaffolds: a therapeutic design to facilitate tendon regeneration.* Materials Technology, 2018. **33**(1): p. 29–37.
20. Martino, M.M., et al., *Heparin-binding domain of fibrin(ogen) binds growth factors and promotes tissue repair when incorporated within a synthetic matrix.* Proc Natl Acad Sci U S A, 2013. **110**(12): p. 4563–8.

21. Soreide, E., et al., *Fibrin glue mediated delivery of bone anabolic reagents to enhance healing of tendon to bone.* Journal of Cellular Biochemistry, 2018. **119**(7): p. 5715–5724.
22. Sarrigiannidis, S.O., et al., *A tough act to follow: collagen hydrogel modifications to improve mechanical and growth factor loading capabilities.* Materials Today Bio, 2021. **10**: p. 100098.
23. MacEwan, S.R. and A. Chilkoti, *Applications of elastin-like polypeptides in drug delivery.* J Control Release, 2014. **190**: p. 314–30.
24. Gigante, A., et al., *Effects of elastin-derived peptide on Achilles' tendon healing: An experimental study.* Journal of Materials Science: Materials in Medicine, 2003. **14**(8): p. 717–720.
25. Wang, J.H.C., *Mechanobiology of tendon.* Journal of Biomechanics, 2006. **39**(9): p. 1563–1582.
26. Nichols, A.E.C., K.T. Best, and A.E. Loiselle, *The cellular basis of fibrotic tendon healing: challenges and opportunities.* Translational Research, 2019. **209**: p. 156–168.
27. Müller, S.A., et al., *Tendon healing: an overview of physiology, biology, and pathology of tendon healing and systematic review of state of the art in tendon bioengineering.* Knee Surgery, Sports Traumatology, Arthroscopy, 2015. **23**(7): p. 2097–2105.
28. Thomopoulos, S., et al., *Mechanisms of tendon injury and repair.* Journal of Orthopaedic Research, 2015. **33**(6): p. 832–839.
29. Jokela, A., et al., *A systematic review of surgical intervention in the treatment of hamstring tendon ruptures: current evidence on the impact on patient outcomes.* Annals of Medicine, 2022. **54**(1): p. 978–988.
30. Freedman, B.R., et al., *Enhanced tendon healing by a tough hydrogel with an adhesive side and high drug-loading capacity.* Nat Biomed Eng, 2022. **6**(10): p. 1167–1179.
31. Freedman, B.R., et al., *Temporal Healing of Achilles Tendons After Injury in Rodents Depends on Surgical Treatment and Activity.* J Am Acad Orthop Surg, 2017. **25**(9): p. 635–647.
32. van der Eng, D.M., et al., *Rerupture rate after early weightbearing in operative versus conservative treatment of Achilles tendon ruptures: a meta-analysis.* J Foot Ankle Surg, 2013. **52**(5): p. 622–8.
33. Kane, S.F., L.H. Olewinski, and K.S. Tamminga, *Management of Chronic Tendon Injuries.* Am Fam Physician, 2019. **100**(3): p. 147–157.
34. Dideriksen, K., *Muscle and tendon connective tissue adaptation to unloading, exercise and NSAID.* Connective Tissue Research, 2014. **55**(2): p. 61–70.
35. Ghourichay, M.P., et al., *Formulation and Quality Control of Orally Disintegrating Tablets (ODTs): Recent Advances and Perspectives.* BioMed Research International, 2021. **2021**: p. 6618934.
36. McPherson, M.L. and N.M. Cimino, *Topical NSAID Formulations.* Pain Medicine, 2013. **14**(suppl_1): p. S35–S39.
37. Dean, B.J.F., et al., *The risks and benefits of glucocorticoid treatment for tendinopathy: A systematic review of the effects of local glucocorticoid on tendon.* Seminars in Arthritis and Rheumatism, 2014. **43**(4): p. 570–576.
38. Nourissat, G., F. Berenbaum, and D. Duprez, *Tendon injury: from biology to tendon repair.* Nature Reviews Rheumatology, 2015. **11**(4): p. 223–233.
39. Prabhath, A., et al., *Growth factor delivery strategies for rotator cuff repair and regeneration.* International Journal of Pharmaceutics, 2018. **544**(2): p. 358–371.
40. Ren, X., et al., *Growth Factor Engineering Strategies for Regenerative Medicine Applications.* Front Bioeng Biotechnol, 2019. **7**: p. 469.
41. Bolt, P., et al., *BMP-14 Gene Therapy Increases Tendon Tensile Strength in a Rat Model of Achilles Tendon Injury.* JBJS, 2007. **89**(6).
42. Zhao, Z., A.C. Anselmo, and S. Mitragotri, *Viral vector-based gene therapies in the clinic.* Bioengineering & Translational Medicine, 2022. **7**(1): p. e10258.
43. Bulcha, J.T., et al., *Viral vector platforms within the gene therapy landscape.* Signal Transduction and Targeted Therapy, 2021. **6**(1): p. 53.
44. Gantenbein, B., et al., *Non-viral Gene Delivery Methods for Bone and Joints.* Front Bioeng Biotechnol, 2020. **8**: p. 598466.

45. Zhou, Y., et al., *Nanoparticle-mediated delivery of TGF-β1 miRNA plasmid for preventing flexor tendon adhesion formation.* Biomaterials, 2013. **34**(33): p. 8269–8278.
46. Yang, Q.Q., et al., *Therapeutic strategies for flexor tendon healing by nanoparticle-mediated co-delivery of bFGF and VEGFA genes.* Colloids and Surfaces B: Biointerfaces, 2018. **164**: p. 165–176.
47. Zhou, Y. and J.H. Wang, *PRP Treatment Efficacy for Tendinopathy: A Review of Basic Science Studies.* Biomed Res Int, 2016. **2016**: p. 9103792.
48. Andia, I., J.I. Martin, and N. Maffulli, *Advances with platelet rich plasma therapies for tendon regeneration.* Expert Opin Biol Ther, 2018. **18**(4): p. 389–398.
49. Galatz, L.M., et al., *Tendon regeneration and scar formation: the concept of scarless healing.* Journal of orthopaedic research, 2015. **33**(6): p. 823–831.
50. Keene, D.J., et al., *Platelet rich plasma injection for acute Achilles tendon rupture: PATH-2 randomised, placebo controlled, superiority trial.* BMJ, 2019. **367**: p. l6132.
51. Lamplot, J.D., et al., *Distinct Effects of Platelet-Rich Plasma and BMP13 on Rotator Cuff Tendon Injury Healing in a Rat Model.* The American Journal of Sports Medicine, 2014. **42**(12): p. 2877–2887.
52. Wang, A., et al., *Autologous tenocyte injection for the treatment of severe, chronic resistant lateral epicondylitis: a pilot study.* Am J Sports Med, 2013. **41**(12): p. 2925–32.
53. Wang, A., et al., *Evidence for the Durability of Autologous Tenocyte Injection for Treatment of Chronic Resistant Lateral Epicondylitis: Mean 4.5-Year Clinical Follow-up.* Am J Sports Med, 2015. **43**(7): p. 1775–83.
54. Conrad, S., et al. *Stem Cell Therapy for Tendon Regeneration: Current Status and Future Directions.* in *Tissue Engineering and Regenerative Medicine.* 2019. Cham: Springer International Publishing.
55. Goldberg, A.J., et al., *Autologous Stem Cells in Achilles Tendinopathy (ASCAT): protocol for a phase IIA, single-centre, proof-of-concept study.* BMJ Open, 2018. **8**(5): p. e021600.
56. Gao, R., et al., *Small extracellular vesicles from iPSC-derived mesenchymal stem cells ameliorate tendinopathy pain by inhibiting mast cell activation.* Nanomedicine, 2022. **17**(8): p. 513–529.
57. Chun, S.-W., et al., *A randomized controlled trial of stem cell injection for tendon tear.* Scientific Reports, 2022. **12**(1): p. 818.
58. Dolgin, E., *First GM pigs for allergies. Could xenotransplants be next?* Nature Biotechnology, 2021. **39**(4): p. 397–400.
59. Karabekmez, F.E. and C. Zhao, *Surface Treatment of Flexor Tendon Autograft and Allograft Decreases Adhesion Without an Effect of Graft Cellularity: A Pilot Study.* Clinical Orthopaedics and Related Research®, 2012. **470**(9).
60. Woon, C.Y.L., et al., *Human Flexor Tendon Tissue Engineering: Revitalization of Biostatic Allograft Scaffolds.* Tissue Engineering Part A, 2012. **18**(23–24): p. 2406–2417.
61. Thorfinn, J., et al., *Bioreactor optimization of tissue engineered rabbit flexor tendons in vivo.* Journal of Hand Surgery (European Volume), 2012. **37**(2): p. 109–114.
62. Lee, K.W., et al., *Tendon-bone interface healing using an injectable rhBMP-2-containing collagen gel in a rabbit extra-articular bone tunnel model.* J Tissue Eng Regen Med, 2017. **11**(5): p. 1435–1441.
63. Gentleman, E., et al., *Development of Ligament-Like Structural Organization and Properties in Cell-Seeded Collagen Scaffolds in vitro.* Annals of Biomedical Engineering, 2006. **34**(5): p. 726–736.
64. Wang, W., et al., *Aligned nanofibers direct human dermal fibroblasts to tenogenic phenotype in vitro and enhance tendon regeneration in vivo.* Nanomedicine, 2016. **11**(9): p. 1055–1072.
65. Kwon, S.-Y., et al., *Silk and collagen scaffolds for tendon reconstruction.* Proceedings of the Institution of Mechanical Engineers, Part H: Journal of Engineering in Medicine, 2014. **228**(4): p. 388–396.
66. Zhang, W., et al., *Weft-knitted silk-poly (lactide-co-glycolide) mesh scaffold combined with collagen matrix and seeded with mesenchymal stem cells for rabbit Achilles tendon repair.* Connective tissue research, 2015. **56**(1): p. 25–34.

67. Shoseyov, O., Y. Posen, and F. Grynspan, *Human collagen produced in plants: more than just another molecule.* Bioengineered, 2014. **5**(1): p. 49–52.
68. Seror, J., et al., *The Potential Use of Novel Plant-Derived Recombinant Human Collagen in Aesthetic Medicine.* Plastic and Reconstructive Surgery, 2021. **148**(6S).
69. Uehara, K., et al., *Effect of Fibrin Formulation on Initial Strength of Tendon Repair and Migration of Bone Marrow Stromal Cells in Vitro.* J Bone Joint Surg Am, 2015. **97**(21): p. 1792–8.
70. Pawelec, K.M., et al., *The effects of scaffold architecture and fibrin gel addition on tendon cell phenotype.* Journal of Materials Science: Materials in Medicine, 2015. **26**(1): p. 13.
71. Wani, S.U.D., et al., *Silk fibroin as a natural polymeric based bio-material for tissue engineering and drug delivery systems-A review.* International Journal of Biological Macromolecules, 2020. **163**: p. 2145–2161.
72. Qian, S., et al., *A Collagen and Silk Scaffold for Improved Healing of the Tendon and Bone Interface in a Rabbit Model.* Med Sci Monit, 2019. **25**: p. 269–278.
73. Chacon, E.L., et al., *Collagen-chitosan-hydroxyapatite composite scaffolds for bone repair in ovariectomized rats.* Scientific Reports, 2023. **13**(1): p. 28.
74. Deepthi, S., et al., *Layered chitosan-collagen hydrogel/aligned PLLA nanofiber construct for flexor tendon regeneration.* Carbohydrate polymers, 2016. **153**: p. 492–500.
75. Arvinius, C., et al., *Enhancement of in vivo supraspinatus tendon–to-bone healing with an alginate-chitin scaffold and rhBMP-2.* Injury, 2021. **52**(1): p. 78–84.
76. Yucel, T., M.L. Lovett, and D.L. Kaplan, *Silk-based biomaterials for sustained drug delivery.* Journal of Controlled Release, 2014. **190**: p. 381–397.
77. Hamedi, H., et al., *Chitosan based hydrogels and their applications for drug delivery in wound dressings: A review.* Carbohydrate Polymers, 2018. **199**: p. 445–460.
78. Park, G.Y., J.P. Yoon, and J.H. Choi, *Alginate Nonwoven-Sponge Composite Scaffold for Rotator Cuff Tendon Repair.* Fibers and Polymers, 2020. **21**(9): p. 1879–1887.
79. Wawrzyńska, E. and D. Kubies, *Alginate matrices for protein delivery–a short review.* Physiological research, 2018. **67**: p. S319–S334.
80. Oliva, F., et al., *The Impact of Hyaluronic Acid on Tendon Physiology and Its Clinical Application in Tendinopathies.* Cells, 2021. **10**(11): p. 3081.
81. Huang, G. and H. Huang, *Application of hyaluronic acid as carriers in drug delivery.* Drug delivery, 2018. **25**(1): p. 766–772.
82. Deng, D., et al., *Repair of Achilles tendon defect with autologous ASCs engineered tendon in a rabbit model.* Biomaterials, 2014. **35**(31): p. 8801–8809.
83. Zhang, J., et al., *Research progress of biodegradable polymers in repairing Achilles tendon injury.* Frontiers in Materials, 2022. **9**: p. 6.
84. Silva, M., et al., *Biodegradable polymer nanocomposites for ligament/tendon tissue engineering.* Journal of Nanobiotechnology, 2020. **18**(1): p. 23.
85. Laranjeira, M., et al., *3D Mimicry of Native-Tissue-Fiber Architecture Guides Tendon-Derived Cells and Adipose Stem Cells into Artificial Tendon Constructs.* Small, 2017. **13**(31).
86. Jayasree, A., et al., *Bioengineered Braided Micro–Nano (Multiscale) Fibrous Scaffolds for Tendon Reconstruction.* ACS Biomaterials Science & Engineering, 2019. **5**(3): p. 1476–1486.
87. Shapiro, E., D. Grande, and M. Drakos, *Biologics in Achilles tendon healing and repair: a review.* Current Reviews in Musculoskeletal Medicine, 2015. **8**(1): p. 9–17.
88. Hortensius, R., L. Mozdzen, and B. Harley, *Chapter 13 – Biomaterial Scaffolds for Tendon Tissue Engineering*, in *Tendon Regeneration*, M.E. Gomes, R.L. Reis, and M.T. Rodrigues, Editors. 2015, Academic Press: Boston. p. 349–380.
89. Kim, S.E., J.G. Kim, and K. Park, *Biomaterials for the Treatment of Tendon Injury.* Tissue Eng Regen Med, 2019. **16**(5): p. 467–477.
90. Adjei-Sowah, E., D.S. Benoit, and A.E. Loiselle, *Drug Delivery Approaches to Improve Tendon Healing.* Tissue Engineering Part B: Reviews, 2023.
91. Hou, J., et al., *Biomaterials strategies to balance inflammation and tenogenesis for tendon repair.* Acta Biomaterialia, 2021. **130**: p. 1–16.

92. Kim, D.H., et al., *Biocompatibility and bioactivity of an FGF-loaded microsphere-based bilayer delivery system.* Acta Biomaterialia, 2020. **111**: p. 341–348.
93. Chen, Y., et al., *Controlled-release curcumin attenuates progression of tendon ectopic calcification by regulating the differentiation of tendon stem/progenitor cells.* Materials Science and Engineering: C, 2019. **103**: p. 109711.
94. Zhang, M., et al., *Advanced application of stimuli-responsive drug delivery system for inflammatory arthritis treatment.* Mater Today Bio, 2022. **14**: p. 100223.
95. Zhao, X., et al., *Opportunities and challenges of hydrogel microspheres for tendon–bone healing after anterior cruciate ligament reconstruction.* Journal of Biomedical Materials Research Part B: Applied Biomaterials, 2022. **110**(2): p. 289–301.
96. Parchi, P.D., et al., *Nanoparticles for tendon healing and regeneration: literature review.* Frontiers in aging neuroscience, 2016. **8**: p. 202.
97. Barua, S. and S. Mitragotri, *Challenges associated with Penetration of Nanoparticles across Cell and Tissue Barriers: A Review of Current Status and Future Prospects.* Nano Today, 2014. **9**(2): p. 223–243.
98. Kim, G.W., et al., *Ultrasonographic Imaging and Anti-inflammatory Therapy of Muscle and Tendon Injuries Using Polymer Nanoparticles.* Theranostics, 2017. **7**(9): p. 2463–2476.
99. Liu, S., et al., *Tendon healing and anti-adhesion properties of electrospun fibrous membranes containing bFGF loaded nanoparticles.* Biomaterials, 2013. **34**(19): p. 4690–4701.
100. Chen, C.-H., et al., *Dual functional core–sheath electrospun hyaluronic acid/polycaprolactone nanofibrous membranes embedded with silver nanoparticles for prevention of peritendinous adhesion.* Acta Biomaterialia, 2015. **26**: p. 225–235.
101. Choi, H.J., et al., *Enhanced tendon restoration effects of anti-inflammatory, lactoferrin-immobilized, heparin-polymeric nanoparticles in an Achilles tendinitis rat model.* Carbohydrate Polymers, 2020. **241**: p. 116284.
102. Zhou, Y.L., et al., *Nanoparticle-coated sutures providing sustained growth factor delivery to improve the healing strength of injured tendons.* Acta Biomaterialia, 2021. **124**: p. 301–314.
103. Younesi, M., et al., *Effects of PDGF-BB delivery from heparinized collagen sutures on the healing of lacerated chicken flexor tendon in vivo.* Acta Biomaterialia, 2017. **63**: p. 200–209.
104. Liu, X., et al., *PDGF-loaded microneedles promote tendon healing through p38/cyclin D1 pathway mediated angiogenesis.* Materials Today Bio, 2022. **16**: p. 100428.
105. Yao, Z., et al., *Parathyroid Hormone-Loaded Microneedle Promotes Tendon Healing Through Activation of mTOR.* Advanced Therapeutics, 2020. **3**(8): p. 2000025.
106. Choi, S., et al., *Therapeutic Efficacy of Intratendinous Delivery of Dexamethasone Using Porous Microspheres for Amelioration of Inflammation and Tendon Degeneration on Achilles Tendinitis in Rats.* BioMed Research International, 2020. **2020**: p. 5052028.

Chapter 6
Drug Delivery Systems for Muscles

Mitsuru Naito and Mao Hori

Contents

M. Naito (✉) · M. Hori
Department of Materials Engineering, Graduate School of Engineering, The University of Tokyo, Tokyo, Japan
e-mail: naito@bmw.t.u-tokyo.ac.jp

S. Murab (ed.), *Drug Delivery Systems for Musculoskeletal Tissues*,
https://doi.org/10.1007/978-3-031-55653-1_6

1 Structural and Molecular Basis of Muscles in the Musculoskeletal System

Muscle tissues in the human body are anatomically categorized into three muscle types: skeletal, smooth, and cardiac. Each muscle tissue type has a distinct structure and function. Skeletal muscle accounts for 30–40% of total body weight and is attached to skeletal bones by tendons. Smooth muscle is found in the walls of hollow organs, such as the stomach, intestines, and uterus, as well as in the walls of blood and lymph vessels. Cardiac muscle is found solely in and accounts for the bulk of the heart. Skeletal and cardiac muscles are also categorized as striated muscles, which feature repeating functional units composed of myosin and actin, named sarcomeres. The skeletal muscles control voluntary movements and posture. They also serve as a storage site for glycogen, one of the primary energy sources in the human body. Smooth and cardiac muscles, which are categorized as involuntary muscles, function under the control of the autonomic nervous system. This section focuses on the anatomy, biology, disorders, and therapeutics related to skeletal muscle, which is a component of the musculoskeletal system.

1.1 Anatomy and Histology of Muscles of the Musculoskeletal System

Skeletal muscle is mainly composed of long cylindrical multinucleated muscle fibers, each of which contains several myofibrils composed of myosin filaments, actin filaments, and mitochondria (Fig. 6.1). Each muscle fiber is wrapped by a cellular membrane (sarcolemma), which is further surrounded by a polysaccharide layer in contact with the basal lamina. Along the basal lamina, blood capillaries run parallel to the muscle fibers, and muscle satellite cells (myosatellite cells) reside between the basal lamina and muscle fibers. Muscle fibers mature during the developmental process. Proliferating myoblasts exit the cell cycle (differentiate) and fuse together to form myotubes, and these primary myotubes further fuse with myoblasts

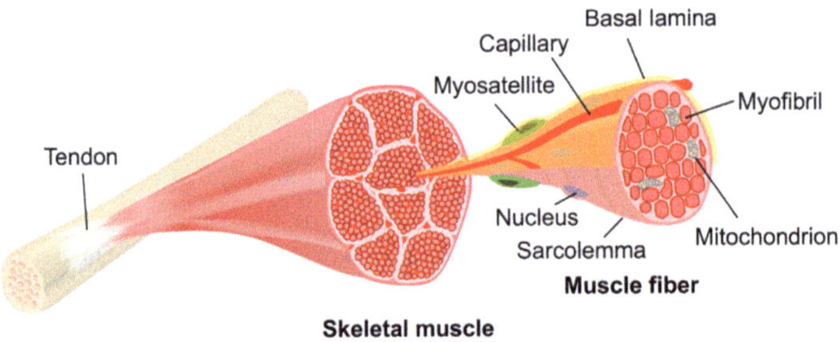

Fig. 6.1 Hierarchical structure of skeletal muscle, muscle fibers, and associated tissues

to form mature muscle fibers. Although the number of muscle fibers remains constant after development, the size of muscle fibers and the myonuclei content increase when differentiated myoblasts are fused with existing muscle fibers. The number of myoblasts decreases dramatically with growth, and myoblasts are usually absent in adults. Instead of myoblasts, myosatellite cells, also known as muscle stem cells, repair damaged muscles in adults [1, 2]. Myosatellite cells, which normally remain dormant, are activated by muscle damage signals and reenter the cell cycle, initiating proliferation. Proliferated myosatellites fuse with damaged muscle fibers for repair. Some proliferating cells then become myosatellites in a dormant state once again. It has been reported that the capacity of myosatellites to regenerate muscle fibers is lost with aging or diseases, such as amyotrophic lateral sclerosis (ALS) and muscular dystrophy [3–5].

1.2 Molecular Biology of Muscles of the Musculoskeletal System

The interaction between actin and myosin filaments in muscle fibers induces muscle contractions and is regulated by calcium ion signals secreted from the ends of motor neurons. There are several types of skeletal muscles, known as slow-twitch type 1, fast-twitch type 2a, and fast-twitch type 2b/x, which are categorized depending on the myosin isoform. Type 1 has a relatively high blood capillary density and is rich in mitochondria. Type 1 muscle fibers are also rich in myoglobin protein, which gives them a reddish appearance. Type 1 is used in aerobic exercises such as long-distance running, and their muscle contraction speed is relatively slow. In contrast, type 2 has fewer mitochondria and is mainly used for anaerobic exercise. Type 2 muscle contracts faster and more powerfully, although it fatigues more rapidly than type 1 muscle. It is subdivided into types 2a and 2b/x, with type 2b/x muscles contracting the fastest and most strongly. The ratio of type 1/2 composition in the body is not constant and generally changes with external stimuli such as exercise, and type 2 muscle fibers are gradually lost with aging [6].

The sarcolemma regulates the influx and outflow of substances across the membrane. Although water, oxygen, and carbon dioxide diffuse freely through the sarcolemma, the diffusion of other molecules is restricted. Several glycoproteins, transporters, and receptors that take up nutrients in the sarcoplasm are found in the sarcolemma (Fig. 6.2). For example, the insulin/insulin-like growth factor (IGF) receptor, transferrin receptor, and glucose transporter 4 (GLUT4) are expressed in the sarcolemma. Insulin/IGF receptors are highly expressed in skeletal muscle, adipocytes, the liver, and the brain, and when insulin (and IGF) binds to these receptors, increased GLUT4 expression is triggered in the sarcolemma by altering the intracellular distribution of GLUT4 in skeletal muscles. This change enhances glucose uptake in skeletal muscles and adipocytes, resulting in decreased blood glucose levels after a meal [7, 8].

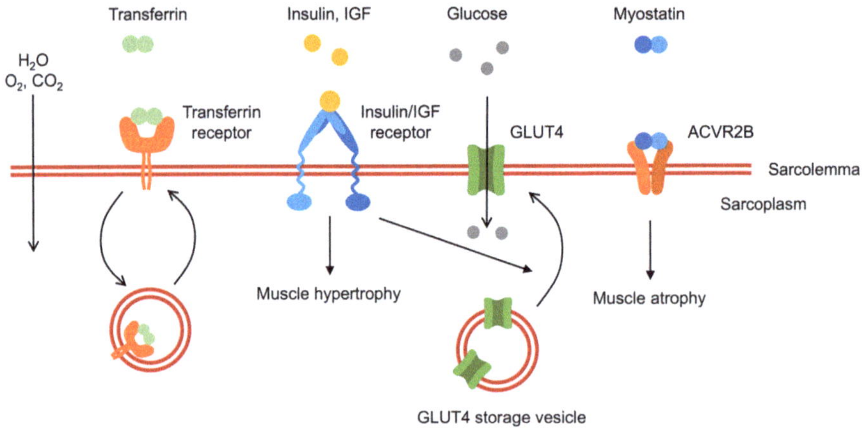

Fig. 6.2 Representative receptors and transporters expressed on the sarcolemma

Molecular mechanisms underlying muscle atrophy and hypertrophy have also been reported (Fig. 6.2). Myostatin, a member of the transforming growth factor-β (TGF-β) family, is secreted from skeletal muscle and acts as a negative regulator of muscle growth. Secreted myostatin binds to activin receptor type 2B (ACVR2B) in the sarcolemma and inhibits muscle protein synthesis by inhibiting the Akt pathway. In addition, myostatin activates ubiquitin ligases, such as atrogin-1 and muscle RING finger-1 (MuRF1), and promotes the proteasomal degradation of muscle proteins, resulting in muscle atrophy [9]. Myostatin also inhibits the proliferation and differentiation of myosatellites, thereby suppressing muscle regeneration. In contrast, the insulin/IGF pathway activates Akt signaling, promotes protein synthesis in the skeletal muscles, and phosphorylates forkhead box protein O1 (FoxO1), resulting in muscle hypertrophy [10, 11]. Exercise activates the IGF pathway to induce muscle protein synthesis and hypertrophy [12].

Recently, skeletal muscle-specific miRNAs have been identified, and certain miRNAs, including miR-1, miR-133a, miR-133b, miR-200, miR-208b, miR-486, and miR-499, are abundantly expressed in skeletal muscle, commonly referred to as myomiRs [13, 14]. MyomiRs are thought to be closely involved in physiological phenomena such as muscle hypertrophy, atrophy, and aging; however, few reports have been published on their detailed functions. As described above, a variety of molecules and signals are involved in the maintenance of skeletal muscle function and are potential therapeutic targets in skeletal muscle diseases.

1.3 Muscle-Related Diseases

Understanding the pathophysiology of muscle dysfunction will allow us to identify molecular targets that regulate disease and develop therapeutic strategies. Muscle-related diseases are roughly categorized into those resulting from defects in muscle function and those resulting from non-muscular disorders. Muscle dystrophy is a general term for muscle diseases that result in muscle weakness and are caused by muscle atrophy and necrosis of muscle fibers [15]. Muscular dystrophies are subdivided into several types, such as Duchenne, Becker, Fukuyama, and facioscapulohumeral, according to the symptoms and causative genes. In recent years, the causative genes of each disorder have been identified. For example, Duchenne muscular dystrophy (DMD) is caused by a dystrophin gene mutation on the X chromosome [16]. DMD is one of the most severe and fatal muscle dystrophies. DMD usually only affects boys and is the most common muscular dystrophy in childhood, with an incidence of approximately 1 in 3500 newborn boys. In DMD, the dystrophin protein lining the sarcolemma is completely defective due to a nonsense mutation, making the skeletal muscle weaker and more easily damaged during muscle contractions [17]. The symptoms of DMD begin to appear in early childhood (approximately 3–5 years of age), and muscle weakness is characterized by its appearance near the center of the body, involving the proximal muscles. Several pre-mRNA mutations have been identified in DMD, and antisense oligonucleotide (ASO)-mediated therapeutics have been developed for DMD treatment (see Sect. 1.4). The gene responsible for Becker muscular dystrophy (BMD) is also the dystrophin gene; this mutated (or truncated) version of dystrophin is expressed in skeletal muscles but is not associated with the complete dystrophin deficits seen in DMD [18]. BMD is also found mainly in boys, though its associated symptoms tend to manifest later in life compared to cases of DMD, appearing during childhood and adulthood. In addition, symptoms of muscle weakness tend to be less severe in BMD than in DMD. Fukuyama muscular dystrophy is mainly found in Japanese children, and the gene responsible for this disease is *fukutin*, which is found on chromosome 9 [19]. Fukutin is necessary for muscle maintenance, and mutated fukutin results in poor muscle function and gradual muscle weakness. Symptoms appear in both males and females from newborns to infancy. Facioscapulohumeral muscular dystrophy (FSHD) affects muscles in the face, shoulders, and upper arms and often develops at 10–20 years of age [20]. It is characterized by an asymmetric progression of muscle atrophy. The causative gene is *double homeobox 4* (*DUX4*), which is located on chromosome 4. Myotonic dystrophy 1 (DM1) is caused by a mutation in *dystrophia myotonica protein kinase* (*DMPK*), which is located on chromosome 19 [21, 22]. DM1 is the most common muscular dystrophy in adults, with an incidence of approximately 7 per every 100,000 individuals. The primary symptoms of this disease are progressive myotonia and muscle weakness. It is an autosomal dominant genetic disease affecting approximately equal numbers of males and females. Similar dystrophy to DM1 in symptoms, known as myotonic dystrophy 2 (DM2), is caused by a mutation in *cellular nucleic acid-binding protein* (*CNBP*) on chromosome 3 [22]. Pompe disease is a fatal disease that causes progressive weakness of the heart and skeletal muscles [23]. It is caused by the recessive inheritance of a mutation in *acid α-glucosidase* (*GAA*) on chromosome 7, which results in the accumulation of glycogen in intramuscular lysosomes. Enzyme replacement therapy is currently used to improve symptoms by reducing stored glycogen levels (see Sect. 1.4). Some muscle-related diseases, such as type 2 diabetes, do not affect motor function [24]. In type 2 diabetes, insulin insensitivity in skeletal muscle induces elevated blood glucose levels.

As previously mentioned, several muscle-related diseases can also be caused by non-muscular tissues, such as neuromuscular junctions (NMJs) or the peripheral nervous system, including ALS, spinal muscular atrophy (SMA), and Friedreich's ataxia [25–28]. These diseases also impair motor control functions and some result in serious symptoms. The NMJ is a synapse connecting the terminal of the motor nerve and muscle fiber cells; each muscle fiber is connected to a single motor nerve in the center, where a high density of acetylcholine (ACh) receptors (AChRs) is expressed. ACh released from the terminals of motor neurons binds to AChRs and induces muscle contractions. NMJ dysfunction-related diseases are known as myasthenia, which are classified into two categories: myasthenia gravis (MG), which is caused by autoimmune disorders, and congenital myasthenic syndromes, which are caused by an abnormality in NMJ-related genes [29]. The characteristic symptoms include generalized weakness, fatigue, and droopy eyelids, with severe cases presenting with dysphagia and dyspnea. MG is primarily caused by the production of autoantibodies against AChRs, resulting in the dysfunction of Ach signaling. Symptomatic treatment with cholinesterase inhibitors can inhibit Ach degradation; however, steroids and immunosuppressive agents are required to suppress autoantibody production. ALS usually develops in middle-aged individuals or later in life, affecting approximately 2 per every 100,000 people [26]. It is characterized by progressive muscle atrophy and weakness, mainly due to the organization and loss of upper and lower motor neurons. Approximately over 90% of all cases are sporadic ALS, and the rest are classified as hereditary familial ALS. Usually, ALS begins with muscle weakness in one upper limb, followed by muscle weakness in the other limbs, respiratory paralysis, and rapid progression of symptoms over a period of 2–5 years. Although no fundamental treatment has been developed, glutamate inhibitors (or antagonists) and antioxidant drugs have been shown to slow disease progression. Approximately 20 genes, including *superoxide dismutase 1* (*SOD1*), *chromosome 9 open reading frame 72* (*C9orf72*), and *fused-in sarcoma* (*FUS*), have been identified as causative genes of familial ALS [30]. SMA is caused by a genetic loss-of-function mutation in exon 7 or 7–8 of *survival motor neuron 1* (*SMN1*), which encodes the SMN protein, with an incidence of 10 per 100,000 people [27]. These loss-of-function mutations induce cell death in the motor neurons, leading to progressive muscle weakness and atrophy. *SMN1* has a paralogous gene, *SMN2*, which encodes the same SMN protein, though *SMN2* usually only produces approximately 10% of functional SMN protein because of a C-to-T substitution in exon 7. Many variations in the *SMN2* copy number determine the severity of SMA symptoms. The higher the copy number of *SMN2*, the less severe the symptoms. The disease is classified into the most severe type I (1–2 copies), intermediate type II (3 copies), mild type III (3–4 copies), and type IV (over 5 copies) [31]. Types I–III develop during childhood, while type IV develops later. Currently, three medications are available for the treatment of SMA (see Sect. 1.4).

Aging also induces a loss of muscle mass, a phenomenon known as sarcopenia, and is associated with impaired physical mobility and decreased quality of life. In a study in the United States, approximately 20% of individuals aged 65–70 years and more than 50% of those aged over 80 years had sarcopenia, indicating that the frequency of sarcopenia increases with age [32, 33]. To date, there is no cure for sarcopenia, and only habitual resistance exercise and proper nutrition can delay its progression. Although the cause of sarcopenia is unclear, reduced muscle regenerative capacity has been identified as one of the responsible factors [34]. Recently, it has been reported that the dysfunction of myosatellite cells and mesenchymal progenitor cells is associated with sarcopenia, which may have future applications in the treatment and prevention of sarcopenia [35].

1.4 Current Treatment Strategies

For most muscle-related diseases, no effective drugs have been established, and supportive therapies, such as rehabilitation and steroid therapy, are usually applied. Exon-modulating medications and adeno-associated virus (AAV)-mediated gene therapeutics have been approved for the treatment of several serious muscle diseases caused by certain genetic mutations, such as DMD [16] and SMA [27, 36]. Steroid (corticosteroid) medications provide short-term benefits for muscular dystrophies and inhibit nuclear factor kappa B (NF-κB) to suppress myonecrosis and inflammation by activating the calcincurin/nuclear factor of activated T cells (NFAT) signaling pathway [37, 38]. However, long-term steroid therapy also has adverse effects, such as obesity and osteoporosis.

DMD is an X-linked genetic disease caused by a dystrophin gene mutation on the X chromosome that results in complete loss of the dystrophin protein [16]. Dystrophin is a membrane-lining protein found in all muscle tissues and neuronal cells. Dystrophin is composed of an N-terminus, a C-terminus, and a rod domain in the middle; the rod domain is composed of multiple spectrin-like repeats [39] (Fig. 6.3a). Dystrophin forms a dystrophin-associated protein complex (DAPC), which stabilizes the sarcolemma structure by linking cytoplasmic actin on the N-terminus of dystrophin to membrane proteins, including α-dystroglycan, on the C-terminus of dystrophin. The DAPC is further connected to the basal lamina via the extracellular matrix, such as laminin α2. Therefore, complete loss of dystrophin leads to weakening of the sarcolemma, resulting in muscle fiber degeneration and fibrosis, and the N-terminal and C-terminal domains are considered to be relatively important for muscle stabilization compared to the rod domain [17, 39]. In patients with DMD, mutations are found in the rod domain sequence that induces nonsense out-of-frame mRNA expression. ASO-mediated therapeutics have been developed to induce exon skipping in mutated dystrophin pre-mRNAs [40] (Fig. 6.3b). Such therapeutics are designed to bind the dystrophin pre-mRNA to shift the reading frame from out-of-frame to in-frame mutation and induce the expression of partially functional dystrophin proteins with a shortened rod structure. Currently, several ASO therapeutics composed of 21- to 30-mer phosphorodiamidate morpholino oligomers (PMOs), which are non-natural and chemically synthesized non-charged

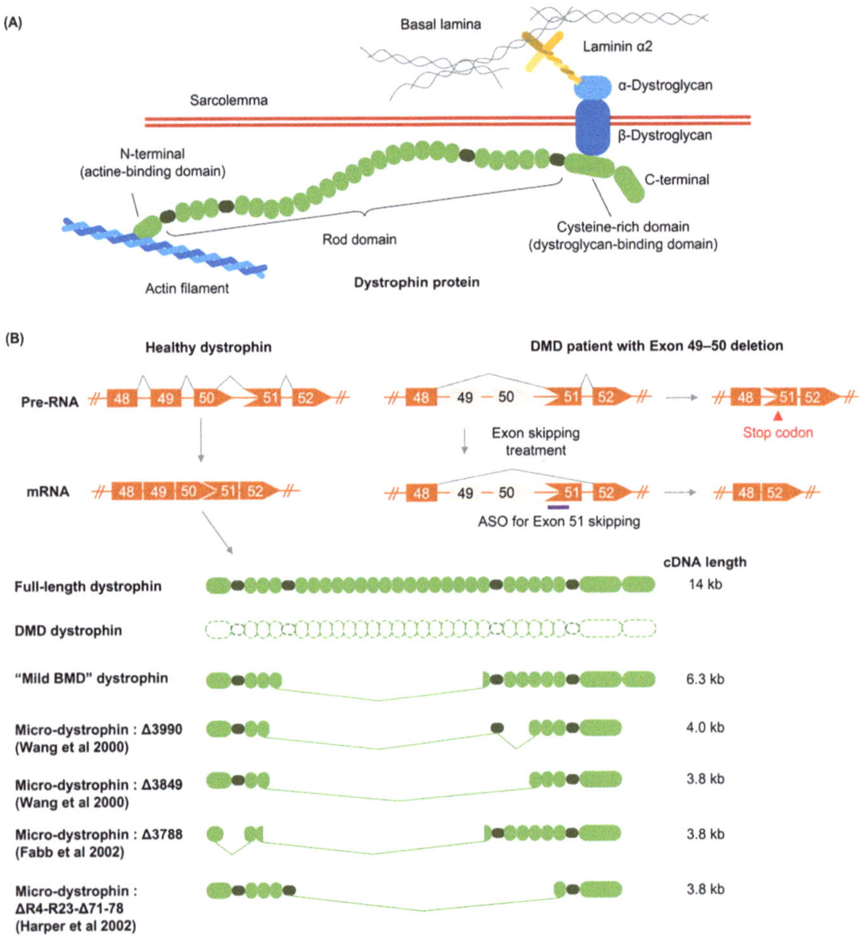

Fig. 6.3 (**a**) Schematic illustration of dystrophin and dystrophin-associated proteins. (**b**) Dystrophin pre-mRNA, mRNA, and protein, and therapeutic strategy of DMD through the exon skipping with ASO

oligonucleotide analogs, have been approved depending on the exon mutation subtype, targeting exon 51 (eteplirsen) [41], exon 53 (golodirsen and viltolarsen) [42, 43], and exon 45 (casimersen) [44]. These PMOs are administered as weekly intravenous injections at 30–80 mg/kg. Other ASO therapeutics are currently being developed for other exon mutations.

In Pompe disease therapy, the defective GAA enzyme is replaced by intravenously administered genetically modified GAA enzymes, alglucosidase alfa [45], and avalglucosidase alfa [46], resulting in the breakdown of glycogen stores and the alleviation of symptoms. Although adverse side effects such as headaches and vomiting have been reported, this therapeutic strategy is the only effective treatment for Pompe disease to date.

Three medications are currently available for SMA treatment. The first is AAV9-mediated gene therapy, onasemnogene abeparvovec, which restores SMN protein expression [47]. This AAV9 vector contains a plasmid with the SMN1 gene and is used for single-dose intravenous injections in children less than 2 years of age, with a recommended dosage of 1.1×10^{14} vg/kg. While a high dose of AAV9 may cause acute liver failure, continuous production of SMN protein in non-dividing motor neurons and high therapeutic performance have been achieved with this method. The second gene therapy for SMA, nusinersen, is an ASO-mediated exon inclusion that alters pre-mRNA splicing of SMN2, similar to the aforementioned DMD treatment strategy [48]. The ASO containing 18-mer chemically modified RNA (with 2′-*O*-methoxyethyl and phosphorothioate modifications) is designed to bind to intron 7 in *SMN2* and enhance exon 7 inclusion of *SMN2* pre-mRNA, resulting in the enhanced production of full-length SMN protein. This ASO is intrathecally administered with four initial loading doses over the initial two months and then three maintenance doses per year, with 12 mg of each dosage. The third gene therapy for SMA, risdiplam, is an orally deliverable small-molecule medication discovered by high-throughput screening to induce exon 7 inclusion in *SMN2* pre-mRNA [49]. It is administered once daily. Both exon-modulating medications are currently approved for the treatment of SMA in patients of all ages.

2 Muscle-Targeted Drugs

As described above (see Sect. 1.4), currently clinically available drugs are limited to muscle-related diseases. However, in recent years, pathogenic genes have been identified and therapeutics have been developed for the treatment of individual mutations. This section focuses on medications that are currently under investigation.

2.1 Low-Molecule and Peptide-Based Drugs

Currently, the development of nucleic acid-based therapeutics (ASO-mediated exon-modulating therapy and AAV-mediated gene therapy) for intractable muscle diseases is mainstream. However, ASO therapeutics, which are hydrophilic macromolecules with molecular weights exceeding 5000, have low permeability through the sarcolemma, and AAVs show some immunogenicity and hepatotoxicity at high doses, limiting their applications. These limitations have increased the importance of developing smaller-molecule medications. In this context, a small-molecule drug that induces exon inclusion for orally administered SMA has already been successfully developed via high-throughput screening [49]. Similarly, small molecules that induce read-through of mutated dystrophin have been developed for DMD therapy, as gentamicin has been shown to restore defective dystrophin in DMD model mice (*mdx* mice) [50, 51]. However, gentamicin is an aminoglycoside antimicrobial agent that is associated with auditory and renal toxicity. Therefore, safer compounds with read-through activity have since been developed, including the clinical trials of orally administered ataluren, which has been licensed in the European Union and the United Kingdom [52, 53]. Small molecules that induce exon skipping, such as PMOs, have also been developed [54]. However, there are more mutation types in DMD than in SMA, and compounds that induce effective exon modulation for each mutation have not yet been developed.

In addition to restoring mutations, other therapeutic strategies that induce muscle hypertrophy are being investigated. For example, inhibitors of myostatin and its receptor (ACVR2B), which induce muscle atrophy, are effective targets for muscle mass recovery. Antibodies targeting myostatin [55, 56] and ACVR2B [57] are currently being developed, and several myostatin antibodies are currently being investigated in clinical trials. As for small-molecule drugs, fast muscle troponin activators [58] and selective androgen receptor modulators [59], both of which contribute to muscle contraction, are underway in clinical trials of ALS and SMA.

2.2 Gene Therapy

For diseases in which therapeutic target genes and mutations have been identified, gene expression by gene therapy using viral vectors (as in SMA) and exon modulation by exon-modulating ASOs, such as PMOs (as in DMD), have been developed. In addition, "gapmer"-type ASOs and small interfering RNAs (siRNAs) have both been found to induce downregulation of target mRNA. Therapeutic miRNAs have been developed as well.

Gene transduction by viral vectors is capable of long-term gene expression in non-dividing cells, such as muscle fibers, with a single administration because they have a long cellular lifespan. Several types of viral vectors, such as AAV [60] and adenovirus (AdV) [61, 62], have shown the potential for gene transduction into skeletal muscles. AdVs have a large capacity, which enables them to encode large exogenous transgenes (~35 kb) [63]. AdV vectors are capable of gene transduction in a wide range of cell types and have a low selectivity for skeletal muscles. However, AdVs generally induce high immunogenicity, limiting the range and duration of transgene expression in vivo after repeated administration. In contrast, AAVs are of interest as viral vectors for gene therapy owing to their relatively low immunogenicity [64]. In fact, various AAV serotypes have been identified with different tissue tropisms, and serotypes 1, 2, 6, 7, and 8 have been reported to have tropisms in skeletal muscles. Unfortunately, one of the limitations of AAVs is their small size, which enables them to encode only a maximum of ~4.7 kb of exogenous DNA [65]. For this reason, it is not possible to encode full-length dystrophin with a total length of 14 kb, so micro-dystrophins (3.8–4.0 kb) with a shortened rod domain packaged into AAVs have been developed for DMD treatment [66–68] (Fig. 6.3b). Increased

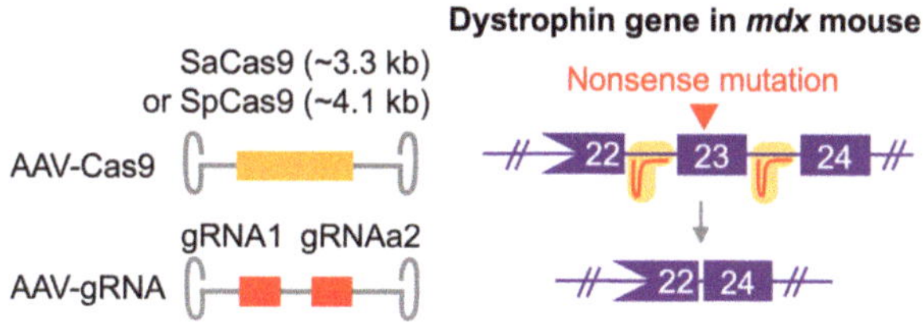

Fig. 6.4 Therapeutic strategy of AAV-mediated gene editing utilizing CRISPR-Cas9 system for DMD treatment

expression of such shortened dystrophin proteins has been observed in animal studies.

In addition to viral gene transduction, other approaches that are being investigated include suppression of the myostatin gene by RNA-cleavable ASOs and siRNAs and gene editing therapy using the CRISPR-Cas9 system for fundamental treatment. CRISPR-Cas9 technology provides an opportunity to directly correct disease-causing mutations in DNA [69]. Recently, the systemic or local administration of two AAVs reportedly restored dystrophin protein expression in *mdx* mice [70–72]: One AAV encoded the Cas9 protein (using either small Staphylococcus aureus Cas9 [SaCas9] or Streptococcus pyogenes Cas9 [SpCas9]) and the other encoded two guide RNAs designed to bind before and after the targeted mutated exon 23 (Fig. 6.4). The recovered dystrophin protein in skeletal muscle was maintained for a year after a single dose [73]. Although these approaches are expected to be next-generation therapies, they require a relatively high dosage to function in skeletal muscles throughout the body, and further improvements are required.

Although PMO-based therapies have been approved for several DMD mutations, challenges remain, particularly in the case of weak exon-skipping activity in the heart. In general, oligonucleotide therapeutics (siRNAs, ASOs, and PMOs) have poor pharmacokinetic profiles and low blood retention because they are excreted through the kidneys, with the highest accumulation in the liver and a small amount reaching muscle tissues (see Sect. 3). To maximize their therapeutic potential, PMOs conjugated with membrane-permeable peptides and octa-guanidine peptides have been clinically developed to increase their cellular uptake and exon-skipping activity, especially in the heart [74, 75]. For siRNAs and ASOs, hydrophobic groups such as cholesterol and fatty acids are conjugated to improve their blood retention time by binding to lipoproteins, such as low-density and high-density lipoproteins, in the blood [76, 77]. Docosanoic acid-conjugated siRNAs [78] or palmitic acid-conjugated ASOs [79] demonstrated higher knockdown efficacy in both the heart and skeletal muscles.

2.3 *Cell Therapy*

Satellite cells exhibit regenerative activity in skeletal muscle tissues, and their self-repairing activity in skeletal muscle is higher than that in the heart and other tissues [2]. Dormant satellite cells can transform into myoblasts that repair myofibers by differentiating and fusing into myofibers. Therefore, a satellite cell-based therapeutic strategy has been developed to regenerate muscle tissues [3–5]. It has been shown that transplanting isolated satellite cells into muscle tissues can improve muscle function in *mdx* and aged mice [80, 81]. However, early clinical trials in humans using this strategy in patients with DMD have failed because of low therapeutic efficacy [82]. This is because isolated human satellite cells lose their self-renewal function during the culture process, resulting in low survival rates at the injection site. Thus, it is necessary to optimize and establish isolation and culture methods to maintain the self-renewal ability of human satellite cells for muscle regeneration therapy [83, 84].

3 Drug Delivery Strategies for Muscle

Although local injections are applicable for localized muscle diseases, systemic administration such as intravenous injections is required for systemic muscle diseases. However, most systemically administered drugs, especially low-molecular-weight drugs and oligonucleotide therapeutics, are mainly eliminated via rapid renal filtration and their accumulated amount in muscle tissues is negligible [85, 86]. In this regard, drug delivery strategies or systems that suppress renal filtration, enhance extravasation from capillaries (particularly continuous endothelial capillaries) into muscle tissues, and enhance uptake into muscle fibers are required to maximize therapeutic effects. In this section, drug delivery systems and strategies for targeting muscle tissues are discussed.

3.1 Targeting Ligands

To achieve selective delivery into muscle tissues, skeletal muscle-targeting ligands that selectively bind to specific molecules expressed on the surface of the sarcolemma can be utilized. Some receptors and transporters, including the myostatin receptor, transferrin receptor, insulin receptor, and GLUT4, are suitable target molecules. Currently, both peptides and antibodies have been developed as muscle-targeting ligands.

Although target molecules are difficult to identify, phage display-based peptide selection methods have been used to determine muscle-selective peptide ligands. As an example of a peptide-based muscle-targeting ligand, a 7-residue amino acid sequence (ASSLNIA) has already been selected through phage display [87]. Insertion of this peptide ligand into the capsid protein of AAV2 increases the transduction efficiency of differentiated mouse myoblast C2C12 cells [88]. In in vivo experiments, intravenous administration of AAV2 achieved gene transduction with particularly high tropism to the heart while reducing tropism to organs other than muscle tissues, such as the liver, spleen, and lungs. However, tropism to skeletal muscles was not as high. In addition, a 12-residue amino acid sequence (SKTFNTHPQSTP) was identified through in vivo screening using *mdx* mice as a peptide ligand that binds to skeletal and cardiac muscles but not to the liver [89]. Conjugation of this peptide ligand with exon skipping-inducing PMOs successfully increased dystrophin protein expression in skeletal muscles throughout the bodies of *mdx* mice. In addition to peptide-based ligands, nucleic acid-based aptamers have been developed using an in vitro selection method, and SELEX and RNA aptamers that induce internalization in C2C12 cells have been identified [90]. Such aptamers have been shown to effectively migrate into muscle fibers after topical administration.

The utilization of receptor-mediated endocytosis could be a more useful strategy to increase intracellular delivery because some of the aforementioned ligands cannot be efficiently internalized into muscle fibers. Transferrin receptors and insulin/IGF1 receptors expressed on the myocyte surface have been investigated as promising target receptors. For example, the F(ab') fragment of an anti-transferrin antibody conjugated to siRNA showed approximately 70% knockdown of the target gene in skeletal muscle via intravenous injection, with little knockdown found in other organs [91]. The combination of antibodies (or their fragments) and siRNAs or ASOs has been clinically developed in humans for the treatment of DM1, DMD, and FSHD.

3.2 Nanoparticle-Based Delivery

In addition to the ligand-installed strategy, avoiding renal filtration and extravasation into muscle tissues are essential strategies for skeletal muscle delivery. To achieve these functions, it is important to control the apparent molecular size within an appropriate range. There is a clear size dependency in renal filtration, and small molecules less than approximately 8–10 nm are rapidly eliminated from the bloodstream into the urine [85, 86]. This indicates that small-molecule drugs and oligonucleotide therapeutics of approximately 20,000 Da in the blood are rapidly cleared after systemic administration. To extend their blood retention time, conjugation of drugs to other macromolecules or encapsulation into nanoparticles is a widely used technique to increase their apparent size to over 10 nm. This approach is widely applicable to cancer- and liver-targeted delivery systems because considerably large pores (or fenestra) are found mainly in the tumor and liver capillary walls [86]. This allows larger molecules (30–100 nm) to enhance extravasation to target tissues (in cancer and the liver) [92, 93]. However, larger-sized molecules are restricted from extravasating effectively in skeletal muscle because only smaller pores are found in the capillary walls, which are known as continuous capillaries. Recent research has revealed that small nanoparticles (10–20 nm in size) can gradually extravasate to the muscle tissue, and smaller nanoparticles showed higher accumulation in a healthy mouse model after systemic administration [94]. This indicates the importance of controlling the apparent size of drug delivery systems to within 10–20 nm to avoid renal filtration and enhance extravasation in skeletal muscles. In this regard, antibody therapeutics (~15 nm), conjugation of drugs to antibodies, and utilization of AAV vectors (~22 nm) are rational designs in terms of size tuning for muscle-targeted delivery. Moreover, in the case of diseased muscle tissues, such as in *mdx* mice, cellular uptake efficiency and extravasation were enhanced due to the weakened sarcolemma and inflammatory response upon chronic muscle damage, increasing the accumulation of relatively large nanoparticles (>30 nm) in skeletal muscle while decreasing accumulation in other organs [94].

3.3 Utilization of External Stimuli

As alternatives to local administration, ligand-based and nanoparticle-based delivery methods and external stimuli-induced muscle-specific delivery methods, such as electroporation, ultrasound, and microbubble, have been investigated [95].

Electroporation has long been a widely used method for gene transfer in vitro by applying electrical pulses to create transient pores in the cellular membrane and is applicable to transfer genes to muscles in vivo by attaching electrodes directly to target muscles [96, 97]. This method does not require any specific delivery system for gene transfer, and gene transfer efficiency is not affected by age or other individual factors. However, this method has the disadvantages of muscle damage caused by electric pulses and the fact that the site of gene transfer is limited to the location where the electric pulse is applied.

Microbubbles with ultrasound are another site-specific delivery approach that involves external stimulation [98, 99]. When a few microbubbles are administered into the bloodstream and irradiated with ultrasound, they expand, contract, and rupture, causing cavitation at the irradiated site. This impacts locally and temporarily disrupts the vessel walls and muscle cell membranes, facilitating tissue penetration and intracellular trafficking and allowing relatively large non-viral nanoparticles and other particles to migrate from the blood vessels into the muscle tissue. In addition, this method is promising for selective drug delivery to muscle tissues because the site of ultrasound irradiation can be easily controlled and is relatively minimally invasive. This method was applied to systemic pDNA delivery; pDNA-loaded bubble liposomes were administered intravenously and immediately irradiated with ultrasound to the hind limb in mice, and increased gene expression at the irradiated site was reported.

4 Perspectives

There are many barriers to the development of drug delivery systems and therapeutics for the treatment of muscle-related diseases. As for therapeutics, however, many kinds of medications are currently available, such as small molecules, protein therapeutics, oligonucleotide therapeutics, and viral-based gene therapy agents; next-generation therapeutics are also undergoing clinical or preclinical studies. Additionally, many efforts have been made to identify disease-related genes not only for genetic defect-related muscle diseases but also for age-related muscle diseases, which will pave the way for a treatment strategy for intractable muscle-related diseases. However, muscle-targeted drug delivery systems have not yet been established. The aforementioned findings and techniques reviewed in this chapter regarding the use of ligands, size tuning, and external stimuli methods will help construct effective drug delivery systems for both healthy and diseased skeletal muscles and aged muscles. The integration of each technology for the treatment of muscle-related diseases is expected to overcome the current challenges in the near future.

References

1. Hill M, Wernig A, Goldspink G, Muscle satellite (stem) cell activation during local tissue injury and repair. ***J. Anat.***, 203(1): 89–99 (2003). https://doi.org/10.1046/j.1469-7580.2003.00195.x
2. Yin M, Price F, Rudnicki MA, Satellite cells and the muscle stem cell niche. ***Physiol. Rev.***, 93(1): 23–67 (2013). https://doi.org/10.1152/physrev.00043.2011
3. Hall JK, Banks GB, Chamberiain JS, Olwin B, Prevention of muscle aging by myofiber-associated satellite cell transplantation. ***Sci. Transl. Med.***, 2(57): 57ra83 (2010). https://doi.org/10.1126/scitranslmed.3001081
4. Chang NC, Chevalier FP, Rudnicki MA, Satellite cells in muscular dystrophy – Lost in polarity. ***Trends Mol. Med.***, 22(6): 479–496 (2016). https://doi.org/10.1016/j.molmed.2016.04.002
5. Pradat P-F, Barani A, Wanschitz J, Dubourg O, Lombès A, Bigot A, Vincent M, Bruneteau G, Salachas F, Lenglet T, Meininger V, Butler-Browne G, Abnormalities of satellite cells function in amyotrophic lateral sclerosis. ***Amyotroph. Lateral Scler.***, 12(4): 264–271 (2011). https://doi.org/10.3109/17482968.2011.566618
6. Lexell J, Henriksson-Larsen K, Winblad B, Sjostrom M, Distribution of different fiber types in human skeletal muscles: effects of aging studied in whole muscle cross sections. ***Muscle Nerve.***, 6: 588–595 (1983). https://doi.org/10.1002/mus.880060809
7. Watson RT, Pessin JE, Intracellular organization of insulin signaling and GLUT4 translocation. ***Recent Prog. Horm. Res.***, 56: 175–193 (2001). https://doi.org/10.1210/rp.56.1.175
8. Leto D, Saltiel AR, Regulation of glucose transport by insulin: Traffic Control of glut4. ***Nat. Rev. Mol. Cell. Biol.***, 13:383–396 (2012). https://doi.org/10.1038/nrm3351
9. Rodriguez J, Vernus B, Chelh I, Cassar-Malek I, Gabillard JC, Sassi AH, Seiliez I, Picard B, Bonnieu A, Myostatin and the skeletal muscle atrophy and hypertrophy signaling pathways. ***Cell. Mol. Life Sci.***, 71(22): 4361–4371 (2014). https://doi.org/10.1007/s00018-014-1689-x
10. Sandri M, Sandri C, Gilbert A, Skurk C, Calabria E. Picard A, Walsh K, Schiaffino S, Lecker SH, Goldberg AL, Foxo transcription factors induce the atrophy-related ubiquitin ligase atrogin-1 and cause skeletal muscle atrophy. ***Cell***, 117(3): 399–412 (2004). https://doi.org/10.1016/s0092-8674(04)00400-3
11. Kousteni S, FoxO1, transcriptional chief of staff of energy metabolism. ***Bone***, 50(2): 437–443 (2012). https://doi.org/10.1016/j.bone.2011.06.034
12. Bolster DR, Jefferson LS, Kimball SR, Regulation of protein synthesis associated with skeletal muscle hypertrophy by insulin-, amino acid- and exercise-induced signaling. ***Proc. Nutr. Soc.***, 63(2): 351–356 (2004). https://doi.org/10.1079/PNS2004355

13. Güller I, Russell AP, MicroRNAs in skeletal muscle: their role and regulation in development, disease and function. ***J. Physiol.***, 588(Pt 21): 4075–4087 (2010). https://doi.org/10.1113/jphysiol.2010.194175
14. Horak M, Novak J, Bienertova-Vasku J, Muscle-specific microRNAs in skeletal muscle development. ***Dev. Biol.***, 410(1): 1–13 (2016). https://doi.org/10.1016/j.ydbio.2015.12.013
15. Emery AEH, The muscular dystrophies. ***Lancet***, 359(9307): 687–695 (2002). https://doi.org/10.1016/S0140-6736(02)07815-7
16. Duan D, Goemans N, Takeda S, Mercuri E, Aartsma-Rus A, Duchenne muscular dystrophy. ***Nat. Ver. Dis. Primers.***, 7(1): 13 (2021). https://doi.org/10.1038/s41572-021-00248-3
17. Gumerson JD, Michele DE, The dystrophin-glycoprotein complex in the prevention of muscle damage. ***J. Biomed. Biotechnol.***, 2011: 210797 (2011). https://doi.org/10.1155/2011/210797
18. Wein N, Alfano L, Flanigan KM, Genetics and emerging treatments for Duchenne and Becker muscular dystrophy. ***Pediatr. Clin. North Am.***, 62(3): 723–742 (2015). https://doi.org/10.1016/j.pcl.2015.03.008
19. Toda T, Kobatashi K, Kondo-Iida E, Sasaki J, Nakamura Y, The Fukuyama congenital muscular dystrophy story. ***Neuromuscul. Disord.***, 10(3): 153–159 (2000). https://doi.org/10.1016/s0960-8966(99)00109-1
20. Tawil R, van der Maarel S, Facioscapulohumeral muscular dystrophy. ***Muscle Nerve.***, 34(1): 1–15 (2006). https://doi.org/10.1002/mus.20522
21. Kaliman P, Llagostera E, Myotonic dystrophy protein kinase (DMPK) and its role in the pathogenesis of myotonic dystrophy 1. ***Cell. Signal.***, 20(11): 1935–1941 (2008). https://doi.org/10.1016/j.cellsig.2008.05.005
22. Meola G, Cardani R, Myotonic dystrophies: An update on clinical aspects, genetic, pathology, and molecular pathomechanisms. ***Biochim. Biophys. Acta***, 1852(4): 594–606 (2015). https://doi.org/10.1016/j.bbadis.2014.05.019
23. Raben N, Plotz P, Byrne BJ, Acid a-glucosidase deficiency (glycogenosis type II, Pompe disease). ***Curr. Mol. Med.***, 2(2): 145–166 (2002). https://doi.org/10.2174/1566524024605789
24. DeFronzo RA, Tripathy D, Skeletal Muscle Insulin Resistance Is the Primary Defect in Type 2 Diabetes. ***Diabetes Care***, 32(Suppl 2): S157–S163 (2009). https://doi.org/10.2337/dc09-S302
25. Li L, Xiong W-C, Mei L, Neuromuscular junction formation, aging, and disorders. ***Annu. Rev. Physiol.***, 80:159–188 (2018). https://doi.org/10.1146/annurev-physiol-022516-034255
26. Al-Chalabi A, Hardiman O, The epidemiology of ALS: a conspiracy of genes, environment and time. ***Nat. Rev. Neurol.***, 9(11): 617–628 (2013). https://doi.org/10.1038/nrneurol.2013.203
27. Tiziano FD, Melki J, Simard LR, Solving the puzzle of spinal muscular atrophy: What are the missing pieces? ***Am. J. Med. Genet. A***, 161(11):2836–2845 (2013). https://doi.org/10.1002/ajmg.a.36251
28. Pastore A, Puccio H, Frataxin: a protein in search for a function. ***J. Neurochem.***, 126(s1):43–52 (2013). https://doi.org/10.1111/jnc.12220
29. Conti-Fine BM, Milani M, Kaminski HJ, Myasthenia gravis: past, present, and future. ***J. Clin. Invest.***, 116(11): 2843–2854 (2006). https://doi.org/10.1172/JCI29894
30. Valdmanis PN, Rouleau GA, Genetics of familial amyotrophic lateral sclerosis. ***Neurology***, 70(2): 144–152 (2008). https://doi.org/10.1212/01.wnl.0000296811.19811.db
31. Prior TW, Perspectives and diagnostic considerations in spinal muscular atrophy. ***Genet. Med.***, 12(3): 145–152 (2010). https://doi.org/10.1097/GIM.0b013e3181c5e713
32. Baumagartner RN, Epidemiology of sarcopenia among the elderly in New Mexico. ***Am. J. Epidemiol.***, 147(8):755–763 (1998). https://doi.org/10.1093/oxfordjournals.aje.a009520
33. Iannuzzi-Sucich M, J Gerontol, Prevalence of sarcopenia and predictors of skeletal muscle mass in healthy, older men and women. ***J. Gerontol. A Biol. Sci. Med. Sci.***, 57(12):M772–M777 (2002). https://doi.org/10.1093/gerona/57.12.M772
34. Sousa-Victor P, Gutarra S, García-Prat L, Rodriguez-Ubreva J, Ortet L, Ruiz-Bonilla V, Jardí M, Ballestar E, González S, Serrano AL, Perdiguero E, Muñoz-Cánoves P, Geriatric muscle stem cells switch reversible quiescence into senescence. ***Nature***, 506:316–321 (2014). https://doi.org/10.1038/nature13013

35. Uezumi A, Ikemoto-Uezumi M, Zhou H, Kurosawa T, Yoshimoto Y, Nakatani M, Hitachi K, Yamaguchi H, Wakatsuki S, Araki T, Morita M, Yamada H, Toyoda M, Kanazawa N, Nakazawa T, Hino J, Fukada S, Tsuchida K, Mesenchymal Bmp3b expression maintains skeletal muscle integrity and decreases in age-related sarcopenia. ***J. Clin. Invest.***, 131(1): e139617 (2021). https://doi.org/10.1172/JCI139617
36. Nicolau S, Waldrop MA, Connolly AM, Mendell JR, Spinal muscular artrophy. ***Semin. Pediatr. Neurol.***, 37: 100878 (2021). https://doi.org/10.1016/j.spen.2021.100878
37. Moxley III RT, Pandya S, Ciafaloni E, Fox DJ, Campbell K, Change in Natural History of Duchenne Muscular Dystrophy With Long-term Corticosteroid Treatment: Implications for Management. ***J. Child. Neurol.***, 25(9):1116–1129 (2010). https://doi.org/10.1177/0883073810371004
38. Hoffman EP, Reeves E, Damsker J, Nagaraju K, McCall JM, Connor EM, Bushby K, Novel Approaches to Corticosteroid Treatment in Duchenne Muscular Dystrophy. ***Phys. Med. Rehabil. Chil. N. Am.***, 23(4): 821–828 (2012). https://doi.org/10.1016/j.pmr.2012.08.003
39. Gao Q, McNally EM, The dystrophin complex: structure, function and implications for therapy. ***Compr. Physiol.***, 5(3): 1223–1239 (2015). https://doi.org/10.1002/cphy.c140048
40. Aartsma-Rus A, Fokkema I, Verschuuren J, Ginjaar I, van Deutekom J, van Ommen G-J, den Dunnen JT, Theoretic applicability of antisense-mediated exon skipping for Duchenne muscular dystrophy mutations. ***Hum. Mutat.***, 30(3) 293–299 (2009). https://doi.org/10.1002/humu.20918
41. Syed YY, Eteplirsen: First Global Approval. ***Drugs***, 76(17): 1699–1704 (2016). https://doi.org/10.1007/s40265-016-0657-1
42. Heo Y-A, Goldirsen: First Approval. ***Drugs***, 80(3): 329–333 (2020). https://doi.org/10.1007/s40265-020-01267-2
43. Dhillon S, Viltolarsen: First Approval. ***Drugs***, 80(10): 1027–1031 (2020). https://doi.org/10.1007/s40265-020-01339-3
44. Shirley M, Casimersen: first approval. ***Drugs***, 81(7): 875–879 (2021). https://doi.org/10.1007/s40265-021-01512-2
45. Alglucosidase alfa: Long term use in the treatment of patients with Pompe disease. Beck M, ***Ther. Clin. Risk Manag.***, 5: 767–772 (2009). https://doi.org/10.2147/tcrm.s5776
46. Dhillon S, Avalglucosidase alfa: First Approval. ***Drugs***, 81(15): 1803–1809 (2021). https://doi.org/10.1007/s40265-021-01600-3
47. Hoy SM, Onasemnogene Abeparvovec: First Global Approval. ***Drugs***, 79(11): 1255–1262 (2019). https://doi.org/10.1007/s40265-019-01162-5
48. Holy SM, Nusinersen: First Global Approval. ***Drugs***, 77(4): 473–479 (2017). https://doi.org/10.1007/s40265-017-0711-7
49. Dhillon S, Risdiplam: First Approval. ***Drugs***, 80(17): 1853–1858 (2020). https://doi.org/10.1007/s40265-020-01410-z
50. Barton-Davis ER, Cordier L, Shoturma DI, Leland SE, Sweeney HL, Aminoglycoside antibiotics restore dystrophin function to skeletal muscles of mdx mice. ***J. Clin. Invest.***, 104(4):375-381 (1999). https://doi.org/10.1172/JCI7866
51. Wagner KR, Hamed S, Hadley DW, Gropman AL, Burstein AH, Escolar DM, Hoffman EP, Fischbeck KH, Gentamicin treatment of Duchenne and Becker muscular dystrophy due to nonsense mutations. ***Ann. Neurol.***, 49(6): 706–711 (2001). https://doi.org/10.1002/ana.1023
52. Welch EM, Barton ER, Zhuo J, Tomizawa Y, Friesen WJ, Trifillis P, Paushkin S, Patel M, Trotta CR, Hwang S, Wilde RG, Karp G, Takasugi J, Chen G, Jones S, Ren H, Moon Y-C, Corson D, Turpoff AA, Campbell JA, Conn MM, Khan A, Almstead NG, Hedrick J, Mollin A, Risher N, Weetall M, Yeh S, Branstrom AA, Colacino JM, Babiak J, Ju WD, Hirawat S, Northcutt VJ, Miller LL, Spatrick P, He F, Kawana M, Feng H, Jacobson A, Peltz SW, Sweeney HL, PTC124 targets genetic disorders caused by nonsense mutations. ***Nature***, 447(7140): 87–91 (2007). https://doi.org/10.1038/nature05756
53. Ryan NJ, Ataluren: First global approval. ***Drugs***, 74(14): 1709–1714 (2014). https://doi.org/10.1007/s40265-014-0287-4

54. Nishida A, Kataoka N, Takeshima Y, Yagi M, Awano H, Ota M, Itoh K, Hagiwara M, Matsuo M, Chemical treatment enhances skipping of a mutated exon in the dystrophin gene. ***Nat Commun.***, 2:308 (2011). https://doi.org/10.1038/ncomms1306
55. Latres E, Pangilinan J, Miloscio L, Bauerlein R, Na E, Potocky TB, Huang Y, Eckersdorff M, Rafique A, Mastaitis J, Lin C, Murphy AJ, Yancopoulos GD, Gromada J, Stitt T, Myostatin blockade with a fully human monoclonal antibody induces muscle hypertrophy and reverses muscle atrophy in young and aged mice. ***Skelet. Muscle***, 5: 34 (2015). https://doi.org/10.1186/s13395-015-0060-8
56. Camporez J-PG, Petersen MC, Abudukadier A, Moreira GV, Jurczak MJ, Friedman G, Haqq CM, Petersen KF, Shulmana GI, Anti-myostatin antibody increases muscle mass and strength and improves insulin sensitivity in old mice. ***Proc. Natl. Acad. USA.***, 113(8): 2212–2217 (2016). https://doi.org/10.1073/pnas.1525795113
57. Lach-Trifilieff E, Minetti GC, Sheppard KA, Ibebunjo C, Feige JN, Hartmann S, Brachat S, Rivet H, Koelbing C, Morvan F, Hatakeyama S, Glass DJ, An Antibody Blocking Activin Type II Receptors Induces Strong Skeletal Muscle Hypertrophy and Protects from Atrophy. ***Mol. Cell Biol.***, 34(4): 606–618 (2014). https://doi.org/10.1128/MCB.01307-13
58. Rudnicki SA, Andrews JA, Duong T, Cockroft BM, Malik FI, Meng L, Wei J, Wolff AA, Genge A, Johnson NE, Tesi-Rocha C, Connolly AM, Darras BT, Felice K, Finkel RS, Shieh PB, Mah JK, Statland J, Campbell C, Habib AA, Kuntz NL, Oskoui M, Day JW, Reldesemtiv in Patients with Spinal Muscular Atrophy: a Phase 2 Hypothesis-Generating Study. ***Neurotherapeutics***, 18(2): 1127–1136 (2021). https://doi.org/10.1007/s13311-020-01004-3
59. Da Fonseca GWP, Dworatzek E, Ebner N, Von Haehling S, Selective androgen receptor modulators (SARMs) as pharmacological treatment for muscle wasting in ongoing clinical trials. ***Expert Opin. Investig. Drugs***, 29(8): 881–891 (2020). https://doi.org/10.1080/13543784.2020.1777275
60. Wang D, Zhong L, Nahid MA, Gao G, The potential of adeno-associated viral vectors for gene delivery to muscle tissue. ***Expert Opin. Drug Deliv.***, 11(3): 345–364 (2014). https://doi.org/10.1517/17425247.2014.871258
61. Ragot T, Vincent N, Chafey P, Vigne E, Gilgenkrantz H, Coutont D, Cartaud J, Briand P, Kaplant J-C, Perricaude M, Kahn A, Efficient adenovirus-mediated transfer of a human minidystrophin gene to skeletal muscle of mdx mice. ***Nature***, 361(6413): 647–650 (1993). https://doi.org/10.1038/361647a0
62. Acsadi G, Jani A, Massie B, Simoneau M, Holland P, Blaschuk K, Karpati G, A differential efficiency of adenovirus-mediated in vivo gene transfer into skeletal muscle cells of different maturity. ***Hum. Mol. Genet.***, 3(4): 579–584 (1994). https://doi.org/10.1093/hmg/3.4.579
63. Liu Q, Muruve DA, Molecular basis of the inflammatory response to adenovirus vectors. ***Gene Ther.***, 10(11): 935–940 (2003). https://doi.org/10.1038/sj.gt.3302036
64. Verdera HC, Kuranda K, Mingozzi F, AAV Vector Immunogenicity in Humans: A Long Journey to Successful Gene Transfer. ***Mol. Ther.***, 28(3): 723–746 (2020). https://doi.org/10.1016/j.ymthe.2019.12.010
65. Carter BJ, Adeno-associated virus and the development of adeno-associated virus vectors: a historical perspective. ***Mol. Ther.***, 10(6): 981–989 (2004). https://doi.org/10.1016/j.ymthe.2004.09.011
66. Wang B, Li J, Xiao X, Adeno-associated virus vector carrying human minidystrophin genes effectively ameliorates muscular dystrophy in mdx mouse model. ***Proc. Natl. Acad. Sci. USA.***, 97(25): 13714–13719 (2000). https://doi.org/10.1073/pnas.240335297
67. Harper SQ, Hauser MA, DelloRusso C, Duan D, Crawford RW, Phelps SF, Harper HA, Robinson AS, Engelhardt JF, Brooks SV, Chamberlain JS, Modular flexibility of dystrophin: Implications for gene therapy of Duchenne muscular dystrophy. ***Nat. Med.***, 8(3): 253–261 (2002). https://doi.org/10.1038/nm0302-253
68. Fabb SA, Wells DJ, Serpente P, Dickson G, Adeno-associated virus vector gene transfer and sarcolemmal expression of a 144 kDa micro-dystrophin effectively restores the dystrophin-associated protein complex and inhibits myofibre degeneration in nude/mdx mice. ***Hum. Mol. Genet.***, 11(7): 733–741. https://doi.org/10.1093/hmg/11.7.733

69. Kupatt C, Windisch A, Moretti A, Wolf E, Wurst W, Walter MC, Genome editing for Duchenne muscular dystrophy: a glimpse of the future? ***Gene Ther.***, 28(9): 542–548 (2021) https://doi.org/10.1038/s41434-021-00222-4
70. Long C, Amoasii L, Mireault AA, McAnally JR, Li H, Sanchez-Ortiz E, Bhattacharyya S, Shelton JM, Bassel-Duby R, Olson EN, Postnatal genome editing partially restores dystrophin expression in a mouse model of muscular dystrophy. ***Science***, 351(6271):400–403 (2016). https://doi.org/10.1126/science.aad5725
71. Nelson CE, Hakim CH, Ousterout DG, Thakore PI, Moreb EA, Rivera RMC, Madhavan S, Pan X, Ran FA, Yan WX, Asokan A, Zhang F, Duan D, Gersbach CA, In vivo genome editing improves muscle function in a mouse model of Duchenne muscular dystrophy. ***Science***, 351(6271): 403–407 (2016). https://doi.org/10.1126/science.aad5143
72. Tabebordbar M, Zhu K, Cheng JKW, Chew WL, Widrick JJ, Yan WX, Maesner C, Wu EY, Xiao R, Ran FA, Cong L, Zhang F, Vandenberghe LH, Church GM, Wagers AJ, In vivo gene editing in dystrophic mouse muscle and muscle stem cells. ***Science***, 351(6271): 407–411 (2016) https://doi.org/10.1126/science.aad5177
73. Nelson CE, Wu Y, Gemberling MP, Oliver ML, Waller MA, Bohning JD, Robinson-Hamm JN, Bulaklak K, Rivera RMC , Collier JH, Asokan A, Gersbach CA, Long-term evaluation of AAV-CRISPR genome editing for Duchenne muscular dystrophy. ***Nat. Med.***, 25(3): 427–432 (2019). https://doi.org/10.1038/s41591-019-0344-3
74. Jearawiriyapaisarn N, Moulton HM, Buckley B, Roberts J, Sazani P, Fucharoen S, Iversen PL, Kole R, Sustained dystrophin expression induced by peptide-conjugated morpholino oligomers in the muscles of mdx mice. ***Mol. Ther.***, 16(9): 1624–1629 (2008). https://doi.org/10.1038/mt.2008.120
75. Betts C, Saleh AF, Arzumanov AA, Hammond SM, Godfrey C, Coursindel T, Gait MJ, Wood MJA, Pip6-PMO, A New Generation of Peptide-oligonucleotide Conjugates With Improved Cardiac Exon Skipping Activity for DMD Treatment. ***Mol. Ther. Nucleic Acids***, 1(8): e38 (2012). https://doi.org/10.1038/mtna.2012.30
76. Osborn MF, Khvorova A, Improving siRNA delivery in vivo through lipid conjugation. ***Nucleic Acid Ther.***, 28(3): 128–136 (2018) https://doi.org/10.1089/nat.2018.0725
77. Prakash TP, Mullick AE, Lee RG, Yu J, Yeh ST, Low A, Chappell AE, Østergaard ME, Murray S, Gaus HJ, Swayze EE, Seth PP, Fatty acid conjugation enhances potency of antisense oligonucleotides in muscle. ***Nucleic Acids Res.***, 47(12): 6029–6044 (2019). https://doi.org/10.1093/nar/gkz354
78. Biscans A, Caiazzi J, McHugh N, Hariharan V, Muhuri M, Khvorova A, Docosanoic acid conjugation to siRNA enables functional and safe delivery to skeletal and cardiac muscles. ***Mol. Ther.***, 29(4): 1382–1394 (2021). https://doi.org/10.1016/j.ymthe.2020.12.023
79. Chappell AE, Gaus HJ, Berdeja A, Gupta R, Jo M, Prakash TP, Oestergaard M, Swayze EE, Seth PP, Mechanisms of palmitic acid-conjugated antisense oligonucleotide distribution in mice. ***Nucleic Acids Res.***, 48(8): 4382–4395 (2020). https://doi.org/10.1093/nar/gkaa164
80. Arpke RW, Darabi R, Mader TL, Zhang Y, Toyama A, Lonetree C, Nash N, Lowe DA, Perlingeiro RCR, Kyba1 M, A new immuno- dystrophin-deficient model, the NSG-mdx4Cv mouse, provides evidence for functional improvement following allogeneic satellite cell transplantation. ***Stem Cells***, 31(8): 1611–1620 (2013). https://doi.org/10.1002/stem.1402
81. Cosgrove BD, Gilbert PM, Porpiglia E, Mourkioti F, Lee SP, Corbel SY, Llewellyn ME, Delp SL, Blau1 HM, Rejuvenation of the muscle stem cell population restores strength to injured aged muscles. **Nat. Med.,** 20(3): 255–264 (2014). https://doi.org/10.1038/nm.3464
82. Tedesco FS, Dellavalle A, Diaz-Manera J, Messina G, Cossu G. Repairing skeletal muscle: regenerative potential of skeletal muscle stem cells. ***J. Clin. Invest.***, 120:11–19 (2010). https://doi.org/10.1172/JCI40373.
83. Bareja A, Billin AN, Satellite cell therapy – from mice to men. ***Skelet. Muscle***, 3: 2 (2013). https://doi.org/10.1186/2044-5040-3-2

84. Rinaldi F, Perlingeiro RCR, Stem cells for skeletal muscle regeneration: Therapeutic potential and roadblocks. ***Transl. Res.***, 163(4): 409–417 (2014). https://doi.org/10.1016/j.trsl.2013.11.006
85. Choi HS, Liu W, Misra P, Tanaka E, Zimmer JP, Ipe BI, Bawendi MG, Frangioni JV, Renal clearance of quantum dots. ***Nat. Biotechnol.***, 25(10): 1165–1170 (2007). https://doi.org/10.1038/nbt1340
86. Kang H, Rho S, Stiles WR, Hu S, Baek Y, Hwang DW, Kashiwagi S, Kim MS, Choi HS, Size-Dependent EPR Effect of Polymeric Nanoparticles on Tumor Targeting. ***Adv. Healthc. Mater.***, 9(1): e1901223 (2020). https://doi.org/10.1002/adhm.201901223
87. Samoylova TI, Smith BF, Elucidation of muscle-binding peptides by phage display screening. ***Muscle Nerve***, 22(4): 460–466 (1999). https://doi.org/10.1002/(sici)1097-4598(199904)22:4<460::aid-mus6>3.0.co;2-l
88. Yu C-Y, Yuan Z, Cao Z, Wang B, Qiao C, Li J, Xiao X, A muscle-targeting peptide displayed on AAV2 improves muscle tropism on systemic delivery. ***Gene Ther.***, 16(8): 953–962 (2009). https://doi.org/10.1038/gt.2009.59
89. Gao X, Zhao J, Han G, Zhang Y, Dong X, Cao L, Wang Q, Moulton HM, Yin H, Effective Dystrophin Restoration by a Novel Muscle-Homing Peptide–Morpholino Conjugate in Dystrophin-Deficient *mdx* Mice. ***Mol. Ther.***, 22(7): 1333–1341 (2014). https://doi.org/10.1038/mt.2014.63
90. Philippou S, Mastroyiannopoulos NP, Makrides N, Lederer CW, Kleanthous M, Phylactou LA, Selection and identification of skeletal muscle targeted RNA aptamers. ***Mol. Ther. Nucleic Acids***, 10: 199–214 (2018). https://doi.org/10.1016/j.omtn.2017.12.004
91. Sugo T, Terada M, Oikawa T, Miyata K , Nishimura S, Kenjo E, Ogasawara-Shimizu M, Makita Y, Imaichi S, Murata S, Otake K, Kikuchi K, Teratani M, Masuda Y, Kamei T, Takagahara S, Ikeda S, Ohtaki T, Matsumoto H, Development of antibody-siRNA conjugate targeted to cardiac and skeletal muscles. ***J. Control. Release***, 237:1–13 (2016). https://doi.org/10.1016/j.jconrel.2016.06.036
92. Maeda H, Wu J, Sawa T, Matsumura Y, Hori K, Tumor vascular permeability and the EPR effect in macromolecular therapeutics: a review. ***J. Control. Release***, 65(1–2): 271–284 (2000). https://doi.org/10.1016/s0168-3659(99)00248-5
93. Zhang Y-N, Poon W, Tavares AJ, McGilvray ID, Chan WCW, Nanoparticle–liver interactions: Cellular uptake and hepatobiliary elimination. ***J. Control. Release***, 240: 332–348 (2016). https://doi.org/10.1016/j.jconrel.2016.01.020
94. Naito M, Watanuki Y, Toh K, Yum J, Kim BS, Taniwaki K, Ogura S, Ishida H, Cho M, Chaya H, Miyajima K, Yamasaki Y, Osada K, Minegishi K, Aoki Y, Miyata K, Size-tunable PEG-grafted copolymers as a polymeric nanoruler for passive targeting muscle tissues. ***J. Control. Release***, 347: 607–614 (2022). https://doi.org/10.1016/j.jconrel.2022.05.030
95. Li Y, Wang J, Satterle A, Wu Q, Wang J, Liu F, Gene transfer to skeletal muscle by site-specific delivery of electroporation and ultrasound. ***Biochem. Biophys. Res. Commun.***, 424(2): 203–207 (2012). https://doi.org/10.1016/j.bbrc.2012.06.090
96. Aihara H, Miyazaki J, Gene transfer into muscle by electroporation in vivo, ***Nat. Biotechnol.***, 16(9): 867–870 (1998). https://doi.org/10.1038/nbt0998-867
97. McMahon JM, Wells DJ, Electroporation for Gene transfer to skeletal muscles, ***BioDrugs***, 18(3): 155-165 (2004). https://doi.org/10.2165/00063030-200418030-00002
98. Hernot S, Klibanov AL, Microbubbles in ultrasound-triggered drug and gene delivery. ***Adv. Drug Deliv. Rev.***, 60(10): 1153–1166, (2008). https://doi.org/10.1016/j.addr.2008.03.005
99. Negishi Y, Endo-Takahashi Y, Matsuki Y, Kato Y, Takagi N, Suzuki R, Murayama K, Aramaki Y, Systemic delivery systems of angiogenic gene by novel bubble liposomes containing cationic lipid and ultrasound exposure. ***Mol. Pharm.***, 9(6): 1834–1840 (2012). https://doi.org/10.1021/mp200554c

Chapter 7
Challenges in Clinical Translation of Musculoskeletal Drug Delivery Systems

Ankita Negi, Sumit Murab, and Patrick W. Whitlock

Contents

A. Negi
School of Biosciences & Bioengineering, IIT Mandi, Mandi, Himachal Pradesh, India

S. Murab
School of Biosciences & Bioengineering, IIT Mandi, Mandi, Himachal Pradesh, India

Indian Knowledge System and Mental Health Applications Centre, IIT Mandi, Mandi, Himachal Pradesh, India

BioX Centre, IIT Mandi, Mandi, Himachal Pradesh, India

Advanced Materials Research Centre, IIT Mandi, Mandi, Himachal Pradesh, India

P. W. Whitlock (✉)
Department of Biomedical Engineering, College of Engineering and Applied Sciences, University of Cincinnati, Cincinnati, OH, USA

Division of Pediatric Orthopaedic Surgery, Cincinnati Children's Hospital Medical Center, Cincinnati, OH, USA

Department of Orthopaedic Surgery, College of Medicine, University of Cincinnati, Cincinnati, OH, USA
e-mail: patrick.whitlock@cchmc.org

S. Murab (ed.), *Drug Delivery Systems for Musculoskeletal Tissues*,
https://doi.org/10.1007/978-3-031-55653-1_7

1 Introduction

For the treatment of many musculoskeletal problems, including infections, tumors, and cancer, drug delivery has been used extensively and has drawn attention as a viable therapeutic technique. Several advanced drug delivery systems (DDS) have been developed with the goal of treating musculoskeletal diseases of different tissues such as bone, cartilage, tendons, ligaments, and skeletal muscles [1]. To achieve controlled and on-demand drug delivery for musculoskeletal treatments, several new material systems are being studied. There are a variety of drug molecules such as low molecular weight, peptide-based, biomacromolecules, and other hybrid drug formulations, which are incorporated into drug delivery systems to improve their delivery in musculoskeletal tissues [1, 2]. The development of delivery systems is the result of the tremendous advancements in material chemistry and drug delivery that have led to the development of novel systems such as micro/nanoparticles, 3D-printed scaffolds, hydrogels, and coated implants [3]. While the field is growing, the clinical translation of early-stage research concepts is crucial.

However, complex designs for drug delivery systems make clinical translation more difficult in terms of quality control and manufacturing processes. In addition to this, multiple clinical trials for current musculoskeletal drug delivery systems have been put on hold because of potential safety risks, low efficacy, and financial concerns. As shown by the few FDA-approved treatments, the journey of a commercially successful DDS is long and challenging [4]. Several challenges at every stage of clinical translation could result in their abrupt failure, thus delaying the commercial translation of DDS. Hence, the future success of drug delivery systems depends on the understanding of the challenges in clinical translation. Therefore, in this chapter, we will discuss the challenges on the way to clinical translation of drug delivery systems for musculoskeletal tissues.

2 Regulatory Challenges

Advanced drug delivery systems have the potential to boost the pharmaceutical industry by enhancing health benefits, but there is a major gap between the development and regulation of such products. For example, the commercialization of nanomedicine is strongly influenced by several governmental regulations regarding patent protection, quality assurance, manufacturing procedures, and safety [5]. Nano-based products are effective for early diagnosis and therapy, but they also have some detrimental effects on the environment, ecology, and human health [6]. Therefore, it is crucial to establish stringent laws and regulations for the monitoring of products based on nanotechnology to protect human health. In view of this, several nations have already started establishing regulatory agencies and applying rules, plans, and guidelines with the purpose of safe and effective regulation of nanoproducts, while many other countries are lagging [7]. Hence, the absence of defined regulatory and safety criteria can hamper the clinical translation of nano-based products. Factors such as social, legal, and ethical concerns also result in delaying the translation of these products.

One of the most extensively researched and versatile systems for tissue engineering and drug delivery applications is hydrogels [8]. There are numerous examples of injectable hydrogels that have been both clinically and commercially successful. Bbut for the musculoskeletal system, the number is low. *EUFLEXXA®* is an example of hyaluronic acid-based hydrogel that is injected intra-articularly for the treatment of knee osteoarthritis, and it is FDA- and EMA-approved. In addition to this, there are injectable hydrogel systems that are currently being investigated in clinics, such as *Hymovis Viscoelastic Hydrogel*, Promedon, and *HYADD® 4 Hydrogel* for osteoarthritis and GelStix® Nucleus Augmentation Device for degenerative disc disease [9]. Despite several commercially available products and ongoing clinical studies, the translation of hydrogels is quite challenging. Moreover, the variety of cross-linking agents and biomaterials used to create hydrogel scaffolds makes their approval and regulation difficult. In accordance with Section 201(g) of the FD&C Act, hydrogel falls within the "devices" category. This section defines the term "device" as "an instrument, implant or in vitro reagent which is intended to use for the treatment, diagnosis, and prevention of diseases in humans or other animals or intended to affect the structure or any function of the body and which does not achieve its primary intended purposes by chemical action within or with the body." In addition, most hydrogel-based products must pass extra FDA evaluations of a 510(k) Pre-Market Notification submission to achieve marketing permission in the United States, with a few exceptions. Hydrogel scaffolds containing drugs or cells that secrete drugs are classified as combination products and take 7–10 years for the regulatory approval process, further limiting their commercial viability.

3 Challenges in the Scale-Up Process

A major barrier in clinical translation is scaling up the production of DDS on an industrial scale. The majority of DDS are fabricated/synthesized in small batches at the preclinical stages. When operations are carried out on a larger scale, batch variability increases and it is difficult to maintain quality, safety, and efficiency. Thus, strong process development teams are needed to develop methods for scaling up. In addition to this, the structural and physicochemical complexity of formulations slows down the clinical translation of drug delivery systems as they involve complex/laborious synthesis processes [10, 11].

For example, most nanoparticles are synthesized in batch mode through traditional laboratory synthesis techniques, which produces variations in the product quality and particle sizes between batches. Hence, it is challenging to implement the manufacturing of nanoparticles at the industrial level for large-scale production [12]. However, the biggest problem arises when the design of a nanoparticle-based drug delivery system becomes more complex by adding surface modification through coating/ligands, using multiple targeting elements/multiple therapeutic agents, etc. This can add multiple steps in the production process as multiple components are integrated into a single nanosized carrier. This can raise the production cost and create difficulties in large-scale good manufacturing production as well as make quality control and assurance of such products more challenging [13]. Hence, access to a manufacturing process that is easy, quick, and reproducible is a crucial necessity for clinical translation. This can facilitate the production of nanoparticles of uniform sizes with a high degree of quality. After the manufacturing process, the stability of the delivery system should also be maintained during clinical administration and long-term storage in order to avoid burst release of drug or aggregation at the site of action. In the case of a hydrogel-based DDS, the production, sterilization, and storage procedures are even more difficult due to the high-water content of hydrogel [9].

As we know, clinical trials are expensive and require substantial financial investments. Academic inventions, in contrast to those developed by huge pharmaceutical corporations, lack significant funding and need help from outside sources. Thus, for the benefits of commercial and clinical translation, the involvement of investors and industrial partners is highly required. The efficacy and effectiveness of the drug delivery system are determined through clinical trials. Before FDA approval, drug delivery systems must demonstrate tolerability and efficacy in the preclinical trial. While it is challenging to cover all aspects in clinical trials, side effects must still be monitored and reported even after successful trials and subsequent commercial launches. There are various reasons for the failure of drug delivery systems in clinical trials, but lack of funding to support expensive trials is the main reason [4]. Therefore, manufacturing-related challenges such as poor-quality control, high material/manufacturing cost, low yield, batch variability, and product stability limit the scale-up process.

4 Biological Challenges

The successful clinical translation of drug delivery systems depends on an understanding of biology and technology. It is important to study the effect of disease pathophysiology on the distribution, retention, accumulation, and efficacy of drug delivery systems [4, 14, 15]. Most of the time, the design of the delivery system is formulated/synthesized from the standpoint of chemists or engineers. Hence, the behavior of the delivery system and its interactions with biological systems and the environment receive less attention. The advancement of drug delivery systems for the benefit of patients requires collaboration between engineers, chemists, and biologists. Biologists can help study the in vivo behavior of DDS in animal and human systems for their successful translation.

In the case of inorganic nanoparticles such as AuNPs and iron oxide NPs, the biological effects need to be studied because it is well known that the size and form of nanocarriers can significantly change biological function and fate [16, 17]. Many preclinical studies utilize small animal models, but this complicates the further process as small animals cannot correctly mimic humans. Nonetheless, later clinical trials are conducted on larger animals. Even at the preclinical level, general conclusions are challenging to draw. Therefore, to uncover fundamental knowledge about nanoparticle (AuNP) biocompatibility, toxicity, and biodistribution in order to maximize possibilities for therapeutic success, these characteristics need to be thoroughly researched at both the preclinical and clinical levels, where applicable.

Other than nanocarriers, an implant-based system for drug delivery has been found to cause a series of immunogenic reactions because of tissue trauma after implantation and foreign body response. Foreign body response (FBR) mainly involves an acute initial inflammatory phase followed by a chronic inflammatory phase. FBR heavily depends on the characteristics of materials used for implant fabrication and their degradation mechanism. Therefore, in-depth understanding of the immune respones against the material system is required to modulate material properties to make them compatible with a specific tissue. This can lead us to the successful clinical outcome of the implantable system [18–20].

Therefore, these kinds of studies can be facilitated by the development of new in vitro approaches such as FBR-on-a-chip, medical devices on a chip, and more sophisticated cell culture systems that stimulate the in vivo environment. Biological reactions to biodegradable materials and devices should be studied using systems of this type, which can track the materials' rates of degradation and provide a native microenvironment for in vitro research on interactions with relevant tissues. These technologies can be implemented into the common evaluation procedure for biocompatibility testing and utilized for routine immunological testing of material or a device [21–23].

5 Material-Based Challenges

Various kinds of designs for drug delivery have been developed such as nondegradable implants, biodegradable systems, or hydrogels. Drugs can be trapped (micro/nanoencapsulated) in biodegradable polymer, where drug release is regulated by diffusion of the drug through the polymer matrix, matrix erosion, and biodegradation, as well as osmotic and swelling events of the polymer. Nondegradable formulations are usually macroscopic removable implants. Implants can support high drug loads and help in prolonged drug release. Typically, these formulations regulate release by allowing drugs to diffuse through a polymer matrix. Lastly, water-based systems used in hydrogel-based formulations can help with drug stability and biocompatibility [24].

There are different kinds of materials present in all the abovementioned drug delivery systems. Most common examples are polycaprolactone (PCL), poly(glycolic acid) (PGA), poly(lactic acid) (PLA), and poly(lactic-co-glycolic acid) (PLGA), and through various studies, it has been found that polymers have received much attention as a platform for the design of the delivery system [21, 24,

25]. But their safety and effectiveness are greatly influenced by factors such as their conjugation chemistry, molecular weight, molecular structure, and polydispersity [7]. Hence, there is a growing need for a suitable regulatory framework so that an increased number of polymer materials can be evaluated for drug delivery systems. Each DDS system is different; hence, it is crucial to evaluate them separately based on the dosages, administration methods, proposed therapeutic use, and frequency of dosing.

Controlled drug delivery system design and biomaterial applications for implants and tissue engineering have been completely transformed by biodegradable polymers. The advantage of using biodegradable polymers is that they do not require removal after usage. Nevertheless, degradation products from the polymer must not be toxic to the patient and drug that is being encapsulated. Importantly, PLGs and PLGAs produce acidic byproducts, which can lower the pH of the surrounding medium and the formulation and can degrade acid-sensitive drugs, especially proteins. Other than synthetic FDA-approved polymers, natural degradable polymers, such as protein-based polymers such as protein-based polymers such as gelatin, albumin, and collagen, and polysaccharides such as alginate, agarose, dextran, hyaluronic acid, chitosan, and dextran, are also used to synthesize biodegradable-based formulation. Due to natural abundance and biocompatibility, naturally degradable materials have been employed in the pharmaceutical or medical field. However, the natural polymer exhibits a wide molecular weight distribution. Controlled and precise drug release can be hampered as they cannot be modulated with additional synthesis. In addition to this, natural polymers can also have precarious outcomes due to batch-to-batch variations [26].

Therefore, selecting, designing, or engineering an appropriate biomaterial is one major challenge in developing drug delivery systems. The ability of a material to encapsulate a bigger drug payload with optimal biocompatibility, degradability, and bioresorbability is one of the important factors that collectively determine a drug delivery system's long-term efficacy and safety. Moreover, the material's ability to integrate with tissue and mechanical strength also needs to be determined, as any kind of failure in these properties can cause serious consequences [27]. Postoperative implant failure is the most common and simple example for understanding the poor integration between biological tissues and the material system, which causes local inflammation and fibrosis in the surroundings, as discussed in the above section. Hence, to prevent inflammation, researchers are functionalizing the implant surface with immunomodulatory drugs and also introducing surface functionalization strategies, including PEGylation, biomolecule coating (such as heparin) [28]. Yet, controlling postoperative complications including fibrosis and inflammation is still challenging.

Nowadays, the development of an advanced polymeric-controlled delivery system is in progress to adjust the temporal release of drugs. There are a majority of traditional controlled-release carriers that provide treatment with a consistent and sustained release for a longer duration. Although these systems have increased retention time and decreased the frequency of administration, they cannot be employed for the optimal treatment of the disease that requires a variable pattern of drug release [29]. The fabrication of a patient-specific customized design for a drug delivery system is another challenge for researchers. Researchers are moving toward personalized therapy as several biodegradable polymers with customizable physiochemical properties regulate the fabrication of a polymeric delivery system that releases drugs at controlled rates. To achieve this, 3D printing has been found to be a potential solution, but such an approach requires rigorous optimization of implant design with respect to the patient. This technique has been adopted widely in healthcare businesses such as the pharmaceutical industry [30]. But it has to overcome a number of challenges as it is the most advanced manufacturing technique in this field.

In 2017, the Food and Drug Administration (FDA) released regulatory guidance on medical devices produced by additive manufacturing techniques. However, regulatory guidelines and laws governing 3D printing products with drug delivery capabilities were not covered. Although, it is anticipated to be published in the future. Other than this, there are additional important factors that need to be overlooked in order to use 3D printing in health care such as standardization of 3D printers to meet quality control, regulatory criteria, and software-based process control. In addition to this, several 3D printers are being utilized in the research sector; hence, such printers are not designed for the manufacturing of pharmaceutical products and, as a result, do not adhere to the Good Manufacturing Practice (GMP) rules. Therefore, some safety concerns and regulatory shortcomings must yet be resolved before these technologies are used in healthcare [30].

Another advanced type of 3D printing technology is 3D bioprinting, which utilizes cells and biomaterials together as printing material. This technique is used to fabricate drug-loaded scaffolds. But logistic requirements for products containing living cells are particularly challenging since they must be produced timely and ecologically. Before printing, cells must be isolated and expanded in a GMP facility according to GMP guidelines. Because of the lengthy processing of such a technique, there is a delay between the isolation, printing, and actual implantation of printed constructs, which can limit the usage of this technique [31].

6 Conclusion

Drug delivery systems for musculoskeletal tissues hold the potential for improving the treatment clinical strategies, which will both improve the outcomes and ease the symptoms for the patients. The clinical translation of these drug delivery systems encounters a lot of challenges, which need to be addressed. Apart from the technical challenges discussed in the chapter, lack of GMP facilities, lack of insurance coverage for the patients during clinical trials, and lack of interdisciplinary teams of scientists, engineers, and physicians in academic settings are some practical hurdles that slow down the clinical translation of these drug delivery systems. Overcoming these challenges as research intensifies in the field will provide us with more commercially available products in the future.

References

1. S. Zhang, M. Xing, B. Li, Recent advances in musculoskeletal local drug delivery, Acta Biomater. 93 (2019) 135–151. https://doi.org/10.1016/j.actbio.2019.01.043.
2. D.S.W. Benoit, C.T. Overby, K.R. Sims, M.A. Ackun-Farmmer, Drug Delivery Systems, Fourth Edition, Elsevier, 2020. https://doi.org/10.1016/B978-0-12-816137-1.00078-7.
3. M. Farokhi, F. Mottaghitalab, M.A. Shokrgozar, K.L. Ou, C. Mao, H. Hosseinkhani, Importance of dual delivery systems for bone tissue engineering, J. Control. Release. 225 (2016) 152–169. https://doi.org/10.1016/j.jconrel.2016.01.033.
4. A.Anselmo, S. Mitragotri, An Overview of Clinical and Commercial Impact of Drug Delivery Systems, J. Control Release. 176 (2019) 139–148. https://doi.org/10.1016/j.jconrel.2014.03.053.An.
5. V. Sainz, J. Conniot, A.I. Matos, C. Peres, E. Zupančič, L. Moura, L.C. Silva, H.F. Florindo, R.S. Gaspar, Regulatory aspects on nanomedicines, Biochem. Biophys. Res. Commun. 468 (2015) 504–510. https://doi.org/10.1016/j.bbrc.2015.08.023.
6. J. Zhao, V. Castranova, Toxicology of nanomaterials used in nanomedicine, J. Toxicol. Environ. Heal. - Part B Crit. Rev. 14 (2011) 593–632. https://doi.org/10.1080/10937404.2011.615113.
7. S. Hua, M.B.C. de Matos, J.M. Metselaar, G. Storm, Current trends and challenges in the clinical translation of nanoparticulate nanomedicines: Pathways for translational development and commercialization, Front. Pharmacol. 9 (2018) 1–14. https://doi.org/10.3389/fphar.2018.00790.
8. C.A. Dreiss, Hydrogel design strategies for drug delivery, Curr. Opin. Colloid Interface Sci. 48 (2020) 1–17. https://doi.org/10.1016/j.cocis.2020.02.001.
9. A. Mandal, J.R. Clegg, A.C. Anselmo, S. Mitragotri, Hydrogels in the clinic, Bioeng. Transl. Med. 5 (2020) 1–12. https://doi.org/10.1002/btm2.10158.
10. S. Svenson, Clinical translation of nanomedicines, Curr. Opin. Solid State Mater. Sci. 16 (2012) 287–294. https://doi.org/10.1016/j.cossms.2012.10.001.
11. S. Tinkle, S.E. Mcneil, M. Stefan, R. Bawa, G. Borchard, Nanomedicines : addressing the scientific, (2014) 1–22. https://doi.org/10.1111/nyas.12403.
12. S. Soares, J. Sousa, A. Pais, C. Vitorino, Nanomedicine : Principles , Properties , and Regulatory Issues, 6 (2018) 1–15. https://doi.org/10.3389/fchem.2018.00360.
13. A. Kad, A. Pundir, S.K. Arya, N. Bhardwaj, M. Khatri, An Elucidative Review to Analytically Sieve the Viability of Nanomedicine Market, (2022) 249–265.
14. A.C. Anselmo, S. Mitragotri, A Review of Clinical Translation of Inorganic Nanoparticles, AAPS J. 17 (2015) 1041–1054. https://doi.org/10.1208/s12248-015-9780-2.
15. J.I. Hare, T. Lammers, M.B. Ashford, S. Puri, G. Storm, S.T. Barry, Challenges and strategies in anti-cancer nanomedicine development: an industry perspective Jennifer, Adv. Drug Deliv. Rev. (2016). https://doi.org/10.1016/j.addr.2016.04.025.
16. J.A. Champion, Y.K. Katare, S. Mitragotri, Particle shape : A new design parameter for micro- and nanoscale drug delivery carriers, 121 (2007) 3–9. https://doi.org/10.1016/j.jconrel.2007.03.022.
17. P. Decuzzi, B. Godin, T. Tanaka, S. Lee, C. Chiappini, X. Liu, M. Ferrari, Size and shape effects in the biodistribution of intravascularly injected particles, J. Control. Release. 141 (2010) 320–327. https://doi.org/10.1016/j.jconrel.2009.10.014.
18. J. Shen, D.J. Burgess, Long Acting Injections and Implants, Long Act. Inject. Implant. (2012) 73–91. https://doi.org/10.1007/978-1-4614-0554-2.
19. D.M. Higgins, R.J. Basaraba, A.C. Hohnbaum, E.J. Lee, D.W. Grainger, M. Gonzalez-Juarrero, Localized immunosuppressive environment in the foreign body response to implanted biomaterials, Am. J. Pathol. 175 (2009) 161–170. https://doi.org/10.2353/ajpath.2009.080962.
20. Y. Chandorkar, K. Ravikumar, B. Basu, The Foreign Body Response Demystified, ACS Biomater. Sci. Eng. 5 (2019) 19–44. https://doi.org/10.1021/acsbiomaterials.8b00252.

21. C. Li, C. Guo, V. Fitzpatrick, A. Ibrahim, M.J. Zwierstra, P. Hanna, A. Lechtig, A. Nazarian, S.J. Lin, D.L. Kaplan, Design of biodegradable, implantable devices towards clinical translation, Nat. Rev. Mater. 5 (2020) 61–81. https://doi.org/10.1038/s41578-019-0150-z.
22. F. Sharifi, S.S. Htwe, M. Righi, H. Liu, A. Pietralunga, O. Yesil-celiktas, S. Maharjan, B. Cha, S.R. Shin, A Foreign Body Response-on-a-Chip Platform, 1801425 (2019) 1–10. https://doi.org/10.1002/adhm.201801425.
23. M. Jannasch, S. Gaetzner, T. Weigel, H. Wal, T. Schmitz, A comparative multi-parametric in vitro model identifies the power of test conditions to predict the fibrotic tendency of a biomaterial, (2017) 1–17. https://doi.org/10.1038/s41598-017-01584-9.
24. W. Li, J. Tang, D. Lee, T.R. Tice, S.P. Schwendeman, M.R. Prausnitz, Clinical translation of long-acting drug delivery formulations, Nat. Rev. Mater. 7 (2022) 406–420. https://doi.org/10.1038/s41578-021-00405-w.
25. G.S. Kwon, D.Y. Furgeson, Biodegradable polymers for drug delivery systems, Woodhead Publishing Limited, 2007. https://doi.org/10.1533/9781845693640.83.
26. D. Dippold, A. Cai, M. Hardt, A.R. Boccaccini, R.E. Horch, J.P. Beier, D.W. Schubert, Investigation of the batch-to-batch inconsistencies of Collagen in PCL- Collagen nano fi bers, Mater. Sci. Eng. C. 95 (2019) 217–225. https://doi.org/10.1016/j.msec.2018.10.057.
27. A. Kar, N. Ahamad, M. Dewani, L. Awasthi, R. Patil, R. Banerjee, Wearable and implantable devices for drug delivery: Applications and challenges, Biomaterials. 283 (2022) 121435. https://doi.org/10.1016/j.biomaterials.2022.121435.
28. N. Ahamad, A. Kar, S. Mehta, M. Dewani, V. Ravichandran, P. Bhardwaj, S. Sharma, R. Banerjee, Immunomodulatory nanosystems for treating inflammatory diseases, Biomaterials. 274 (2021) 120875. https://doi.org/10.1016/j.biomaterials.2021.120875.
29. A. Geraili, M. Xing, K. Mequanint, Design and fabrication of drug-delivery systems toward adjustable release profiles for personalized treatment, View. 2 (2021) 20200126. https://doi.org/10.1002/viw.20200126.
30. I. Seoane-Viaño, P. Januskaite, C. Alvarez-Lorenzo, A.W. Basit, A. Goyanes, Semi-solid extrusion 3D printing in drug delivery and biomedicine: Personalised solutions for healthcare challenges, J. Control. Release. 332 (2021) 367–389. https://doi.org/10.1016/j.jconrel.2021.02.027.
31. S. Murab, A. Gupta, M.K. Włodarczyk-Biegun, A. Kumar, P. van Rijn, P. Whitlock, S.S. Han, G. Agrawal, Alginate based hydrogel inks for 3D bioprinting of engineered orthopedic tissues, Carbohydr. Polym. 296 (2022). https://doi.org/10.1016/j.carbpol.2022.119964.

Index

S. Murab (ed.), *Drug Delivery Systems for Musculoskeletal Tissues*, https://doi.org/10.1007/978-3-031-55653-1

MIX
Papier aus verantwortungsvollen Quellen
Paper from responsible sources
FSC® C105338

FSC
www.fsc.org

If you have any concerns about our products,
you can contact us on
ProductSafety@springernature.com

In case Publisher is established outside the EU,
the EU authorized representative is:
Springer Nature Customer Service Center GmbH
Europaplatz 3, 69115 Heidelberg, Germany

Printed by Libri Plureos GmbH
in Hamburg, Germany